Arts in Place

This interdisciplinary book explores the role of art in placemaking in urban environments, analysing how artists and communities use arts to improve their quality of life. It explores the concept of social practice placemaking, where artists and community members are seen as equal experts in the process. Drawing on examples of local level projects from the USA and Europe, the book explores the impact of these projects on the people involved, on their relationship to the place around them, and on city policy and planning practice.

Case studies are Art Tunnel Smithfield, Dublin, an outdoor art gallery and community space in an impoverished area of the city; The Drawing Shed, London, a contemporary arts practice operating in housing estates and parks in Walthamstow, London; and Big Car, Indianapolis, an arts organisation operating across the whole of this Midwest city.

This book offers a timely contribution, bridging the gap between cultural studies and placemaking. It will be of interest to scholars, students and practitioners working in geography, urban studies, architecture, planning, sociology, cultural studies and the arts.

Cara Courage is an arts and urban researcher, curator and commentator. She completed her PhD at the University of Brighton, UK, and has an urban arts career spanning over fifteen years, with her own arts-based placemaking practice and specialism in public engagement in the built environment.

Routledge Research in Culture, Space and Identity
Series editor: Dr. Jon Anderson, School of Planning and Geography, Cardiff University, UK

The *Routledge Research in Culture, Space and Identity Series* offers a forum for original and innovative research within cultural geography and connected fields. Titles within the series are empirically and theoretically informed and explore a range of dynamic and captivating topics. This series provides a forum for cutting edge research and new theoretical perspectives that reflect the wealth of research currently being undertaken. This series is aimed at upper-level undergraduates, research students and academics, appealing to geographers as well as the broader social sciences, arts and humanities.

For a full list of titles in this series, please visit www.routledge.com/Routledge-Research-in-Culture-Space-and-Identity/book-series/CSI

Memory, Place and Identity
Commemoration and remembrance of war and conflict
Edited by Danielle Drozdzewski, Sarah De Nardi and Emma Waterton

Surfing Spaces
Jon Anderson

Violence in Place, Cultural and Environmental Wounding
Amanda Kearney

Arts in Place
The Arts, the Urban and Social Practice
Cara Courage

Arts in Place
The Arts, the Urban and Social Practice

Cara Courage

LONDON AND NEW YORK

First published 2017
by Routledge

2 Park Square, Milton Park, Abingdon, Oxfordshire OX14 4RN
52 Vanderbilt Avenue, New York, NY 10017

Routledge is an imprint of the Taylor & Francis Group, an informa business

First issued in paperback 2019

Copyright © 2017 Cara Courage

The right of Cara Courage to be identified as author of this work has been asserted by her in accordance with sections 77 and 78 of the Copyright, Designs and Patents Act 1988.

All rights reserved. No part of this book may be reprinted or reproduced or utilised in any form or by any electronic, mechanical, or other means, now known or hereafter invented, including photocopying and recording, or in any information storage or retrieval system, without permission in writing from the publishers.

Notice:
Product or corporate names may be trademarks or registered trademarks, and are used only for identification and explanation without intent to infringe.

British Library Cataloguing-in-Publication Data
A catalogue record for this book is available from the British Library

Library of Congress Cataloging in Publication Data
Names: Courage, Cara, author.
Title: Arts in place : the arts, the urban and social practice / Cara Courage.
Description: Abingdon, Oxon ; New York, NY : Routledge, [2017] | Series: Routledge research in culture, space and identity | Includes bibliographical references and index.
Identifiers: LCCN 2016041910| ISBN 9781138962637 (hbk) |
ISBN 9781315659299 (ebk)
Subjects: LCSH: Arts and society. | Community arts projects. |
Sociology, Urban.
Classification: LCC NX180.S6 C68 2017 | DDC 701/.03—dc23
LC record available at https://lccn.loc.gov/2016041910

ISBN: 978-1-138-96263-7 (hbk)
ISBN: 978-0-367-21909-3 (pbk)

Typeset in Times New Roman
by diacriTech, Chennai

What is public space? For me now, it's an ongoing exploration.

–Big Car artist

Contents

List of figures xi
Preface xiii
Acknowledgements xv
Explanation of terms xvii

1 Introduction: social practice placemaking in a global context 1
   1.1 The context of the research project 1
   1.2 Theoretical framework 3
   1.3 Research project design and methods 4
   1.4 Definition of the field of investigation 5
        1.4.1 Locating in the city 5
        1.4.2 Locating in space 6
        1.4.3 Experiencing the city ambiguously 9
        1.4.4 Shaping urban space and place 11
        1.4.5 Introducing place attachment 13
   1.5 The demographic context of the case studies 17
        1.5.1 Dublin and Ireland 18
        1.5.2 Indianapolis and the USA 19
        1.5.3 London and the UK 19
   1.6 The reflexive journey into the research project 20
   1.7 Note 21
   1.8 Bibliography 21
   1.9 List of URLs 28

2 Arts in place 29
   2.1 Moving art into the urban realm 29
   2.2 Moving art towards participation 32
   2.3 Relational and dialogical aesthetics 35
   2.4 Emplaced arts 37
        2.4.1 The aesthetic dislocation of emplaced art 38
        2.4.2 The aesthetic third 39

2.5 Social practice art(s) 40
    2.5.1 Social practice arts practice 41
    2.5.2 Social practice arts process 43
    2.5.3 The re-materialised art object as output 45
    2.5.4 Outcomes of social practice art 46
    2.5.5 'Social practice arts as social work' 47
2.6 Art in the public realm and placemakings schematic 47
2.7 Bibliography 48
2.8 List of URLs 52

3 A typology of placemaking   53

3.1 Understanding placemaking 53
3.2 Understanding creative placemaking 55
3.3 Aims and benefits of placemaking and creative placemaking 56
3.4 Architecture and placemakings 58
3.5 Expanding the understanding of placemaking 60
    3.5.1 Public Realm placemaking 61
    3.5.2 Participatory placemaking 62
3.6 Social practice placemaking 65
    3.6.1 Social practice placemaking practice 66
    3.6.2 Social practice placemaking process 68
    3.6.3 Aims and outcomes of social practice placemaking 70
3.7 The placemaking typology 72
    3.7.1 Presenting the placemaking typology 73
3.8 The political implications of social practice placemaking 76
    3.8.1 'The neoliberal dilemma' and political implications of social practice placemaking 77
    3.8.2 Right to the city, agonism and progressive politics 79
    3.8.3 Citizenship 80
    3.8.4 Ambivalent pragmatism 81
3.9 Notes 83
3.10 Bibliography 83
3.11 List of URLs 92

4 The Drawing Shed, London   93

4.1 Introducing The Drawing Shed 93
4.2 Facets of social practice placemaking practice and process 94
    4.2.1 The informal aesthetic 94
    4.2.2 The relational and the dialogical 94
    4.2.3 Performativity 96
    4.2.4 Duration 99
4.3 The position of the artist 100
    4.3.1 The special dispensation of the artist 100
    4.3.2 The embedded artist position 101
    4.3.3 Problematising the artist position 102

4.4 The place of the art 105
    4.4.1 The gallery 105
    4.4.2 The housing estate 108
    4.4.3 Public space 111
4.5 Degrees of participation 112
    4.5.1 Non-artist participation 113
    4.5.2 Problematising participation 115
4.6 Reflexive and transformative outcomes 118
4.7 Bibliography 121
4.8 List of URLs 123

5 Art Tunnel Smithfield, Dublin   125
5.1 Introducing Art Tunnel Smithfield 125
5.2 The art practice and process of Art Tunnel Smithfield 126
    5.2.1 Social horticulture, the informal aesthetic and beautification 127
    5.2.2 Issues of and around participation 129
    5.2.3 Place attachment and place loss 131
    5.2.4 Space/place thresholds and boundary crossing 135
5.3 'The vacant land issue', active citizenship and generative planning 138
    5.3.1 Active citizenship 138
    5.3.2 Temporary land use and 'Dublin new urbanism' 140
5.4 Note 149
5.5 Bibliography 149
5.6 List of URLs 152

6 Big Car, Indianapolis   153
6.1 Introducing Big Car 153
6.2 The artist as leader and the Staff Artist 153
6.3 The arts practice and process of Big Car 154
    6.3.1 Collaborative participation and the temporary aesthetic 155
    6.3.2 Access to arts and social justice 156
    6.3.3 The artist disposition and role 156
6.4 Social practice placemaking and social cohesion 157
6.5 Social practice placemaking and place identity 160
    6.5.1 Travel and the car 160
    6.5.2 Territory crossing and the city terroir 160
6.6 Social practice placemaking as arts-led regeneration 164
    6.6.1 Arts-led regeneration and the material and cultural 164
    6.6.2 The role of the artist and arts organisation 168
6.7 Note 170
6.8 Bibliography 170
6.9 List of URLs 172

7 Conclusion: towards a deeper understanding of the arts in placemaking 173

7.1 Syntheses of findings 173
- 7.1.1 Thematic one: the practice and process of social practice placemaking and its effect on the emplaced arts experience 173
- 7.1.2 Thematic two: reinterpreting the urban public realm through the arts in (social practice) placemaking 181
- 7.1.3 Thematic three: social practice placemaking and social cohesion and active citizenship 191

7.2 Limitations and expansions 198
- 7.2.1 Limitations of the text 198
- 7.2.2 Unanswered and alternative research questions uncovered in the research project 199
- 7.2.3 New knowledge contribution 202
- 7.2.4 Implications of the research 203
- 7.2.5 Directions for future research 204
- 7.2.6 Recommendations of future actions 208

7.3 Closing remarks 208

7.4 Bibliography 211

7.5 List of URLs 219

Appendix: demographic and statistical information 220

A.1 Global demographic information 220

A.2 Dublin, Republic of Ireland 220

A.3 London, UK 221

A.4 Indianapolis, Indiana, USA 222

A.5 Bibliography 222

Index 224

Figures

2.1	Schematic of art in the public realm and social practice art and placemakings.	48
3.1	Placemaking typology.	74
3.2	Strategic Public Realm placemaking matrix illustration.	74
3.3	Tactical creative placemaking matrix illustration.	75
3.4	Opportunistic participatory placemaking matrix illustration.	75
3.5	Tactical social practice placemaking matrix illustration.	75
4.1	The Drawing Shed (2009–2015) TL, CW, BR, *The Drawing Shed, Printbike, Go-Kart, Typing Pool*, London, 2014.	95
4.2	The Drawing Shed (2014), *LiveElse[W]Here Live Lunch* [event], Attlee Estate, Walthamstow, London, 2014.	98
4.3	The Drawing Shed (2014), *IdeasFromElse[W]Here* [residency], Winns Gallery, Lloyd Park, London, 2014.	107
4.4	Ed Woodham (2014), *Danger Deep Water* [performance], during The Drawing Shed (2014) *IdeasFromElse[W]Here* [residency], Winns Gallery, Lloyd Park, London, 2014.	113
5.1	TL, CW, BL, Art Tunnel Smithfield, view of *Art Tunnel* and *Art Platform*, Dublin, 2013, Benburb Street graffiti, Dublin, 2015, Smithfield, Dublin, 2014.	128
5.2	L, Image of Art Tunnel Smithfield, Dublin, after closure, March 2015;R, An impromptu planning meeting between volunteers at Mary's Abbey garden, Dublin, September 2014.	133
5.3	Images of Art Tunnel Smithfield, Dublin, fence signage, July 2013.	134
6.1	Big Car (2014), *Galería Magnifica* [gallery and workshop], Superior Market, North High School Road and West 38th Street, Indianapolis, 2014.	159
6.2	L, Big Car (2014), *Indy Do Day* [n.9] mural painting [event], Indianapolis, 2014. R, Example of interstate infrastructure bisecting pedestrian route that linked two retail parades, South Meridian, Indianapolis (2014), image taken during Big Car (2014) *Indy Do Day* mural painting.	162

xii *Figures*

6.3 TR, CW, BR, Big Car (2014) *Showroom* (centre, left, TR) [storefront gallery, events and workshop space] and *Listen Hear* (centre, right, TR) [sound art production, exhibition, installation and performance space], Lafayette Place, Indianapolis, 2014. 163

6.4 TL, Exterior aspect of one of *The Tube* warehouses (pre-renovation) from Cruft Street, Indianapolis, 2014. TR, Interior aspect of one of *The Tube* warehouses (pre-renovation) from Cruft Street, Indianapolis, 2014. BR, Example house for renovation around *The Tube* site (pre-renovation), Indianapolis, 2014. BL, Site of Garfield Park *Listen Hear* sound art gallery (pre-renovation), Indianapolis, 2014. 166

6.5 Big Car and Ed Woodham (2014), *Art In Odd Places Indianapolis* [event], Indianapolis, 2014. 170

Preface

There is so much stuff happening around us and we can't focus on it to see it for what it is. Art can focus on something that can help people see things that they couldn't normally see (Rick Lowe, Project Row Houses [n.1], *keynote lecture, at Open Engagement* [n.2] *2015, Pittsburgh)*

This text, *Arts in Place: the Arts, the Urban and Social Practice*, results from a PhD research project focused on a practice of placemaking informed by performative and social practice artforms.

Operating at the intersection of the arts, placemaking and urban theory and based on case study research, this project investigated participation in social practice and performative arts–based placemaking projects in the urban realm. It has a special focus on outcomes of place attachment and civic participation and on inter-disciplinary practice and learning across the arts and placemaking practices.

In response to the question 'How are urban places made and remade through performative arts practices?', the research was concerned with grassroots arts-led interventions in the urban realm, participated in by citizens with an aim to improve the urban lived experience and to form and cultivate connections between people, place and community. This came to be termed in the course of the research *social practice placemaking*, a practice observed in the placemaking sector as an approach that is informed by social practice arts and an attention on these arts as a means of urban revitalisation.

The research used a comparative case study approach based on participant observation and interviews at three sites: Art Tunnel Smithfield [n.3], Dublin, an outdoor art gallery and garden social practice arts; The Drawing Shed [n.4], London, a social arts practice predominantly operating in housing estates in Walthamstow and Wandsworth; and Big Car [n.5], Indianapolis, an arts organisation operating across the whole of this Midwest USA city.

Findings were along three themes. First, of the art practice and process of social practice placemaking, revealing the collaborative social practice placemaking art experience. Second, of urban social practice arts and place and social practice placemaking as a means of reinterpreting both social practice arts and cultural activities of the city. Third, of place attachment and social practice placemaking and its role in citizenship conscientisation and the politics of social practice placemaking activity in the urban public realm. These three themes are interwoven through the case study chapters and synchronised in the concluding chapter.

The research presents an original typology of practice for the placemaking sector, examines the practice, process and role of arts in placemaking and positions social practice placemaking in the social practice arts field. Significantly, the presentation of data includes the voice of the artist and non-artist protagonists.

The research has various implications for the sector. First, for creative and urban professionals and communities, by revealing how social practice placemaking can deepen an understanding of the relative agencies of the various modes of arts in place. Second, how this practice may advance placemaking practice as a whole by its use to better understand differences and similarities between placemakings within the placemaking sector, and from this, better communicate its practices to constituent stakeholders in the creative, urban design and community sectors. Third, to inform the understanding of collective progressive citizenship in the urban realm and inform generative planning practices.

Acknowledgements

The author would like to thank the institution of University of Brighton and its Arts and Humanities Centre for Research and Development and its staff; the PhD supervisory team of Alan Boldon, Professor Andrew Church and Dr. Lesley Murray, and also Professor Steve Miles, Manchester Metropolitan University, formerly of University of Brighton; and the PhD readers, Professor David Cotterell, University of Brighton, and Professor Doina Petrescu, University of Sheffield. Thanks are also due to the University and College Union and the Students Union University of Brighton representatives.

The author would also like to thank their friends and family for their support, the writing retreats provided in Berlin, Germany, and Ripe, East Sussex, UK, and Richard Wolfstrome for artworking the diagrams.

Special thanks are given to all those involved as part of the case studies and to the global cohort of social practice arts and placemaking sectors that have supported and welcomed this research, as well as the Routledge team.

Explanation of terms

As participatory art is problematised in the text it is insincere to perpetuate in this discourse the use of the term 'participant' in all instances. Where the term is used, it is in direct reference to participatory art thinking or practice; when an all-encompassing term for those involved in a social practice placemaking or other intervention is required, generally, those 'being part of something'; or when directly quoted from academic texts or interviewees. With the latter, in the conversational vernacular language of a project, artists would commonly (though not always) refer to all non-artists as 'participants', regardless of actual degree of participation. As appropriate, the terms 'collaborator', 'co-producer' or 'protagonist' (O'Neill, 2014, p.201) or co-creator will be used otherwise.

In the art sector vernacular for arts in the social practice realm, it is predominantly *social practice art* in the UK and *socially engaged art* in the US; the meaning in the field has come to mean one and the same common denominator of activity (Bishop, 2011) and this text is not explicitly concerned with teasing out the debates in nuances of this terminology. The term *social practice art(s)* will be used throughout this text. While social practice arts practice and process can, and are here, differentiated, inherent in the art form is the collapse of the two, where the process becomes the practice, an 'approach not an output' (Hoskins, 1999, p.287) – which can lead to an (understandable) conflation of the terms in both theory and in practice. Froggett et al. (2011, pp.6–8) determine a total of eleven dimensions of social practice arts, which are of both practice and process. Those relating to practice are: experimentation and diversity; aesthetic of engagement; philosophy, civic mission and politics; innovation and ethical practice; authoring and participation; and transformative practice. Those relating to process are: modes of engagement; personalisation; local and the global; intensity and duration; and partnerships and collaboration. This text adopts a similar categorisation pertaining to data findings. Any demarcation however can only be illustrative as practice-process is fluid and operational through an affective spacetime (Munn, 1996, in Low, 2014, p.20).

American English spellings will be used throughout the text where presented as direct quotes from text and field data.

The term 'Indy', for Indianapolis, will also be found in this text in direct quotes, as the vernacular for the city name.

Bibliography

Bishop, C. (2011) 'Participation and spectacle – where are we know', from Living As Form lecture series, Creative Time 18 May 2011 [Online]. Available at: http://creativetime.org/programs/archive/2011/livingasform/talks.htm (Accessed: 13th February 2016).

Froggett, L., Little, R., Roy, A. and Whitaker, L. (2011) New Model of Visual Arts Organisations and Social Engagement University of Central Lancashire Psychosocial Research Unit [Online]. Available at: http://clok.uclan.ac.uk/3024/1/WzW-NMI_Report%5B1%5D.pdf (Accessed: 12th August 2015).

Hoskins, M. W. (1999) 'Opening the door for people's participation' in White, S. (ed.) *The art of facilitating participation: releasing the power of grassroots communication*. New Delhi: Sage Publications.

Low, S. (2014) 'Spatializing culture: an engaged anthropological approach to space and place' in Gieseking, J. and Mangold, W. (eds.) *The people, place, and space reader*. New York: Routledge.

O'Neill, P. (2014) 'The curatorial constellation – durational public art, cohabitattion time and attentiveness' in Quick, C., Speight, E. and van Noord, G. (eds.) *Subplots to a city: ten years of In Certain Places*. Preston: In Certain Places.

List of URLs

1 Project Row Houses: http://projectrowhouses.org/
2 Open Engagement: http://openengagement.info/
3 Art Tunnel Smithfield: http://arttunnelsmithfield.com/
4 The Drawing Shed: www.thedrawingshed.org/
5 Big Car: www.bigcar.org/

1 Introduction
Social practice placemaking in a global context

Cities are the new black (Legge, 2013, p.7)

1.1 The context of the research project

The purpose of the research project was to understand and reconceptualise the practice and process of placemaking through a participative and social practice arts lens, as *social practice placemaking*, a term created and introduced in the research project, and to explore its production of agency for protagonists in relation to place. The research project worked to move placemakings' axiomatic abstracted appreciation of the arts in its practice to a granular and nuanced understanding of differences between and through practices. Social practice placemaking is informed by performative social practice arts and operates as an extension of this artform enacted in the urban built environment with an intentional material outcome, which may be absent from social practice arts practices.

Contemporary everyday urban life plays out at the intersection of multiple forms of conviviality (Bourriaud, 1998/2006, p.165), mundane interactions, lifestyles, belonging and placemaking (Connerton, 1989, in Sen and Silverman, 2014, p.4; Kester, 2004, pp.77–8; Low, 2014, in Gieseking and Mangold, 2014, p.35; Tonkiss, 2013, p.10). As located in place, social practice placemaking acts as a process of place conscientisation for the protagonists, who may become (more) civically minded and active. Cities offer dynamic social and cultural experiences and chances of interactions with diverse groups across social, creative, intellectual and political milieus. City-making is a social process, a relationship between social and physical shaping of cities, between how people use, create and live in social spaces, and the formal and informal material and embodied production of urban environments (Tonkiss, 2013, p.1). Cities also take a majority and growing share of global population [A1.1] with rapidly increasing urban populations which is consequently impacting on urban form and lived experience, which is 'central to the study of human settlements and social arrangements' (ibid.). This research project was thus concerned with how social practice placemaking contributes to the strategies urban dwellers use to understand and enact their lived experience, often through informal design processes, with an aim to inform intra-city communities' dialogue i.e. the conversation between citizens at the grassroots and with those in decision-making positions.

Challenges posed to urban thinking/theory development come from its 'loose disciplinary fit' across social science, architecture, urban design, planning, engineering and environmental science (Tonkiss, 2013, p.3) – and this text includes the category of art in this listing too. Diverse arts-based placemaking practices at this time are grouped under the term *creative placemaking* (Markusen and Gadwa, 2010 [a, b]); this text states that the popular use of this term renders its meaning unsatisfactory in describing a proportion of arts-based placemaking and delineates between practices in a placemaking typology [3.7]. In the context of placemaking, *any* explicitly arts-based approach to this process is commonly termed creative placemaking. Though useful in particular regards, the definition of creative placemaking is inadequate, too broad and fiscally concerned to describe the observed practice of social practice placemaking. The research project was a reaction to research gaps in the realm of *emplaced arts* [2.4] and its modes and mechanisms of participation; the practice and process of arts in placemaking; and the processes and outcomes of such activity in relation to place attachment and citizenship. This text problematises this nomenclature and practice and offers an alternative and deeper understanding through the contextual review and case studies of difference types of arts-*led*, not arts-*based*, placemaking. Types of placemaking are presented in a placemaking typology [3.7.1], adding to a growing body of critical discourse on contemporary placemaking activities (Kester, 2004; McKeown, 2015).

As such, this text has an intended outcome of addressing the gap in knowledge in the placemaking sector pertaining to an understanding of the practice and process of a social practice arts-informed placemaking – social practice placemaking – which would notionally be otherwise termed creative placemaking, and in doing so, the text delineates between different types of placemakings, along arts and participation lines. The text further intends to bring placemaking knowledge into the social practice arts and wider arts sector, to galvanise the social practice arts field of knowledge of working in place and with material processes and outcomes, to inform the wider arts sector of this ever-growing practice and to extend its knowledge of arts in place and in the public realm. The text also intends in its outcomes to communicate to decision makers in city administrations, planners and funders the detail of the practice of social practice placemaking to which they may not be aware, to aid dialogue between parties and to help inform policy making and grant giving. In this latter regard, where the micro of the social practice placemaking project meets the meso of city administrations and funders and the macro of policy, the text also aims to contribute to debates on the role of art in place relating to an instrumentalised use of the arts in the public realm or as social work (Bishop, 2006; Froggett et al., 2011; Hamblen, 2014; Jackson, 2011; Miles, 1997) [2.5.5] and its position vis-à-vis a neoliberal administration (Brown, 2014; McAllister, 2014; Sennett, 2012; Zukin, 1995), what this text terms *the neoliberal dilemma* [3.8.1] of emplaced arts and the neoliberal rhetoric of social inclusion and the co-opting of emplaced arts to the ends of market forces and administrative institutions (Kwon, 2004). It also aims to address the gaps in knowledge pertaining to 'a host of overlapping and poorly defined terms' (Carmona et al., 2008, p.4) used in the arts and placemaking sectors by both interrogating those terms and presenting a placemaking typology.

By mapping over and then extrapolating from the research gaps, the research question and its three aims were formed. The central question was 'How are urban places made and remade through performative arts practices?' The aims were threefold: firstly, to examine the practice and process of performative arts–informed placemaking and its effect on the emplaced arts experience; secondly, to investigate what existing social practice arts and place thinking can contribute to performative arts–informed placemaking, and this artform as a means of reinterpreting the urban public realm; and thirdly, to explore the role of emplaced performative arts practice in shaping social cohesion, arts and civic participation and citizenship.

1.2 Theoretical framework

The research endeavour situated itself in an ambiguous regard to both *adverse* and *positive urban experience*: it contends that cities can be psychologically fragmented, anonymous and cohesive, emancipatory and communal [1.4.3]. It is framed in the re-imagining of the city that is provoking a reconsideration of expansion of categories of inclusion of city actors and urban form and design (Marcuse, 2012) and to place spotlight on the range of actors involved in placemaking (Tonkiss, 2013, p.10) from both professional and non-professional categories. The text further places itself in, and focuses on, the social practice turn in the arts, social sciences and humanities, 'which puts social practice arts and place at the center of analysis of culture and history' (Sen and Silverman, 2014, p.2). It further views the culturisation of cities as a strategy and tactic of urban development and concurrent in the increase in the symbolic economy of art (Zukin, 2010, p.xiii) and positions this vis-à-vis the urban lived experience as a motivation to reappropriate space and place by urban dwellers into their own making.

Special focus should be given here to the conceptual framework position with regards to the category of community. Firstly, the conceptual framework used is in agreement with Bourdieu (1986) that an individual's presence in a place changes the physicality of it and that people live within their own mini-city habitus (Lee, 1997) and contends that the social life of urban form – the urban lived experience – pertains to 'how cities are structured as social environments around and through social relations, practices and divisions' (Tonkiss, 2013, p.16). Secondly, the text understands and uses the term *neighbourhood* as a social concept and – where used – *community* as a functional concept (Nicholson, 1996, p.116). The neighbourhood and the community may be synonymous, operating 'a multidimensional spectrum from a pure network model to a pure neighbourhood model' (ibid., p.118) but the terms as used here will not transcend one another (Agnew and Duncan, 1989). It further aids the research's conceptualisation of community as one that can both deny and repress difference and be excluding and not unified (Young, 1990, in Gieseking and Mangold, 2014, p.247). Amin's (2008) concept of the *micropublic* presents a community of people, emplaced or not, as a diverse collection of individuals with shared interests and with a need to galvanise collectively in pursuit of mutual goals, closer to a network model based on social, work or ethnic links for example (Nicholson, 1996, p.113). But, as Nicholson also signals, 'everyone lives somewhere' and that place has to be negotiated and managed – which is part of a social turn (ibid.) in the arts and urban design that the research project is concerned with. The micropublic that is focused on in the research project will galvanise around a place, formed thus of interest and geographic affiliations. This understanding also encompasses a generalised 'sort of belonging' (ibid., p.138) that is commonly associated with a place and the informal relations within it, a tipping point into concepts of place attachment that underpin the research. Together, these positions have been taken to guide the examination of the key issues identified to further the understanding of social practice arts and social practice placemaking that will address the research question and aims, and thus address the stated research gaps.

1.3 Research project design and methods

This research project conjoined in transdisciplinary study arts theory and urban and placemaking theory, with a reference to the thinking of place attachment, to offer a critical discourse on arts and placemaking. It did this through empirical and qualitative case study research, interrogating social practice placemaking projects as a placemaking practice informed by social practice arts, signalling to the placemaking sector how arts practice can be understood in place as having a rich heritage of practice and a vital contemporary practice. It did this by examining three case studies of projects working in this arena, and collecting data from participant observation and interviews with artist and non-artist protagonists in the project and other stakeholders as identified in the field as having a relation with the project.

Case studies had to involve an arts-led practice and process and be situated in an urban neighbourhood or community setting. It was expected at the outset of the research project that the case study portfolio would include both emergent and longstanding projects which would offer a comparison of the formation, impact and sustainability of such work. Whilst the issue(s) each project addressed would be contextual to its time and place, at the heart of social practice placemaking is the concern with the urban built environment and its social and cultural experience and the potential for participation to strengthen or create a sense of community (to any or what degree would be interrogated in the research process); case studies thus had to offer scope for such investigation whether community building was a stated aim or not. The final selected case studies were: Art Tunnel Smithfield, Dublin, Ireland [n.1]; Big Car, Indianapolis, Indiana, USA [n.2]; and The Drawing Shed, Walthamstow and Wandsworth, London, UK [n.3]. Qualitative and comparative case study methodology, using interviews and participant observation as its main methods (Andrews and Drass, 2016; Gillman, 2002) was chosen for the research project. This offered the opportunity to gain a holistic phenomenological view of social practice placemaking by the study of projects in context through an inductive process (DeWalt and DeWalt, 2002, in Kawulich, 2005, p.4; Yin, 2003, 2009). These methodologies involve explanation building based on participant observation, and allow the simultaneous collection of data and analysis; the pursuit of emergent themes from a simultaneous process; the inductive construction of data categories that explain social processes as observed; and the integration of these categories into a theoretical framework (Yin, ibid.).

1.4 Definition of the field of investigation

The following section of this chapter will locate the research project within various fields of theory, relating to the city, space and place, the lived experience of the city and of place attachment. This is presented here by way of an introduction to the wider field of study and the theories applied in further depth and extrapolated from in subsequent chapters.

1.4.1 Locating in the city

Following urban studies from the turn of the twentieth century (amongst others, Mead, 1934; Park, 1925; Simmel, 1903; Wirth, 1930) this text takes the position that as much as cities can be defined by legal, territorial, political and economic terms, they are also social, cultural and environmental entities operating as a whole and complex urban ecology, based on individual and collective meaning-making and memory (Sepe, 2013, p.8) operating through extended networks and managed by diverse actors (Tonkiss, 2013., pp.4–14). The category of the city is thus expansive (Hubbard, 2006, p.1). Both media and academic publications have made much of the trend of urbanisation and city population increase and the megacity [A1.1] (Goldbard, 2011; Haase et al., 2010; Legge, 2012, p.5; Rennie Short, 2006, p.1): where city size intersects with this research project is in how urban dwellers meet their functional social, cultural and material functions at a localised scale in the city and with accompanying interest in how place is made and who is making it and the use of aesthetic power in the urban discourse (Zukin, 1995, p.7). The term *liveability* (Legge, 2012, p.5; Project for Public Spaces with UN-HABITAT, 2012, p.1; Sepe, 2013, p.xiv) is used often, and often approximately, to denote the 'lively city' that precipitates functions over and above necessary quotidian activities to optional and spontaneous social and collective activities (Gehl, 1996).

A common critique of such approaches is that they are symptomatic of a 'culturisation' (Zukin, 2010, p.3) of city life and form, which has created a homogenised public culture of identity and citizenship (ibid., 1995, p.264) and privatised and alienated (Hooper and Boyle, 2008, p.267) public space that 'transcends', 'crowds out' and 'destroys' (to paraphrase) localised urban cultures (Sepe, 2013, p.xiv) – the destruction of the 'urban terroir' as Zukin (2010, p.xi) terms it, where neoliberal urban economics brands neighbourhood identity in a process of mutually destructive gentrification. In the neoliberal culturised city, spaces are 'produced for us rather than by us' (Mitchell, 2003, p.18); any sense of frustration with this is compounded by there being 'too much process and not enough doing' (Lydon and Garcia, 2015, p.83) in planning and urban design and a malaise with regards to political administrations (ibid., p.63). Thus, the site and context of urban struggle is over space, what space means, whom it is for and how it functions, a 'struggle over place making [sic]' (Lepofsky and Fraser, 2003, p.129). In this, it is concerned with the Lefebvrian (1984) *right to the city* political discourse that articulates the human, common and collective right of the citizenry to make change to the city (Harvey, 2008).

This text draws on this thinking by focusing on its aspect of arts in place and individual and collective meaning-making, concerned as it is with an 'authentic' 'cultural phase' in which people are re-determining space use on their own terms and a 'spatial identity phase' in which public spaces are interconnected in new meaning-making networks (Sepe, 2013, p.81). It does this through the lens of social practice placemaking as an emplaced artform [2.4] which operates as a cultural, dialogic and embodied production of people and place. Contemporary urban design, just as urban political discourse, is asking who is, can and should be designing the urban form and the nature of the dominant designs and their consequences. This questioning is increasingly located at the hyperlocal, community level and enacted on a neighbourhood basis. As Silberberg (2013, p.7) states, 'The contemporary challenge to placemakers is to address the pressing needs of our cities in a way that transcends physical place and empowers communities to address these challenges on an ongoing basis'. The research project thus took place in, and was observing, social practice placemaking practice, process and outcomes in the contemporary, culturised city, where arts and non-arts actors were questioning the function of space and becoming active in the creation of space, acting as a force contra to processes of culturisation and in turn, becoming citizen placemakers. The following section addresses this concern.

1.4.2 Locating in space

With the emplaced arts of the research located most commonly in *countersites* (Holsten, 1998, p.54, in Watson, 2006, p.170), this section will focus on theories pertaining to these liminal and often informal (Tonkiss, 2013, p.1) urban spaces as a multidimensional process of urbanisation. Countersites is a useful term to utilise in the text as it pertains to notions of the liminal site, where 'people, symbols and objects are encountered outside cultural frames of reference and normal instrumental relations' (Stevens, 2007, p.74; Whybrow, 2011) and an informal meanwhile use of space (Thompson, 1984, in Sime, 1986, p.54). Such non-traditional spaces have a socialisation function (Carmona et al., 2008, p.14) which in the polylogic of people, site and process in the urban realm, acts as the parasitical third which marks and disrupts the subject/object position in the urban realm (Serres, 1980/2007).

Prefacing the subsequent section and to further locate the text in an alignment with ambiguous urban thinking, an affective dimension to the production of space/place though *spacetimes* (McCormack, 2013, p.2) is now introduced, the 'symbolic nexus of relations produced out of interactions between bodily actors and terrestrial spaces' (Munn, 1996, in Low, 2014, p.20) as a socialised and contextualised reality as 'perceived, conceived and lived' (paraphrasing Schmid, 2014, p.31). Spacetimes operate in the Lefebvrian 'generative relation' of affective spaces (McCormack, 2013, p.2), produced through a process of assemblage and experienced through a variety of sensory registers with varying scope and intensity. As an embodied participation in affective spaces, the generative relation 'demands particular attention to the affective qualities of these spaces combined with a commitment to experimenting with different ways of becoming attuned to these qualities' (ibid., p.3). As affective spaces, spacetimes are relational, between bodies and artefacts; processual, of techniques of attention, participation and involvement; and nonrepresentational, the affectual presence is not cognitive for it to be manifestly felt (ibid., p.4). Spacetimes are composed of demarcating 'refrains', 'generating a certain expressive consistency through the repetitions of practices, techniques, and habits' (ibid., p.7). Guattari (2006) terms the refrains as 'existential territories', 'kinaesthetic, conceptual, "material", and gestural', attributes associated also with emplaced arts.

Vacant space is that land or built environment that goes un- or under-used in the city. It functions as 'diverted space', that, from its conceptual indeterminacy is 'vague' (based on Lefebvre, Lilliendahl Larsen, 2014, p.319). While nominally associated with the 'discursively weak', vacant space holds the potential for new productions of space emanating from its 'transducive affinity' with the urban lived experience and its signifying of 'new spatial codes' (ibid., p.336) of perceived, conceived and lived (paraphrasing Schmid, 2014, p.31) spacetimes, a mutually productive process that begets further new sociospatial and political practices and lived and conceived representations. Thus, when considering vacant space, one must adopt a networked local-scale perspective to contextualise the area in question within the wider physical and socioeconomic spatial systems to which it belongs (Foo et al., 2014, p.175). Vacant space then is key to urban morphologies (Sennett, 2007) that when activated in liminal urban spaces of political vacuum or disempowerment, operates as countersite (Holsten, 1998, p.54, in Watson, 2006, p.170), as a 'critical instrument of social and environmental justice, empowering marginalised and disadvantaged communities and neighborhoods' (Moyersoen and Swyngedouw, 2013, p.149).

As a situated urbanity, 'place always involves "appropriation and transformation of space and nature that is inseparable from the reproduction and transformation of society in time and space"' (Pred, 1986, in Low, 2014, p.21). The re-appropriation of space acts as a signifier to, and of, new ways of producing space (Lilliendahl Larsen, 2014, p.336), and is a personalisation of space by the individual or group (Sime, 1986, p.60) that self-defines and self-territorialises through socio-spatial production (Friedmann, 2010, p.154). As a transition from vagueness to form, the re-appropriation of space is a boundary position re-presentation of that space by the public, 'both shaping public space for social interaction and constructing a visual representation of the city' (Zukin, 1995, p.24).

The cultural production anew in the re-appropriation of vacant space is an informal *critical spatial practice* (Rendell, 2006; Tonkiss, 2013, p.107) that breaks the signal-noise of traditional approaches to space (Németh and Langhorst, 2014, p.148). As a critical spatial practice, informality is contextually performative (Whybrow, 2011, p.35; Yoon, 2009), both a site and object of experiencing as well as the means to start a process of embodied reflection and tactical response, the experience constituted too by what the viewer brings to it (Rendell, 2006, p.52; Lippard, 1997, p.267). The transformation from critical spatial practice encompasses physical, psychological, cultural – and more – factors (Lerner, 2014, p.xiii). Critical spatial practice is dialogic and has a civic potential (Lilliendahl Larsen, 2014, p.326) and is 'city-making as an ordinary practice', that embodies a pluralist urban design in the interrelation between urban form and human aim and objectives (Tonkiss, 2013, p.91), a '"public urbanity", combining the insights of radical democratic theory and the perceptions of new urban cultures' (Lilliendahl Larsen, 2014, p.329). Part of the city symbolic economy, critical spatial practice incorporates new visual representations into the city by integrating social and ethnic groups and forming new group identities (Zukin, 1995, p.20) – as seen in Amin's micropublic (2008). Informality challenges the increased/increasing regulation of space (Minton, 2009; Tonkiss, 2013) with 'loose space' (Franck and Stevens, 2007, p.272; Tonkiss, 2013, p.108) and that draws on, knowingly or not, Sennett's (1970) 'uses of disorder', the social value of urban freedom afforded by encounter and informal spaces. This ordinary practice or *everyday urbanism* (Roy, 2005, p.148) has 'provided a platform' for non-commissioned design professionals to enact temporary or improvised space productions (Tonkiss, 2013, p.108), 'eliciting an aesthetic response' (Lynch, 1981, p.21) as both an ongoing or ephemeral phenomenon, which surfaces new urban forms constantly (Schmid, 2014, pp.2–3). An informal critical spatial practice conceives performative primary agency in the city as 'the everyday inhabitants, who make and remake the city in ways of their choosing but not always of their choice in the operational context' (Tonkiss, 2013, p.10), harnessing creative energies that are outside of formal urban design and fuse concepts of the expert/amateur and formal/informal (ibid.). This mode also demands that urban design experts work in co-production in the public realm, which demands the same of public space:

> Urban professionals should learn to listen to inhabitants' place-based knowledges and create the possibilities of place-based forms of stewardship by working with residents and artists before they begin conceptualising, designing, and implementing planning and development projects. Why? Residents know the ways that places are made, the mazeways that sustain them, and the ways places and peoples are connected through various networks (Till, 2014, p.168).

This co-production in urban design – as will be shown below with arts participation and placemakings also – is a 'dialogic space of exchange and social relation that strengthens as active citizenship' (Schneekloth and Shibley, 2000, p.138), and creates a co-produced body of *relative expertism* [2.5.1]. As a public urbanity (Lilliendahl Larsen, 2014, p.329), paradoxically, even though often ephemeral, interventions have a long-lasting affect and effect on the sociocultural aspects of the city (Klanten et al., 2012, p.9). Whilst operating often from the marginal countersite informal critical spatial practice is a 'core means of ordering urban processes at quite different scales of income and urban power' (Tonkiss, 2013, p.111) and it is affecting policy and formal urban planning. The modelling of the temporary use of vacant space offers 'a rich and diverse territory within which to accommodate testing of a wide range of uses and processes and their effects' (Németh and Langhorst, 2014, p.149) and is advocated by the tactical urbanism 'movement' as a key advantage. Such use offers the opportunity to test and pilot urban realm design solutions in a 'continuous editing' process rather than a single transformation event (ibid.). Informal temporary use of vacant spaces can also mediate to a degree against trends of neglect and social and economic devaluation. The process of this will also construct models to include a range of civic stakeholders (Foo et al., 2014, p.181), including citizens as co-author with city administration officials. This adds a fourth dimension to the normative three-dimensional urban planning strategy that works through the generative relation (McCormack, 2013, p.2) and encourages generative urban design (Németh and Langhorst, 2014, p.149). As informality moves processually towards the formal, it can influence policy and become legalised (Tonkiss, 2013, p.107). Thus, critical spatial practice can empower residents in the urban development process (Németh and Langhorst, 2014, p.149), vacant space acting as a bridging tertius (Moyersoen and Swyngedouw, 2013, p.149) between notionally divergent interest groups. Here critical spatial practice functions as a tactical rupture (de Certeau, 1984; Jackson, 2011, p.52) and as a catalysing interactional and informational thirdspace – the Serres (1980/2007) parasite. This can be positive or negative and can manifest as a 'glocal empowerment' bridging the inter-scalar gap between the local everyday life spaces and extra-local cultural political and economic processes (Swyngedouw and Kaïka, 2003).

This section has introduced the concepts of spacetimes and countersites as holding agentive potential in the urban realm and specifically focused on the productive potential of liminal, vacant and re-appropriated urban space. This has been located in urban space and place theory as a critical spatial practice, its informality being its operative aesthetic, which is shared with social practice arts in place. An individual's experience of the city is a motivating factor to participate (or not) in social practice placemaking and the following section addresses this as an ambiguous condition.

1.4.3 Experiencing the city ambiguously

Urban theory commonly positions the urban lived experience as a dualism between negative and positive experiences, a binary though that is neither useful nor accurate in understanding the urban dweller. Through a negative lens, increasing city size and density is an entropic experience (Friedmann, 2010, p.150; Krupat, 1985, p.51), where an individual cannot fathom the concept of the city as a whole, nor its diverse inhabitants (Wirth, in Lin and Mele, 2005, p.37). In this city, emotional life is fragmented, depersonalised and intensified

(Gotteliener and Hutchinson, 2006, p.49; Rennie Short, 2006, p.1; Simmel, 1903/1950, and Wirth, 1930, in Lin and Mele, 2005; Lin and Mele, 2005, p.23–4), which dislocates the individual from their core personality (Bowlby, 1979/2010; Krupat, 1985, p.52; Simmel, 1903, in Gieseking and Mangold, 2014, p.223). This serves to create 'psychosocial distancing' (Hubbard, 2006, p.17), 'civil inattention' (Lofland, 1998, in Hubbard, 2006, p.17), a 'blasé outlook' (Sennett, 2012, p.188; Simmel, 1903/1950, in Gieseking and Mangold, 2014, p.224) and 'reserve' (Simmel 1903/1950, in Lin and Mele, 2005, p.27) where gaze is averted to avoid intimacy (Sennett, 2012, p.8). Here the individual experiences a *mixophobia*, a forgetting of the art of interacting with the other and approaching it with apprehension (Bauman, 2003, in Watson, 2006, p.168) and a consequent disempowering of the body 'as a site of material influences on the city' (Sen and Silverman, 2014, p.6). The outcome of this is an increase in superficial, impersonal and transitory secondary relations (Gotteliener and Hutchinson, 2006; Hubbard, 2006, p.18; Lin and Mele, 2005, p.23–4; Wirth, in Hubbard, 2006, p.21) at the expense of primary relations and defended and closed spatialand social proximity (Krupat, 1985, p.57; Milgram, 1970; Uitermark et al., 2012, p.2546; Valentine, 2014, p.79).

Through a positive lens, cities are experienced as fecund and serendipitous sites of heterogeneous inter-cultural sociality (Sennett, 2012, p.38; Soja, 1997, p.20; Watson, 2006, p.6) which gives potential for cultural, creative and intellectual cross-fertilisation of ideas and where the city is a stage for diverse subcultural groups and 'meaningful social worlds' (Fischer, 1975; Key, in Krupat, 1985, pp.133–4), group identity constructions, interpersonal knowledge exchange (Sassen, 2005; Young, 1990, in Gieseking and Mangold, 2014, p.247–50) and participative relational performances (Anderson and Nielson, 2009, p.305–6). This city embraces two meta-psychological positions: a dynamic one, of psychological forces involved in its phenomenon, and an adaptive one of an inter-relationship of the phenomenon to the environment (Rapaport and Gill, in Bowlby, 1979/2010). This is a volte-face of mixophobia to *mixophilia*: the increase of city scale and density brings diverse groups into contact with one another producing positive affirmations of subcultural identity and mutual and positive cultural exchange (Fischer, 1975; Lefebvre, 1970/2003, p.96; Mitchell, 2003; Watson, 2006, p.6). Here the individual has agency to make myriad personality choices according to their personal preferences and affiliate to groups accordingly (Anderson and Neilson, 2009, p.318; Krupat, 1985, p.53), the variety of people and circumstances presented in the city constructing the individual as versatile, flexible and adaptable (Anderson and Nielson, 2009; Proshansky 1978, in Krupat, 1985, p.54) and giving a space to form their true identity (McAllister, 2014, p.189). An affinity does not necessarily collapse into homogeneity but both disrupts and affirms diversity (Fischer, 1975, p.1319; Iveson, 1998, in Gieseking and Mangold, 2014, p.189) and, enacted in space, engenders a familiarisation through residence and the manifesting of public space as a 'home territory' (Krupat, 1985, p.61). This chimes with the much-contested theory and practice of *new urbanism*, 'a mixed-use method of city design that makes neighbourhoods more self-sufficient and friendly to social interaction' (Brown, 2014, p.176), moving the positive experience of encounter into application through design.

Introduction 11

However, this text contends that to think of the city as being composed solely of fragmented social groups is no longer relevant or constructive (Buck et al., 2005, p.57), and similarly, to think of the lived experience of the city as one of a binary is inaccurate. Rather, an ambiguous position may be someone's holistic experience of city life: the city lived experience has the potential for isolation and integration and a variety in quality and quantity of relations, their meaning, function, form, distribution and formation process (Krupat, 1985, pp.128–9). The ambiguity is an ambivalent dialectic interaction, stemming from the 'human desire to be seen and known' in contrast to 'our equally human wish for anonymity and the freedom it confers' (Goldbard, 2011, p.170): as an individual, the city dweller inhabits one of many ambiguous dimensions, being able to act for self as a free agent (Tönnies, from Loomis, 1963/1987, in Lin and Mele, 2005, pp.17–8) as well as for common good, alone and in a group, and inhabit different types of functional interrelations. Group membership precipitates a *communitas*, a larger sense of self known through encounters with others (Turner, in Tuan, 2014, p.106). However, individualised separate identities coupled with an increase in contact relations can also lead to social antagonism (Tönnies, from Loomis, 1963/1887, in Lin and Mele, 2005, p.19; Valentine, 2014, p.78). For Lefebvre (1970/2003, p.133), human interaction in the urban realm can both strengthen and weaken mutual knowledge of difference, and contact between different social groups alone is not sufficient to engender meaningful social worlds with intra-human contact based on a 'naïve assumption' (Valentine, 2014, p.78) of mutual respect. People may respond differently to the same situation too – for example, the 'density-intensity hypothesis' (Freedman, 1975, in Krupat, 1985, p.54) where a negative outlook person may become more isolated living in a city and a positive outlook person would become more active.

To bring this to the concern of placemaking, as a complex multivariate site, the city can be a site of instability and transitoriness (Sepe, 2013, p.xv) where an embodied sense of place is desired by city protagonists (McClay, 2014, p.3) and where marginal or liminal countersites offer significant creative and cultural roles in the urban lived experience as contingent and open-ended (Watson, 2006, pp.171–2). Thus, the ambiguous urban dweller is both shaped by, and shapes, the urban form and lived experience (Tonkiss, 2013, p.19) and this text places its study of social practice placemaking in that process, with the following section focused on this aspect of space and place thinking.

1.4.4 Shaping urban space and place

As discussed, space is a 'complex ecology' (Lynch, 1981, p.119) with attributed material and socially constructed qualities (Sen and Silverman, 2014, pp.2–3), of heterogeneous inter-relations and interactions (Anderson, 2004, in Brown, 2012, p.6; Beyes, 2010, p.231; Massey, 2005). As an embodied experience, the production of space is generative, with an 'immediate relationship between the body and its space ... each living body is space and has its space: it produces itself in space and it also produces that space'; the individual is a 'spatiotemporal unit ... who creates space as a potentiality for social relations, giving it meaning, form, and ultimately through the patterning of everyday movements, produces place and landscape' (Lefebvre, 1984, p.170). Here, meanings are multi-scalar, flexible and in constant development, emerging from socially, politically and economically interconnected interactions among people, institutions and systems, a dialogical process that aims to, or does, produce the physical creation of the material setting' (Low, in Geiseking and Mangold, 2014, p.35). For Lefebvre (1984), space is produced by a triad of interrelated modes: spatial practice, the representation of space, and, key in this text, space as representation, a political and social expressive and reflective dialectic. This is inter-reactive and interdependent, the social roles of production being both space framing and space contingent (Rendell, 2006, p.17; Soja, 1997). Thus space is socially produced, as well as social relations being spatially produced (Lilliendahl Larsen, 2014, pp.329–30; Schmid, 2014, p.31).

Place is not of a particular and *a priori* fixed and local scale (Massey, 2005; Pierce et al., 2011) and phenomenologically, is the 'environmental locus in and through which individual or group actions, experiences, intentions, and meanings, are drawn together spatially'; place is not separate from the people associated with it, but is an indivisible articulation of the lifeworld of those in place (Foo et al., 2014, p.177; Seamon, 2014, p.11–2; Speight, 2014, p.113). The group transforms the space which it is in, and the space changes them (Halbwachs, 1992, p.54, in Sepe, 2013, p.7). Everyday experience of place differentiates it from other places, forming a 'habit-memory' (Connerton, 1989, in Sen and Silverman, 2014, p.4). To come into 'being', place incorporates generative processes into nodes of social and creative encounter (McClay and McAllister, 2014, p.8), a 'structural imperative', which engenders an 'interiority', the place then becoming 'cherished' by those that inhabit it (Friedmann, 2010, pp.154–6). Thus place becomes a cultural entity that is constitutive of 'identity, memory, language, material culture, and symbolic and affective message' (Magnaghi, 2005, p.37, in Sepe, 2013, p.6). 'Places are termed "places" and not just "spaces" when they are endowed with identity' (Hague and Jenkins, 2005, in Sepe, 2013, p.xiii).

Placemaking literature makes (often unqualified) use of the term *sense of place* as 'an overarching concept encompassing a cognitive (place identity), an affective (place attachment), and a behavioural dimension (place dependence)' (Carrus et al., 2014, p.155). Sense of place includes the 'imageability' (Lynch, 1960, pp.9–10) of place, based on a visual recognition of the structure of a place, and the memory and meaning of it – the material form and the emotional and psychological layering on it – its *genius loci* (Norberg-Schulz, 1989, pp.13–4, in Sepe, 2013, p.4), which will change when the place is physically or materially changed (Sen and Silverman, 2014, pp.1–2). Whilst place has an undoubted physical manifestation, Lefebvre (1984) changed the focus from urban form to urban process, with place comprising a social and imagined element of fluid meanings (Friedmann, 2010; Hou and Rios, 2003; Mihaylov and Perkins, 2014; Relph, 1976, in Sime, 1986; Sen and Silverman, 2014; Sepe, 2013). This text sees a similar value to urban emplaced arts and views the social practice placemaking process and intervention as a 'spatio-temporal event' that locates place in the social and cultural context, over and above the built environment manifestation of space (Massey, 2005, pp.130–1; Sen and Silverman, 2014, p.4). Thus place is of three dimensions: built and natural material form; patterns of social activities; and sets of personal and shared meanings (Relph 1976, in Sime, 1986, p.55). The latter of these dimensions can also be divisible into three further dimensions: the person dimension, of individually or collectively determined meanings; the psychological dimension, of affective, cognitive, and behavioural components of attachment; and the place dimension, of the characteristics of attachment, including spatial level, specificity, and the prominence of social and/or physical elements (Scannell and Gifford, 2010, p.1).

Of note to the research project, and linked to both adverse and positive urban experiences, some sense a feeling of loss at the citizen level over any say in the production of place (Tohme, in Doherty et al., 2015, p.122). However, key to the intellectual framing of this text is the assertion that placing art into the Lefebvrian triad of space and representation activates its agency as a relational practice. It is of networks, as a place of exchange and interaction; of manifold political, social and cultural borders; and of denoting difference in lived experiences (ibid., p.38). The 'creative' in urban planning however is a 'fuzzy' 'buzzword' (Lilliendahl Larsen, 2014, p.330), one that is used systemically through concepts of vitality and vibrancy to articulate how arts and culture change the qualities of place, such as with the Vitality Indices in the US [n.4] and UK [n.5] (Gilmore, 2014, p.20). In the culturised city, culture and 'the arts' become part of the city's symbolic and fiscal economy. Creativity in the city though is also a site of resistance to culturisation, a 'call and response among different social groups' (Zukin, 1995, p.264) to find, create and maintain sites of different cultural value through new city visualisations (Stern, in Lowe and Stern, in Finkelpearl, 2013, p.146) – which will be discussed with regards to an intersectional understanding of space.

The shaping of space and place, as generatively produced by an active social body, has a relation to place attachment; one of space and place's operative dimensions will be the degree to which its inhabitants or users feel attached, or not, to the place in question. The arts in this process too may work with existing, or galvanise new feelings of, place attachment, to the end goal, as in the case of social practice placemaking, of an ecological change to the urban form and function. The following section thus turns to place attachment theory.

1.4.5 Introducing place attachment

Place attachment theory was developed by Altman and Low in 1992, in the eponymously entitled book (Manzo and Devine-Wright, 2014), stemming from Bowlby's (1979/2010) attachment theory, a behavioural theory of the interaction between the individual and social systems of which they are a part (Berzoff, 2011, p.222; Marris, 1991, p.77). In broad terms, place attachment can be defined as 'the emotional bonds between people and a particular place or environment' (Seamon, 2014, p.12), a multifaceted and complex phenomenon that incorporates different aspects of people-place bonding and involves the interplay of affect and emotions, knowledge and beliefs, and behaviours and actions in reference to a place (Carrus et al., 2014; Low, 1992, in Gieseking and Mangold, 2014, p.73; Rollero and De Piccoli, 2010, p.198; Seamon, 2014, p.16). It is a social process and has a community dimension rooted in the individual human desire to feel an attachment to a group (Cozolino, 2006, p.11) through symbolic and shared meanings of place (Scannell and Gifford, 2010), a mechanism which links the political and personal; is a nexus between sociological and psychological understanding; and is a vision of democratic social co-operation, helping the individual find security in their lives through the fostering of close emotional bonds (Goldberg, 2000; Holmes, 1993, p.200). Holmes (ibid., p.202) asserts that social co-operation is dependent on a population that, through a positive emotional foundation, has learnt to love and trust, and that has leaders that are prepared to listen and show that the opinions of the public are valued and respected and that the secure attachment disposition of a public will affect the broad cultural and economic conditions of a society (ibid., p.204).

Bowlby's (1979/2010, p.137) attributes of, and shown in, an 'attached person' – qualities of self-reliance, autonomy, mutual trust/need relationships – are akin to those qualities attributed to a community with high social capital. As a phenomenological experience, place attachment encompasses holistic, dialectic and generative place perspectives (Seamon, 2014, p.12) and 'three senses of "insideness", expressing different aspects of his respondents' affinity with their surroundings': a physical insideness of an embodied knowledge and awareness of place; a social insideness of community and social connection; and autobiographic insideness, an 'idiosyncratic sense of rootedness' (Dixon and Durrheim, 2000, p.32). Place attachment has a behavioural enactment aspect and theory holds a 'general agreement on the idea that it represents a set of positive affective bonds or associations between individuals, groups, communities, and their daily life settings' (Carrus et al., 2014, p.154) – although these can of course also be adverse, as with perspectives on urban experience. The text will use place attachment thinking to provide a textured understanding of people and their relation and bond to place composed of social, emotional, cognitive and behavioural elements, and to provide an explanation for participative motivation to join social practice placemaking projects as well as observed outcomes of social practice placemaking, those of place-care and of civic engagement.

The development of emotional bonds to place is a three-dimensional, person–process–place organising framework, with individually or collectively determined meanings interacting with affective, cognitive, and behavioural components of attachment relating to place identity, self and other and sense of community (Scannell and Gifford, 2010). When there is a positive connection between people and place, it is said to be *topophillic* (Tuan, 2014, p.4). Place attachment operates counter to macro structures of place, operating through symbolic (Scannell and Gifford, 2010, p.26) and psychological processes of orientation and identification (Norberg-Schulz, 1971, in Sime, 1986, p.51). The place attachment process has six components: place interaction, the typical goings-on in a place; place identity, the process whereby people living in place or otherwise associated with a place take up that place as a significant part of their world; place release, an environmental serendipity of unexpected encounters and events, through which people are more 'released' into themselves; place realisation, the palpable presence of place; place creation, human beings being active in place; and place intensification, the independent power of well-crafted policy, design, and fabrication to revive and strengthen place (Seamon, 2014, pp.16–9).

Place is a site of elective belonging and performing identities, *place identity* being a relational cognitive sub-structure of an individual's identity (Hernández et al., 2010, p.281, in Mihaylov and Perkins, 2014, p.66; Proshansky et al., 1958, in Gieseking and Mangold, 2014, p.77). It is the inter-relation of question of *who we are* with *where are we* (Dixon and Durrheim, 2000, p.27; Gieseking and Mangold, 2014, p.73) where place comes to function as an attachment figure (Scannell and Gifford, 2010, p.27), the secure base in a changing world (Manzo, 2005; Proshanky et al., 1983, in Sime, 1986, p.56; Sennett, 2012), underpinning the feeling of unity in an area (Unwin, 1921, in Nicholson, 1996, p.114), akin to Anderson's (1986) territorial specificity, Castells' (1983) spatial meaningfulness and Lynch's (1981) sense of connection. Positive psychological benefits of self-esteem and sense of self are derived from place (Twigger-Ross et al., 2003, in van Hoven and Douma, 2012, p.66), through an interactionist 'body-subject' (Merleau-Ponty, 1962). These may be of the interactionist past (memories) or of interactional potential (future experience anticipation). The degree of meaningfulness of the interactional past is directly related to the degree of current attachment to place (Milligan, 1998, pp.1–2) in a mutually sustaining past-to-future loop (Bulmer, 1969, p.2; Seamon, 2014, p.13). Memory acts as a glue to connect people to place (Giuliani, 2003; Lewicka, 2014, p.51; Rubinstein and Parmelee, 1992; Scannell and Gifford, 2010) with bonds forming with the place itself as well as the people experienced in it (Altman and Low, 1992; Rowles, 1983; Sixsmith, 1986). This is a discursive interaction, placing individual mental processes into the interpersonal space of conversation where place identity is constituted as something that people form, reproduce and modify through conversation as collective practice (Dixon and Durrheim, 2000, p.32).

In common with positive urban experience thinking, individuals do not operate as separate 'social atoms' (Pahl, 1996, p.96) but group activity in place acts as an affective bond for both individual and communal aspects of identity (Brown and Perkins, 1992, p.284, in Mihaylov and Perkins, 2014, p.62; Giddens, 1992; Nowak, 2007, p.98; Wilmott, 1987). It follows that individuals that are place-attached are more likely to engage in social, place-protective (Carrus et al., 2014, p.157; van Hoven and Douma, 2012, p.74), neighbourly and civic behaviour and be more resilient to change, attachedness increasing in intensity with an increase in social activity and stability (Livingston et al., 2008; Shumaker and Taylor, 1983). The qualities required for this behaviour are 'sensitivity, responsiveness, mutual understanding, consistency and ability to negotiate – all of which are found in secure attachment behaviours' (Marris, 1991, in Holmes, 1993, p.205; Cozolino, 2006, p.14) – attributes also ascribed to those of the micropublic. Thus, place attachment has *social cohesion* and *social capital* aspects, acting counter to Durkheim's (1893) *anomie* and an adverse urban experience. Instead, social cohesion acts to maintain moral order, increasing attachment to place (Sennett, 2012, p.257), friendship and other peer-to-peer proximal relations (Ainsworth, 1991, p.38; Gosling, 1996, p.149; Pahl, 1996, p.98; Marris, 1991, p.70). Those that strongly identify with a group will act in its favour, transferring interests from the individual to the collective (Carrus et al., 2014, p.156), engendering an enacted status change from social cohesion to social capital. Social capital is defined as 'bonding ties' of social interactions within place and 'bridging ties', of functional social place-located connections (Mihaylov and Perkins, 2014, pp.69–70) and this secure inhabited position engenders a deeper exploration of citizenship (Holmes, 1993, p.208). This is made visible in community and neighbourhood intra-relations and their galvanisation around common issues or participation in communal events and activities (Ainsworth, 1991, p.43; Carrus et al., 2014, p.155; Friedmann, 2010, p.155) – thus, at the community level, place attachment has a cognitive aspect pertaining to community identity; an affective element pertaining to sense of community; and a behavioural element pertaining to neighbourliness (Manzo and Perkins, 2006, in Mihaylov and Perkins, 2014, p.63).

Place-attached communities are created when people are politically mobilised outside of commercial or state interests, from the grassroots (Gosling, 1996, p.147–9), creating a culturally cross-cutting micropublic. This can only ever affect localised change though: mass mobilisation is not possible this way as it challenges an individual's and community's ability to make sense of a wider citizenship (Gosling, 1996). A community-based place attachment is something that planners need to have an understanding of, for two-fold reasoning. Firstly, to overcome by place design the forces that may fragment the community. Secondly, to offer a new political paradigm based on communal values (Benington, 1996, p.152), and whilst Rekte (2011) states that an aim of the new urbanism walkable city design movement is to create cities also that have a sense of community, this is not automatically consequential. Instead, 'developmental relationships' (Benington, 1996, p.162) are offered as a new model for public-authority interaction, whereby people's positive relation to place is recognised and cultivated and capacity built to give those place constituents voice and authorship – a term in current times that would be understood as new urbanism.

Place identity however is based on what the place is supposed to be like and how self and others are supposed to behave in it (Proshansky et al., 1958, in Geiseking and Mangold, 2014, p.77); any upset in this expectation, a 'mediating change function' (ibid.), engenders a change in relation to place and thus place identity (Relph, 1976, p.55, in Seamon, 2014, p.14). When place identity is threatened it can lead to collective actions or adaptations (Mihaylov and Perkins, 2014, p.71) based on the fear of change in the everyday experience of place (Marris, 1991, p.80). In line with adverse urban experience thinking, a loss of community relationships can lead to feelings of loneliness (Weiss, 1973, in Marris, 1991, p.70), impoverished wellbeing and grief (Twigger-Ross et al., 2003, in van Hoven and Douma, 2012, p.67) and a sense of being othered and excluded from place and community (Rustin, 1996, pp.214–6), an 'emancipation from place' which has psychologically negative consequences (McAllister, 2014, p.191). Gated communities can also manifest an exaggerated sense of inwardness of place, commensurate with strong outwardness and detached feelings to the outside place (Seamon, 2014, p.15). The literature on changes to place and subsequent psychological harm is based on negative change to place assumptions – depopulation, High Street recession for example – and does not consider if 'positive' changes – such as an increase in public services or improvements to housing stock – can have the same psychological impact: one person's duplex living is another's gentrification. As will be seen in the case studies, place attachments are a motivation for people to spend time in places, meeting their neighbours, talking of local matters and fostering ideas for their solutions (Mihaylov and Perkins, 2014, p.61) and will be affected by both general and detailed change in an area (Livingston et al., 2008; Nicholson, 1996, p.118).

1.4.5.1 Place attachment and the arts

Nowak (2007) sees the inclusion of *community art* in city development as the creative act of social capital; the art functions as a further relational glue through which complex social networks can navigate inward–outward opportunity, conflict and institutional interactions, citing community centres and creative programmes as examples of 'workshop[s] of civic engagement', 'serving both as a staging ground for community identity and a source of neighbourhood stability and growth'. Thus, emplaced arts in a process of place-attached social cohesion moves a group from a fragmented group 'serial identity' to a 'group-in-fusion' (Lilliendahl Larsen, 2014, p.326), where group membership precipitates a *communitas*, of a larger sense of self known through encounters with others (Turner, in Tuan, 2014, p.106). This has a political dimension by virtue of being sited in the public realm with the place-attached subject 'capable of insights and taking part in creative action' (Saegert, 2014, in Gieseking and Mangold, 2014, p.397; Carrus et al., 2014, p.156). In this context, emplaced arts are involved in the making of 'authentic places' as a social product of placemaking (Zukin, 2013, p.17). Authenticity in this context is a 'a moral right to the city' to be able to put down roots and a continuous process (ibid., p.6) based on a surety of place, its people and built environment (ibid., 2010, p.6). Thus, authenticity is an expression of fear of change and the desire to counter place-based *anomie* through iterative actions that also connect to wider structural forces (ibid., 2013, p.220). It is said of middle-sized cities in the US – of which case study city Indianapolis is one – that 'We haven't been saddled with the expectations that go with being well known and recognized beyond a local scene.… That may be changing, but it's changing in a good way right now in that its tied to things that are authentic to this place' (Seltzer, in Fletcher and Seltzer, in Finkelpearl, 2013, p.162).

1.5 The demographic context of the case studies

The following section now turns to the location of the case studies and gives the demographic context that the arts projects and organisations were operative in[1].

1.5.1 Dublin and Ireland

Historically, Ireland's population entered decline in 1841 to the 1960s, from 6.5 million to 2.8 million at its lowest in 1961. The years 2002–2006 however saw an 'unparalleled' annual population increase of 80,000 or close to 2 per cent and at the highest variant predictions this rate is set to continue to 2021 (Central Statistics Office [b], pp.6–8) and 2026 (Central Statistics Office [e], p.1). The city saw a population percentage change 2002–2006 of 4.2 per cent and its regional area a change of 7.2 per cent. In 2011 the city had a total population of 1,273,069 (Central Statistics Office [c], pp.9–10), and 97.8 per cent of the population in the aggregate town area (Central Statistics Office [c] p.14). To place this in context, the other four large city and suburban areas of Cork, Limerick, Galway and Waterford had a combined population of 403,083 in 2006, with a 3.8 per cent increase to 418,333 in 2011 (Central Statistics Office [c], p.117). Dublin's 2011 population figure was 527,612 in Dublin City and 1,273,069 in its wider area (Central Statistics Office [d]). The Dublin City area has 208,716 households, the wider Dublin region 468,122 (Central Statistics Office Ireland [a]). Thirty-two per cent of the inner city population is aged 20–34 years, compared to the national average of 24 per cent (Cudden, 2013, p.51). Smithfield – the location of case study project Art Tunnel Smithfield – is located north of the River Liffey/*An Life* – 'the north side' as it is called in the vernacular, an area of the city where 70 per cent of the population of the electoral division are born outside Ireland (Duncan, 2012). Smithfield is located in and around the Arran Quay administrative area of inner city Dublin; this had a total population in 2011 of 15,841, the area experiencing an overall population increase (Cudden, 2013, p.75). Smithfield median property value is €145,521, compared to Dublin's at €265,000 and the rest of Ireland at €175,000; median rental prices are €1,037, €1,018 and €993 respectively (RateMyArea.com).

As vacant space in the Dublin case study arose as a predominant concern from the data [5.3], it is prudent now to turn to literature specific to this. Three hundred vacant sites had been identified by Dublin City Council/*Comhairle Cathrach Bhaile Átha Cliath* (DCC) of an estimated, and thought underestimated, sixty-three hectares (Kearns and Ruimy, 2014, p.66). In the past twenty years, and in marked effect since Ireland's economic crash, public space in Dublin has been privatised, financialised and regulated with independent cultural production commodified, 'Dublin's "great enclosure"' (Bresnihan and Byrne, 2014, pp.1–4). This has seen a push-back in the form of urban commoning of public, in particular vacant, spaces, re-appropriating such spaces to meet collective needs and in the process, redefining forms of ownership, social, cultural and material (re)production and governance (ibid.). A 'the praxis of the commons' emerges from a dissatisfaction with the Dublin urban realm's proscribed cultural production and symbolic economy within it (ibid., p.5). The praxis is a relational social body (ibid., p.12) where the spaces are collectively 'owned' by those that participate in them and financed collectively through crowdfunding and/or non-monetary exchange and circulation, a subversive act 'when we consider how expensive, individualised and disconnected much of the social and cultural activity in the city has become' (ibid., pp.9–10). This emerging Dublin new urbanism (Kearns and Ruimy, 2014, p.48) breaks the 'liveable-city glass ceiling' of a 'bigotry of low expectation' (ibid., p.48) where Dubliners do not believe the city can become a desirable place for people to live and is attempting to address the contemporary Dublin urban difficulty of a successful, liveable inner-city by rendering its response on a cultural and social reimagining (ibid., p.15). These spaces operate independent of, and often do not seek, formal institutions and are led by communities of 'urban pioneers' (ibid., p.98) – read, urban co-creators [3.6.2] – that operate beyond the auspices of public and private management and which have grassroots, DIY ethos. The urban commoning praxis has created a place in Dublin for informal critical spatial practice, integrating people, space and place, materials and knowledges-in-exchange (Bresnihan and Byrne, 2014, p.11) which have produced new ways of 'working, playing, and deciding together, and the production of shared knowledge and resources' (ibid., p7). Open participation in the urban commoning process has been cited as a motivating factor to join and continue in the process in contrast to the boundaried participation of enclosed spaces that alienated them from commodified and normative Dublin culture.

1.5.2 Indianapolis and the USA

In the USA, 80 per cent of Americans live in urbanised areas (Lydon and Garcia, 2015, p.67). Indianapolis has a population estimate of 843,393 for 2014, an estimated increase of 1.7 per cent since previous census. Its population is 86.1 per cent white, compared to 77.4 per cent for the US nationally, with 4.7 per cent 'foreign born', compared to 12.9 per cent nationally. Its median household income is $48,248, with 70 per cent home ownership rate of a median $122,800 property value and 15.3 per cent living below the poverty line (United States Census Bureau State and County QuickFacts). The city is ranked the US's ninth poorest city, with 29.1 per cent of households with income below $25,000, one of the largest national increases in child poverty and income inequality rates (Kennedy, 2015 – who also states the city's population to be 828,841). Indianapolis is seventy-eighth on the Creative Class list in the US, the creative class having a 33 per cent share of the working population (compared to 48.8 per cent of Durham, ranked first), with 43.6 per cent in the service industries (Florida, 2011, pp.404–8) and is in the bottom ten of the 'Openness-to-Experience' ranking, a personality inclination that favours innovation and creativity (ibid., p.250).

1.5.3 London and the UK

London's population stood at 8.3 million in mid-2012, 13 per cent of the total UK population (Office for National Statistics, 2013). The Drawing Shed operated firstly in Walthamstow, in the borough of Waltham Forest, and secondly in Wandsworth. Waltham Forest is London's second most densely populated borough, its population of 265,800 people living in approximately 96,900 households

20 *Introduction*

(Waltham Forest). The ethnic demographic of the area has changed from a white population of 74.4 per cent in 1991 to 50 per cent, with the largest ethnic minority groups being Black Caribbean, Pakistani and Black African. It has the eighth largest Muslim population in England and the fourth largest in London. It is a 'deprived borough', ranked the fifteenth most deprived borough nationally, and London's sixth most deprived, according to the 2010 Index of Multiple Deprivation. Its unemployment rate is 9.1 per cent, below London's average; the proportion of its 'Lower Super Output Areas' in the 30 per cent most deprived in England is 81 per cent (Trust for London [a]). The area's average age of residents is 34.4 years, compared to the UK average of forty years, and average earnings are £24,200, slightly lower than the London average. The average house price in the borough at June 2014 was £333,300, up 28 per cent from 2013 (Waltham Forest Borough Council). Those without educational qualifications stands at 20.8 per cent; 49.9 per cent of its population own or have a mortgage on residential property; and 10.7 per cent live in social housing (I Live Here).

Wandsworth is in London's southwest. Here, the 1960s saw slum clearance and industrial decline and consequent movement of population to other London boroughs and suburbs; the area is subject to regeneration at the time of writing however. The population was 307,000 in 2011; it has a population density of ninety persons per hectare, higher than the Greater London average, though comparatively low for inner London. Seventy-one per cent of residents are white, compared to 60 per cent in London; it has a high proportion of black and minority ethnic (BME) residents compared to the national average although it has the second lowest proportion of BME residents in Inner London; and recent population growth has been attributed to young professional workers, who, whilst bringing wealth into the area, are transient with career progression (Trust for London [b]). The median household income in Wandsworth is above the London median but there are wards in the borough where one in four households earn under £15,000 and approximately 50 per cent of all households with children in Wandsworth are in receipt of Child Tax Credits. Over half the borough population is single or cohabiting, amongst the highest rates in the country (Wandsworth Borough Council).

1.6 The reflexive journey into the research project

The theory of social practice placemaking and this research project originates from the researchers' previous academic, professional and personal observations of a growing number of grassroots-initiated and -located creative projects that themselves originated variously from a desire to improve the immediate local of the project site, from a questioning of the place and/or a belief potential of art in the city context. Guattari (2006) advances that the growing consumption of art in the modern, western world is attributed to the increased uniformity of the urban lived experience and in the course of the researcher's professional practice, they have witnessed a growing number of grassroots, arts-led interventions in the city as a direct response to its lived experience that challenge such uniformity and prescription. In their project management, curation and advocacy roles in the architecture, arts and placemaking sector in the UK and internationally, the researcher was involved in and saw an emerging urban-located movement that was both seemingly precipitous and symptomatic of a new form of urban citizenship that had a citizen-led incremental approach to process of city building and its placemaking that was collaborative and utilised communicative arts interactions in non-traditional art settings.

From this (and from a career forged in the years of a Labour administration that largely supported an instrumental role of the arts in society (Butler and Reiss, 2007) the researcher began to formulate thinking that such projects and interventions acted as an agent in the built environment, contributing to the creation of public life and acting as the adjoining conduit of art, city, and public/community relations. It was with MSc Psychosocial Studies (in the School of Applied Social Studies at University of Brighton) [n.6] learning that the researcher furthered this thinking and the beginnings of a theory of social practice placemaking formed, at that point termed *relocalism*. The MSc dissertation positioned relocalism as emblematic of a proletarian and bottom-up change to society that was community-based and acted without reference to policy or city administrative protocol and was a possible means of psychosocial articulation and a factor in the expressive order. The dissertation positioned it as a movement that reintroduced a creative tension into the city narrative and aimed to 'win back' urban development from the private, profit-orientated and corporate-led agenda and that sought to counter public apathy to allow creativity, dissent and critique to thrive, as affirmed by Chatterton (2006, p.73) for example: 'A healthy civic culture is based on a sense of democracy which is defined through conflict and disagreement'. Herbert and Thomas (1990) assert that there is not enough controlled research into the built environment and behaviour cause-and-effect and whilst much has been advanced since their time of writing, as an emerging movement, the researcher believed that relocalism deserved its own investigation as a contemporary urban movement and was able to undertake this, and question, interrogate and progress thinking to social practice placemaking, through the PhD research project.

1.7 Note

1   Statistics correct at time of writing in 2016.

1.8 Bibliography

Agnew, A. and Duncan, J. (1989) *The power of place*. Boston: Unwin Hyman.
Ainsworth, M. D. (1991) 'Attachments and other affectional bonds across the lifecycle' in Parkes, C. M., Stevenson-Hinde, J. and Marris, P. (eds.) *Attachment across the life cycle*. London: Routledge.
Altman, I. and Low, S. M. (1992) *Place attachment*. New York: Plenum Press.
Amin, A. (2008) 'Collective culture and urban public life' in *City* 12, 1, April.
Anderson, S. (1986) 'Studies towards an ecological model of urban environment' in Anderson, S. (ed.) *On streets*. Cambridge, MA: MIT Press.

Anderson, S. E. and Nielson, A. E. (2009) 'The city at stake: "stakeholder mapping" the city' in *Culture Unbound* 1, 2009, 305–29.

Andrews, G. J. and Drass, E. (2016) 'From the pump to senescence: two musical acts of more-than-representational "acting into" and "building new" life' in Fenton, N. and Baxter, J. *Practicing qualitative research in health geography*. London: Ashgate.

Benington, J. (1996) 'New paradigms and practices for local government capacity building within civil society' in Kraener, S. and Roberts, J. (eds.) *The politics of attachment: towards a secure society*. London: Free Association Books.

Berzoff, J. (2011) 'Relational and intersubjective theories' in Berzoff, J., Melano Flanagan, L. and Hertz, P. (eds.) (3rd edn.) *Inside out and outside in: psychodynamic clinical theory in psychopathy in contemporary multicultural contexts*. Plymouth: Rouman and Littleford Publishers, Inc.

Beyes, T. (2010) 'Uncontained: the art and politics of reconfiguring urban space' in *Culture and Organisation* 16, 3: 229–46.

Bishop, C. (2006) 'The social turn: collaboration and its discontents' in *Artforum*, 178–83 [Online]. Available at: www.gc.cuny.edu/CUNY_GC/media/CUNY-Graduate-Center/PDF/Art%20History/Claire%20Bishop/Social-Turn.pdf. (Accessed: 28th November 2015).

Bourdieu, P. (1986) 'The forms of capital' in Richardson, J. (ed.) *Handbook of theory and research for the sociology of education*. New York: Greenwood.

Bourriaud, N. (1998/2006) 'Relational aesthetics' (1998) in Bishop, C. (ed.) *Participation*. London: Whitechapel Gallery and The MIT Press.

Bowlby, J. (1979/2010) *The making and breaking of affectional bonds*. Abingdon: Routledge.

Bresnihan, P. and Byrne, M. (2014) 'Escape into the city: everyday practices of commoning and the production of urban space in Dublin' in *Antipode* 47, 1: 36–54.

Brown, A. (2012) 'All the world's a stage: venues, settings and the role they play in shaping patterns of arts participation' in *Perspectives on non-profit strategies*. Boston: Wolfbrown.

Brown, B. (2014) 'The rise of localist politics' in McClay, W. M. and McAllister, T. V. (eds.) *Why place matters: geography, identity, and civic life in modern America*. New York: New Atlantis Books.

Buck, N., Gordon, I., Harding A. and Turok, I. (2005) 'Conclusion: moving beyond the conventional wisdom' in Buck, N., Gordon, I., Harding A. and Turok, I. (eds.) *Changing cities*. Basingstoke: Palgrave Macmillan.

Bulmer, H. (1969) *Symbolic interactionism: perspective and method*. Berkeley: University of California Press.

Butler, D. and Reiss, V. (eds.) (2007) *Art of negotiation*. Manchester: Cornerhouse Publications.

Carmona, M., de Magalhães, C., and Hammond, L. (2008) *Public space: the management dimension*. London: Routledge.

Carrus, G., Scopelliti, M., Fornara, F., Bonnes, M. and Bonaiuto, M. (2014) 'Place attachment, community identification, and pro-environmental engagement' in Manzo, L. C. and Devine-Wright, P. (eds.) *Place attachment: advances in theory, methods and applications*. Abingdon: Routledge.

Castells, M. (1983) *The city and the grassroots: a cross-cultural theory of urban movements*. London: Edward Arnold.

Central Statistics Office (Ireland) [a] Number of private households and persons in private households in each Province, County and City. Available at: www.cso.ie/quicktables/GetQuickTables.aspx?FileName=CNA33.asp&TableName=Number+of+private+households+and+persons+in+private+households+in+each+Province+,+County+and+City&StatisticalProduct=DB_CN. (Accessed: 20th August 2013).

Central Statistics Office (Ireland) [b] Population and Labour Force Predictions 2011–2041. Available at: www.cso.ie/en/media/csoie/releasespublications/documents/population/2008/poplabfor_2011-2041.pdf. (Accessed: 20th August 2013).

Central Statistics Office (Ireland) [c] Population Classified by Area. Available at: www.cso.ie/en/media/csoie/census/documents/census2011vol1andprofile1/Census,2011,-,Population,Classified,by,Area.pdf. (Accessed: 20th August 2013).

Central Statistics Office (Ireland) [d] Population of Each Province, County and City, 2011. Available at: www.cso.ie/en/statistics/population/populationofeachprovincecountyandcity2011/. (Accessed: 20th August 2013).

Central Statistics Office (Ireland) [e] Regional Population Projections. Available at: www.cso.ie/en/media/csoie/releasespublications/documents/population/current/poppro.pdf. (Accessed: 20th August 2013).

Chatterton, P. (2006) 'Retrofitting the corporate city: five principles for urban survival ...' in Orta, L. and Orta, J (eds.) *Collective space*. Article Press and ixia PA Ltd.

Cozolino, L. (2006) *The neuroscience of human relationships: attachment and the developing social brain*. New York: WW Norton and Company.

Cudden, J. (2013) Dublin Demographics 2013. Available at: www.slideshare.net/jcudden/dublin-demographics-2013. (Accessed: 23rd September 2015).

de Certeau, M. (1984) *The practice of everyday life*. Berkeley: University of California Press.

Dixon, J. and Durrheim, K. (2000) 'Displacing place-identity: a discursive approach to locating self and other' in *British Journal of Social Phycology* 39: 27–44.

Doherty, C., Eeg-Tverbakk, P. G., Fite-Wassilak, C., Lucchetti, M., Malm, M. and Zimberg, A. (2015) 'Foreword' in Doherty, C. (ed.) *Out of time, out of place: public art (now)*. London: ART/BOOKS with Situations and Public Art Agency Sweden.

Duncan, P. (2012) 'Dublin holds most "people born abroad"' 14th May 2012, The Irish Times [Online]. Available at: www.irishtimes.com/news/dublin-holds-most-people-born-abroad-1.714646. (Accessed: 23rd September 2015).

Durkheim, E. (1893) *The division of labour in society reprint*. Basingstoke: Palgrave Macmillan, 2013.

Finkelpearl, T. (2013) *What we made: conversations on art and social cooperation*. Durham: Duke University Press.

Fischer, C. S. (1975) 'Toward a subcultural theory of urbanism' in *American Journal of Sociology* 80, 6: 1319–41.

Florida, R. (2011) *The rise of the creative class revisited*. New York: Basic Books.

Foo, K., Martin, D., Wool, C. and Polsky, C. (2014) 'Reprint of "The production of urban vacant land: relational placemaking in Boston, MA neighborhoods"' in *Cities* 40: 175–82.

Franck, K. A. and Stevens, Q. (2007) 'Tying down loose space' in Franck, K. A. and Stevens, Q. (eds.) *'Loose' unintended use of space, from impromptu to planned, splash to embedded practice*. Abingdon: Routledge.

Friedmann, J. (2010) 'Place and place-making in cities: a global perspective' in *Planning Theory and Practice* 11, 2: 149–65.

Froggett, L., Little, R., Roy, A. and Whitaker, L. (2011) New Model of Visual Arts Organisations and Social Engagement, University of Central Lancashire Psychosocial Research Unit [Online]. Available at: http://clok.uclan.ac.uk/3024/1/WzW-NMI_Report%5B1%5D.pdf (Accessed: 12th August 2015).

Gehl, J. (1996) *Life between buildings: using public space* (3rd edn.). Skive: Arkitektens Forlag.

Giddens, A. (1992) *The transformation of intimacy*. Cambridge: Polity Press.

Gieseking, J. and Mangold, W. (eds.) (2014) *The people, place, and space reader*. New York: Routledge.

Gillman, B. (2002) *Case study research methods*. London: Continuum.

Gilmore, A. (2014) 'Raising our quality of life: the importance of investment in arts and culture', Centre for Labour and Social Studies [Online]. Available at: http://classon-line.org.uk/docs/2014_Policy_Paper_-_investment_in_the_arts_-_Abi_Gilmore.pdf (Accessed: 12th August 2015).

Giuliani, M. (2003) 'Theory of attachment and place attachment' in Bonnes, M., Lee, T. and Bonainto, M. (eds.) *Psychological theories for environmental issues*. Aldershot: Ashgate.

Goldbard, A. (2011) 'Nine ways of looking at ourselves (looking at cities)' in *Culture and Local Governance* 3, 1–2.

Goldberg, S. (2000) *Attachment and development*. London: Arnold.

Gosling, J. (1996) 'The business of "community"' in Kraener, S. and Roberts, J. (eds.) *The politics of attachment: towards a secure society*. London: Free Association Press.

Gotteliener, M. and Hutchison, R. (2006) *The new urban sociology*. Boulder, CO: Westview Press.

Guattari, F. (2006) 'Chaosmosis: an ethico-aesthetic paradigm' in Bishop, C. (ed.) *Participation*. London: Whitechapel Gallery and The MIT Press.

Haase, A., Kabisch, S., Steinführer, A., Bouzarovski, S., Hall, R. and Ogden, P. (2010) 'Emergent spaces of reurbanisation: exploring the demographic dimension of inner-city residential change in a European setting' in *Population, Space and Place* 16: 443–63.

Hamblen, M. (2014) 'The city and the changing economy' in Quick, C., Speight, E. and van Noord, G. (eds.) *Subplots to a city: ten years of In Certain Places*. Preston: In Certain Places.

Harvey, D. (2008) 'The right to the city' in *New Left Review* 53: 23–40.

Herbert, D. T. and Thomas, C. J. (1990) *Cities in space, city as place*. London: David Fulton Publishers.

Holmes, J. (1993) *John Bowlby and attachment theory*. London: Routledge.

Hooper, L. and Boyle, P. (2008) 'Living city: an experiment in urban design education' in Coults, G. and Jokela, T. (eds.) *Art, community and environment: education perspectives*. Bristol: Intellect.

Hou, J. and Rios, M. (2003) 'Community-driven place making: the social practice of participatory design in the making of Union Point Park' in *Journal of Architectural Education* 57, 1, 19–27.

Hubbard, P. (2006) *City*. London: Routledge.

I Live Here: Britain's Worst Places To Live (2015) Walthamstow, Waltham Forest. Available at: www.ilivehere.co.uk/statistics-walthamstow-waltham-forest-40928.html. (Accessed: 23rd September 2015).

Jackson, S. (2011) *Social works: performing art, supporting publics*. Abingdon: Routledge.

Kawulich, B. (2005) 'Participant observation as data collection method' in *Forum: Qualitative Social Research* 6, 2, Art 43, May 2005.

Kearns, P. and Ruimy, M. (2014) *Beyond Pebbledash and the puzzle of Dublin*. Kinsale: Gandon Editions Kinsale.

Kennedy, B. (2015) 'Indianapolis' in 'America's 11 poorest cities', 8th February 2015, CBS Moneywatch [Online]. Available at: www.cbsnews.com/media/americas-11-poorest-cities/4/. (Accessed: 23rd September 2015).

Kester, G. H. (2004) *Conversation pieces: community and communication in modern art*. Berkeley: University of California Press.

Klanten, R., Ehmann, S., Borges, S, Hübner, M and Feireiss L. (2012) *Going public: public architecture, urbanism and interventions*. Berlin: Gestalten.

Krupat, E. (1985) *People in cities: the urban environment and its effects*. Cambridge: Cambridge University Press.

Kwon, M. (2004) *One place after another: site-specific art and located identity*. Cambridge, MA: The MIT Press.

Lee, M. (1997) 'Relocating location: cultural geography, the specificity of place and the city habitus' in McGuigan, J. (ed.) *Cultural methodologies*. London: Sage.

Lefebvre, H. (1970/2003) *The urban revolution*. Minneapolis: University of Minneapolis.

Lefebvre, H. (1984) *The production of space*. Translated Nicholson-Smith, D. Malden, MA: Blackwell Publishing.

Legge, K. (2012) *Doing it differently*. Sydney: Place Partners.

Legge, K. (2013) *Future city solutions*. Sydney: Place Partners.

Lepofsky, J. and Fraser, J. C. (2003) 'Building community citizens: claiming the right to place-making in the city' in *Urban Studies* 40, 1: 127–42.
Lerner, J. (2014) *Urban acupuncture*. Washington: Island Press.
Lewicka, M. (2014) 'In search of roots: memory as enabler of place attachment' in Manzo, L. C. and Devine-Wright, P. (eds.) *Place attachment: advances in theory, methods and applications*. Abingdon: Routledge.
Lilliendahl Larsen, J. (2014) 'Lefebvrean vagueness: going beyond diversion in the production of new spaces' in Stanek, Ł., Schmid, C. and Moravánszky, Á. (eds.) *Urban revolution now: Henri Lefebvre in social research and architecture*. Farnham: Ashgate Publishing Ltd.
Lin, J. and Mele, C. (eds.) (2005/2011) *The urban sociology reader*. Abingdon: Routledge.
Lippard, L. (1997) *The lure of the local: senses of place in a multi-centered society*. New York: New Press.
Livingston, M., Bailey, N. and Kearns, A. (2008) *People's attachment to place: the influence of neighbourhood deprivation*. Glasgow: Glasgow University.
Low, S. (2014) 'Spatializing culture: an engaged anthropological approach to space and place' in Gieseking, J. and Mangold, W. (eds.) *The people, place, and space reader*. New York: Routledge.
Lydon, M. and Garcia, A. (2015) *Tactical urbanism: short-term action for long-term change*. Washington: Island Press.
Lynch, K. (1960) *The image of the city*. Cambridge, MA: MIT Press
Lynch, K. (1981) *A theory of good city form*. Cambridge, MA: MIT Press.
Manzo, L. (2005) 'For better or worse: exploring the multiple dimensions of place meaning' in *Journal of Environmental Psychology* 25, 1: 67–86.
Manzo, L. C. and Devine-Wright, P. (2014) 'Introduction' in Manzo, L. C. and Devine-Wright, P. (eds.) *Place attachment: advances in theory, methods and applications*. Abingdon: Routledge.
Marcuse, P. (2012) 'Re-imagining the city critically' in Peter Marcuse's Blog [Online]. Available at: http://pmarcuse.wordpress.com/2012/12/14/blog-25-re-imagining-the-city-critically/ (Accessed: 20th August 2013).
Markusen, A. and Gadwa, A. (2010) [a] Creative Placemaking White Paper [Online]. Available at: www.nea.gov/pub/CreativePlacemaking-Paper.pdf (Accessed: 5th October 2013).
Markusen, A. and Gadwa, A. (2010) [b] Creative Placemaking White Paper Executive Summary [Online]. Available at: www.nea.gov/pub/CreativePlacemaking-Paper.pdf (Accessed: 5th October 2013).
Marris, P. (1991) 'The social construction of uncertainty' in Parkes, C. M., Stevenson-Hinde, J. and Marris, P. (eds.) *Attachment across the life cycle*. London: Routledge.
Massey, D. (2005) *For space*. London: Sage Publications.
McAllister, T. V. (2014) 'Making American places: civic engagement rightly understood' in McClay, W. M. and McAllister, T. V. (eds.) *Why place matters: geography, identity, and civic life in modern America*. New York: New Atlantis Books.
McClay, W. M. (2014) 'Introduction: why place matters' in McClay, W. M. and McAllister, T. V. (eds.) *Why place matters: geography, identity, and civic life in modern America*. New York: New Atlantis Books.
McClay, W. M. and McAllister, T. V. (2014) 'Preface' in McClay, W. M. and McAllister, T. V. (eds.) *Why place matters: geography, identity, and civic life in modern America*. New York: New Atlantis Books.
McCormack, D. P. (2013) *Refrains for moving bodies: experience and experiment in affective spaces*. Durham: Duke University Press.
McKeown, A. (2015) *Cultivating permaCultural resilience: towards a creative placemaking critical praxis*. Unpublished PhD thesis. National College of Art and Design, Dublin.
Mead (1934) *Mind, self, and society: from the standpoint of a social behaviourist*. Chicago: University of Chicago Press.
Merleau-Ponty, M. (1962) *The phenomenology of perception*. New York: Humanities Press.

Mihaylov, N. and Perkins, D. D. (2014) 'Community place attachment and its role in social capital development' in Manzo, L. C. and Devine-Wright, P. (eds.) *Place attachment: advances in theory, methods and applications*. Abingdon: Routledge.
Miles, M. (1997) *Art, space and the city: public art and urban futures*. London: Routledge.
Milgram, S. (1970) 'The experience of living in cities' in *Science* 167: 1461–8.
Milligan, M. J. (1998) 'Interactional past and potential: the social construction of place attachment' in *Symbolic Interaction* 21, 1: 1–33.
Minton, A. (2009) *Ground control: fear and happiness in the twenty-first century city*. London: Penguin Books.
Mitchell, D. (2003) *The right to the city*. New York: The Guildford Press.
Moyersoen, M. and Swyngedouw, E. (2013) 'LimiteLimite: cracks in the city, brokering scales, and pioneering a new urbanity' in Nicholls, W., Miller, B. and Beaumont, J. (eds.) *Spaces of contention: spatialities and social movements*. Farnham: Ashgate.
Németh, J. and Langhorst, J. (2014) 'Rethinking urban transformation: temporary uses for vacant land' in *Cities* 40: 143–50.
Nicholson, G. (1996) 'Place and local identity' in Kraener, S. and Roberts, J. (eds.) *The politics of place attachment: towards a secure society*. London: Free Association Press.
Nowak, J. (2007) A Summary of Creativity and Neighbourhood Development: Strategies for Community Investment TRFund [Online]. Available at: www.sp2.upenn.edu/siap/docs/cultural_and_community_revitalization/creativity_and_neighborhood_development.pdf (Accessed: 4th December 2013).
Office of National Statistics (2013) London's population was increasing the fastest among the regions in 2012. Available at: www.ons.gov.uk/ons/rel/regional-trends/region-and-country-profiles/region-and-country-profiles---key-statistics-and-profiles--october-2013/key-statistics-and-profiles---london--october-2013.html (Accessed: 23rd September 2015).
Pahl, A. (1996) 'Friendly society' in Kraener, S. and Roberts, J. (eds.) *The politics of attachment: towards a secure society*. London: Free Association Books.
Park, R. E. (1925) *The city: suggestions for the study of human nature in the urban environment*. Chicago: University of Chicago Press.
Pierce, K. Martin, D. G., and Murphy, J, T. (2011) 'Relational place-making: the networked politics of place' in *Transaction* NS 36: 54–70.
Project for Public Spaces with UN-HABITAT (2012) Placemaking and the future of cities. Available at: www.pps.org/wp-content/uploads/2012/09/PPS-Placemaking-and-the-Future-of-Cities.pdf (Accessed: 21st August 2013).
RateMyArea.com. Smithfield, Dublin, Ireland. Available at: http://dublin.ratemyarea.com/areas/smithfield-162 (Accessed: 23rd September 2015).
Rekte, S. (2011) 'New Urbanism and Pocket Neighborhoods: Developing Stronger Communities', 18th August 2011. Global Site Plans [Online] available at: http://globalsiteplans.com/environmental-design/new-urbanism-and-pocket-neighborhoods-developing-stronger-communities/ (Accessed: 4th December 2013).
Rendell, J. (2006) *Art and architecture: a place between*. London: I. B. Tauris.
Rennie Short, J. (2006) *Urban theory: a critical assessment*. Basingstoke: Palgrave Macmillan.
Rollero, C. and De Piccoli, N. (2010) 'Place attachment, identification and environmental perception: An empirical study' in *Journal of Environmental Psychology* 30: 198–205.
Rowles, G. D. (1983) 'Place and personal identity in old age: observations from Appalachia' in *Journal of Environmental Psychology* 3, 4: 299–313.
Roy, A. (2005) 'Urban informality: toward an epistemology of planning' in *Journal of the American Planning Association* 71, 2.
Rubinstein, R. and Parmelee, O. (1992) 'Attachment to place and the representation of the life course by elderly' in Altman, I. and Low, S. (eds.) *Place attachment*. London: Plenum Press.

Rustin, M. (1996) 'Attachment in context' in Kraener, S. and Roberts, J. (eds.) *The politics of attachment: towards a secure society*. London: Free Association Press.

Sassen, S. (2005) 'Cityness in the urban age' in *Urban Age Bulletin* 2: 1–3.

Scannell, L. and Gifford, R. (2010) 'Comparing the theories of interpersonal and place attachment' in Manzo, L. C. and Devine-Wright, P. (eds.) *Place attachment: advances in theory, methods and applications*. Abingdon: Routledge.

Schmid, C. (2014) 'The trouble with Henri: urban research and the theory of the production of space' in Stanek, Ł., Schmid, C. and Moravánszky, Á. (eds.) *Urban revolution now: Henri Lefebvre in social research and architecture*. Farnham: Ashgate.

Schneekloth, L. H. and Shibley, R. G. (2000) 'Implacing architecture into the practice of placemaking' in *Journal of Architectural Education* 53, 3: 130–40.

Seamon, D. (2014) 'Place attachment and phenomenology: the synergistic dynamism of place' in Manzo, L. C. and Devine-Wright, P. (eds.) *Place attachment: advances in theory, methods and applications*. Abingdon: Routledge.

Sen, A. and Silverman, L. (2014) 'Introduction – embodied placemaking: an important category of critical analysis' in Sen, A. and Silverman, L. (eds.) *Making place: space and embodiment in the city*. Bloomington: Indiana University Press.

Sennett, R. (1970) *The uses of disorder: personal identity and city life*. New Haven: Yale University Press.

Sennett, R. (2007) 'The open city' in Burdett, R. and Sudjic, D. (eds.) *The endless city*. London: Phaidon.

Sennett, R. (2012) *Together: the rituals, pleasures and politics of cooperation*. London: Allen Lane.

Sepe, M. (2013) *Planning and place in the city: mapping place identity*. Abingdon: Routledge.

Serres, M. (1980/2007) *The parasite*. Minneapolis: University of Minnesota Press.

Shumaker, S. and Taylor, R. (1983) 'Toward a clarification of people-place relationships: a model of attachment to place' in Feimer, N. and Geller, E. (eds.) *Environmental psychology*. New York: Praeger.

Silberberg, S. (2013) Places in the making: how placemaking builds places and communities. MIT Department of Urban Studies and Planning [Online]. Available at: http://dusp.mit.edu/cdd/project/placemaking (Accessed: 13th August 2015).

Sime, J. D. (1986) 'Creating places or designing spaces?' in *Journal of Environmental Psychology* 6: 49–63.

Simmel, G. (1903/2014) 'The metropolis and mental life (1903)' in Gieseking, J. and Mangold, W. (eds.) *The people, place, and space reader*. New York: Routledge.

Sixsmith, J. (1986) 'The meaning of home: an exploratory study of environmental experience' in *Journal of Psychology* 6, 4: 281–98.

Soja, E. W. (1997/2005/2011) 'Six discourses on the postmetropolis' in Westwood, S. and Williams, J. (eds.) *Imagining cities; scripts, signs, memory*. London: Routledge.

Speight, E. (2014) 'Subplots, tactics and stories-so-far' in Quick, C., Speight, E. and van Noord, G. (eds.) *Subplots to a city: ten years of In Certain Places*. Preston: In Certain Places.

Stevens, Q. (2007) 'Betwixt and between: building thresholds, liminality and public space' in Franck, K. A. and Stevens, Q. (eds.) *Loose space: possibility and diversity in urban life*. Abingdon: Routledge.

Swyngedouw, E. and Kaïka, M. (2003) 'The making of "glocal" urban modernities: exploring the cracks in the mirror' in *City* 7, 1.

Till, K. E. (2014) '"Art, memory, and the city" in Bogotá: Mapa Teatro's artistic encounters with inhabited places' in Sen, A. and Silverman, L. (eds.) *Making place: space and embodiment in the city*. Bloomington: Indiana University Press.

Tonkiss, F. (2013) *Cities by design: the social life of urban form*. Cambridge: Polity.

Trust for London [a] Waltham Forest. Available at: www.londonspovertyprofile.org.uk/indicators/boroughs/waltham-forest/ (Accessed: 23rd September 2015).

Trust for London [b] Wandsworth. Available at: www.londonspovertyprofile.org.uk/indicators/boroughs/wandsworth/ (Accessed: 23rd September 2015).

Tuan, Y. (2014) 'Place/Space, ethnicity/cosmos: how to be more fully human' in McClay, W. M. and McAllister, T. V. (eds.) *Why place matters: geography, identity, and civic life in modern America*. New York: New Atlantis Books.

Uitermark, J., Nichools, W. and Loopmans, M. (2012) 'Cities and social movements: theorizing beyond the right to the city' in *Environment and Planning A* 44: 2546–54.

United States Census Bureau State and County QuickFacts: Indiana. Available at: http://quickfacts.census.gov/qfd/states/18000.html (Accessed: 23rd September 2015).

Valentine, G. (2014) 'Living with difference: reflections on geographies of encounter' in Paddison, R. and McCann, E. (eds.) *Cities and social change: encounters with contemporary urbanism*. London: Sage Publications.

van Hoven, B. and Douma, L. (2012) '"We make ourselves at home wherever we are" – older people's placemaking in Newton Hall' in *European Spatial Research and Policy* 19, 1: 65–79.

Waltham Forest. *Data and Statistics*. Available at: www.walthamforest.gov.uk/Pages/Services/statistics-economic-information-and-analysis.aspx (Accessed: 23rd September 2015).

Wandsworth Borough Council Statistics and census information. Available at: www.wandsworth.gov.uk/downloads/200088/statistics_and_census_information. (Accessed: 23rd September 2015).

Watson, S. (2006) *City publics: the (dis)enchantments of urban encounters*. Abingdon: Routledge.

Whybrow, N. (2011) *Art and the city*. London: I B Tauris.

Wilmott, P. (1987) *Friendship networks and social support*. London: Policy Studies Institute.

Wirth, L. (1930/2005/2011) 'Urbanism as a way of life' from American Journal of Sociology 1930 in Lin, J. and Mele, C. (eds.) *The urban sociology reader*. Abingdon: Routledge.

Yin, R. K. (2003) *Applications of case study research* (2nd edn.). Thousand Oaks, CA: Sage Publications, Inc.

Yin, R. K. (2009) *Case study research: design and methods* (4th edn.). Thousand Oaks, CA: Sage Publications, Inc.

Yoon, M. J. (2009) 'Projects at play: public works' in Lehmann, S. (ed.) *Back to the city: strategies for informal urban interventions*. Ostfildern, Germany: Hatje Cantz.

Zukin, S. (1995) *The cultures of cities*. Malden, MA: Blackwell Publishers.

Zukin, S. (2010) *Naked city: the death and life of authentic urban place*. Oxford: Oxford University Press.

Zukin, S. (2013) 'Whose culture? Whose city?' in Lin, J. and Mele, C. (eds.) *The urban sociology reader* (3rd ed.) Abingdon: Routledge.

1.9 List of URLs

1. Art Tunnel Smithfield: http://arttunnelsmithfield.com/
2. Big Car: www.bigcar.org/
3. The Drawing Shed: www.thedrawingshed.org/
4. Creative Cities Vitality Index: http://creativecities.org/the-vitality-index/
5. UK Vitality Index: www.lsh.co.uk/campaigns/uk-vitality-index/about
6. School of Applied Social Studies at University of Brighton: www.brighton.ac.uk/about-us/contact-us/academic-departments/school-of-applied-social-science.aspx

2 Arts in place

Arts in the urban context have been introduced thus far: as the arts in placemaking is the central concern of the text, this section pays deeper attention to the moves in the arts sector to the performative and social practice arts domain that social practice placemaking is informed by. The chapter specifically focuses on the practice and process of an emplaced social practice arts and issues of participation. It then goes on to inform Chapter 3 and the delineation of types of placemaking practice based on the degrees of both. This chapter locates this subject first in the situational turn out of the gallery and consequently in site-specific and participative arts, underpinned by relational and dialogical aesthetics.

2.1 Moving art into the urban realm

The situational turn in the arts out of the gallery, predominantly from the 1960s onwards, was borne from an ancillary move away from monologically expressive art where art functioned as a 'repository of values' (Kester, 2004, p.87) and the artist was abstracted from the audience via an intermediary art object (Finkelpearl, 2013, p.27; Kester, 2004, p.89; Reed, 2005, pp.28–9). Artists such as Beuys [n.1] and Kaprow [n.2] and collectives Fluxus [n.3] and The Diggers [n.4] removed the artist ego position and created space for subsequent anti-individualistic and participatory art (Finkelpearl, 2013, p.20). They desired to act as 'radically related', social and communicative and with a focus on the means and ways of interaction and intersubjectivity (Gablik, 1992, pp.2–6), inviting the audience closer to the artwork (Kwon, 2004, p.66). This was reflective of the culture of the community (Hein, 1995/2005, p.436) in which the 'viewer's physical and cognitive interaction is integral to the work itself' (Kester, 2004, p.51) – the *New Situationists* (Doherty, 2004). For Kaprow, the modern concern of the artist was to use art to frame experience and meaning, creating the participatory *art experience* (Kelly, 1993/2003, p.xiii). This endeavour in turn created a 'form', 'Templates for modern experience, they are situational, operational, subject to feedback and open to learning' (ibid., p.xvii). Play, as inventive, instructive and participatory, was invoked and the artist positioned as the 'un-artist' and educator (ibid., p.xxii). As with Dewey's (1958) 'art as experience' and 'doing is knowing', in Kaprow's participatory art, meaning comes from 'not the content in art, but from the art in content' (Kelly, 1993/2003, pp.xxiii–xxiv). Such art operates Freire's (1972) 'problem-posing pedagogy' (Finkelpearl, 2013, p.30); it does not replicate the dominant ideologies or modes of production but operates in a third space between art and its own critique, 'revealing temporality and renewed possibility of society as horizontal and dialogic' (Sherlock, 1998, p.219).

In the public realm specifically, such artists operate to question the proscribed city function and engage local people in a social criticism of this (Miles, 1997, p.188), involving strategies of micro-communities of human interaction (Kester, 2011, p.29). The status of the *art object* changed with the situational turn: in interaction with the audience, the art object was de-materialised, a 'tactic of anti-commodification and a creative moment, increasing the collaborative process' (Lippard, 1997). This was influenced by earlier arts movements such as Dadaism that moved art away from an object-led practice to a process and interventionist art form (Miles, 1997, p.207). It should be noted however that various forms of art in the public realm will hold different intentionality by degree of the prominence given to process and the product – the art object – that form these differing practices (Miles, 1989, p.39). This has emerged as a central concern of the case studies where a re-materialisation of the art object was seen [2.5.3].

The situational turn expresses a site-specific tendency, art work responding to, reflecting and exploring 'the temporal and circumstantial context in which it inhibits' (Klanten et al., 2012, p.131), simultaneously relinquished to the physical context, directed and formed by it, and acting as an ideological critique of the art sector's institutionalised frameworks (Kwon, 2004, p.38; Lippard, 1973, in Suderberg, 2000, pp.1–5; Rendell, 2006, p.25) with *site-specific art* linked to the political through its perception as the 'activist branch' of the art sector (Lippard, 1997, p.274). Site-specific art does not operate in the 'literal site' of art – such as the white-cube gallery space – but in the *functional site*: the functional site is more complex; it may or may not incorporate a physical place, does not privilege the place, is processual, is informational, refuses the intransigence of literal site specificity, is a temporary thing and a chain of meanings, and is a place marked and swiftly abandoned (Meyer, 1995/2009, p.38). By being temporary, the functional site is complicit in its own destruction (ibid.) and by being temporary both site and place change by stimulating interest in people as they move on and/or by reflecting back the changes to place (Lippard, 1997, p.288).

Again, this is a political process whereby artists 'investigate urban topographies as sites of resistance, the human form is configured and employed as ideologically resonant, and spatial rearrangements compel a reassessment of perceptual boundaries' (Kwon, 2004, p.19). Site-specific art thus does not manifest in isolation but is rather 'imbricated' to the social and political process of its time, its aim to find the authentic meaning of a place (Kwon, 2000, pp.40–56); it can be precipitative of uncovering voices silenced and places hidden by the dominant discourse (Kwon, 1997, in Gieseking and Mangold, 2014, p.31). Site-specific art is purposed to provoke a critical acuity; can be anti-visual, informational, textual, expositional, didactic, gestural and eventual; and demands an intellectual and physical 'spatial expansion' of what is considered site, operating across phenomenological, social and discursive paradigms (Kwon, 2000, pp.43–6).

Arts in place 31

'New genre public art' emerged out of a critique of monumental 'Public Art' *(capitalisation intentional)* as a monologically expressive artform of the sole-author artist (Lacy, in Kwon, 2004, p.106). It is an idiom created by Lacy (2008) of an emplaced art that is contextual and site-specific (Brown, 2012; Hein, 1995/2005, p.438; Maksymowicz, 1990, pp.148–9; Rendell, 2006, p.5), community-informed and a form of community cultural development (Lacy, 2008, p.19), shifting the location of value to the interaction between artist and community (Kwon, 2004, p.95) in an ethnographic performance (Rendell, 2006, p.15). Artists will work with issues of local and global significance (Lacy, 2008, pp.19–21) and hold the public contingent to the art practice (Kravagna, 2012, p.254; Wilson, in Doherty et al., 2015, p.89). In new genre public art the artist moves from an 'elevated outsider' to a 'sympathetic facilitator' and 'engaged partner' (Adamek and Lorenz, 2008, p.57), that 'seeks to understand and help people determine their own agenda' (Kiiti and Nielsen, 1999, pp.52–64). The public is active in this process, the artist a catalyst of the creativity of others, the community providing the context and thus a further value of the art (McEwan, 1984, in Miles, 1997, p.93), marking a shift from artist to audience, object to process and production to reception (Kwon, 2004, p.106), the artwork intended for, and possibly also made by, the public (Lippard, 1997, p.272). It is claimed of much new genre public art that it becomes a 'framework of empowerment' and a 'catalyst to forms of sociation' (Miles, 1997, p.172), enabling ongoing community participation in the wider political sphere (Kester, 2004, p.174). An understanding of new genre public art moves the text closer to its concern of social practice arts and then social practice placemaking as a step on a continuum of art practice out of the gallery and with increasing democratisation of cultural practices. It also employs Lefebvre's (1984, p.73) 'art function is urban' thinking, that such art is a political act of access-giving that precipitates an increase in personal cultural capital (in accordance with Bourdieu, 1986) that leads to an autonomous organisation of individual living (Puype, 2004, pp.300–1).

However, the new genre public art term is not adequate to describe the subject of the social practice emplaced art of this text – and whilst used as a tool to ever closer locate the text in social practice arts and social practice placemaking, the text departs from new genre public art here. New genre public art is still of an administered and largely proscribed public realm and funded by governmental schemes such as Percent for Art [n.5] in the UK that have been open for deliberate wide interpretation to avoid public art commissioning strictures (Finkelpearl, 2013, p.40). Whilst new genre public art has the potential to reveal the 'personal-as-political' it can tend towards being an instrumental art that is at the service of a civic agenda (Lacy, 2008, pp.23–4) and accused of being complicit in perpetuating uneven urban development (Finkelpearl, 2013, p.41). New genre public art is effectively othered, at once situated within the community but also its art creation separated from society and this positioned as part of the dominant and privileged form of culture (Miles, 1997, pp.88–9). It's participation model is not necessarily equitable, but rather scaled according to the degree of intervention: firstly, 'community of mythic unity', an overarching grouping of intersectional groups; secondly, 'sited communities', those that share a geographical base or sense of purpose; thirdly, 'temporary invented communities', a new constituted group, gathered around the creative endeavour; and fourthly, 'ongoing invented communities', the same as three, but sustained past the creative endeavour (Kwon, 2004, pp.117–30). The community can be viewed as incapable of a critical historical distance from the work as a cultural comment as the issues involved are too 'alive' for them (Kastner, 1996, p.42).

To critique site-specific art however, firstly, it is not equitable that the site-specific art site is meaningful to the public; it may have an *a priori* function placed upon it and its public realm siting does not mean it gathers audience by virtue of being there (Kaji-O'Grady, 2009, p.109) or that an audience that understands its intended meaning. There is an inherent friction between the acceptance in visual arts theory that the intended meaning of the artist of a piece of work may not be the one interpreted by the audience, and conversely in urban studies, there being an importance given to the construction of meaning by the public (Hall and Smith, 2005, p.176). Secondly, the virtue of being outside of a gallery or museum setting does not automatically divest art of a structural, institutional influence or direct systemic relationship; there may still be a funded and commissioned relationship with site-specific art and, as exemplified with the activity of some institutions, galleries and museums that are actively seeking to work in this sphere and position themselves as thought-leaders in this practice. As a site-specific and performative artform, social practice placemaking shares its concerns and the text is thus informed by this thinking; the notion of site- and audiences-relation, as problematised here, also arose in data collection and will be addressed accordingly throughout the text. The participation of the non-artist is a central conceit of social practice arts and social practice placemaking and the text now turns to inspect the participatory turn in the arts, detailed further in direct relation to social practice placemaking and placemaking.

2.2 Moving art towards participation

Alongside, and further subsequent to, the situational and site-specific art turn is that of the participatory turn in the arts, a move that called for a physical inclusion of the viewer into the artwork to engender reflection and for participatory arts to align and fulfil certain agendas. Firstly, the casual production of an active subject who would, empowered by the participatory art experience, go on to act as politically autonomous. Secondly, ceding authorial control in the arts practice from a co-produced aesthetic (Bishop, 2006, p.12) where the artist and audience are active constituents in creation via a subjective and differentiated experience from one person and instance to another (Brown, 2012, p.1; Grodach, 2010, p.476; Kravagna, 2012, p.254). Thirdly, the concern with the postulated crisis of community and a collective responsibility through restored social bonds (Bishop, 2006, p.12). Whereas Modernist art was seen to abstract the individual from their quotidian functional lived experience, post-Modernist participatory, and situated, art represented a paradigmatic change where the artwork was both experimental and conflicting, transgressing communicatory, participatory and authoring boundaries and discussant of issues of social conflict and marginality (Bosch and Theis, 2012, pp.6–7), and artists were explicitly concerned with social issues, transforming the former's aesthetic art experience into an embodied one of meaning (Kaprow, in Kelly, 1993/2003, p.xviii).

By virtue of taking place in a community platform, participatory arts claims to facilitate intra-group relations and augment social interaction and engagement and in this opening up of interaction between strangers, and via focusing on roles in public space, foster community development (Grodach, 2010). The process is led by an 'artist-as-facilitator' who is a 'trained communicator' and a 'catalysing force' of a two-way co-authored process between facilitator and participants (Gablik, 1992, p.6; McGonagle, 2007; White, 1999, pp.16–7); though it may be the artist that creates the setting or opportunity, it is the participant that creates its actual terms, in a fluid and perpetually open, mutual system (Kaprow, in Kelly, 1993/2003, p.xviii). In this 'socially responsive' (Gablik, 1992, p.6) system, protagonists are affected by each other; protagonists are not observers but 'corporeally and intellectually part of the work' (ibid., p.94), forming an instrument of social change (Madyaningrum and Sonn, 2011, p.358). Community art sits within a participative arts discourse and practice. It encompasses a range of cultural activities as practice involving community members, often geographically defined and operating in notionally socially deprived areas, enabling people to be creative in ways that facilitate new community meanings and relationships (Madyaningrum and Sonn, 2011, p.358; Vaughan Williams, 2005, p.221).

Several criticisms can be levelled at participatory arts however. Participation rests on ontological assumptions that themselves rest on value judgements around the supposed political role that participation can or should have in wider society (Grodach, 2010; Rounthwaite, 2011, p.92) and assumes pluralism and inclusivity and consensus building as an extension of arts hegemony and an artist-audience binary (Beech, 2010, p.15; Kwon, 2004). However, participation is as much about who is participating and who is not – who has been *invited* and who has been *excluded* from the process (Beech, 2010, pp.25–6); participatory arts process cannot include all and this premises an exclusory binary of social participant/non-participant and one where those invited to participate have to accept the parameters of the art project (ibid., p.25) and the artist often maintains implicit authorship (Bishop, 2006, p.10; Colombo et al., 2001, p.462). Participatory art also may not be as emancipating as the practice can purport to be: the practice may liberate the artist from a distanced relation with society but it is not automatic that it liberates community participants from theirs (Bishop, 2006, p.10; Colombo et al., 2001, p.462). Whilst the artist acts as signifier of social issues, participants are not in fact made significant; rather, they are subject to an 'urban colonisation of difference' that strengthens the artist position as one that is speaking for another (Kwon, 2004, pp.139–40; Kravagna, 2012, p.242). Here also there is a semblance of collective action but no actual difference made and the artist/non-artist positions remain differentiated, the artist deployed in 'aesthetically digestible' bites (Kravagna, 2012, pp.240–1). Similarly, whilst participatory art may emerge from a feeling of political impotence and a need to redress this, it does not involve the required multiple affecting factors to influence collective action and is thus not truly interactive or capable of lasting impact (ibid.). Conversely, excessive emphasis is given to individualised transformation generated by an artist as a measure of success: 'The effect of this rhetoric can be to elide any analysis of the systematic causes of poverty and to put in its place a closed circuit of creative personal transformation presided over by the artist' (Kester, 2004, p.138). Recruitment and motivation to participate is obscured by thinking of the community as homogeneous (ibid., p.130) and even if this is understood, members of a community may be apathetic to participating (Froidevaux, 2013, p.189), the supposed 'catalysing force' of the artist only activated by the self-actualisation of the community (White, 1999).

When addressing community art in particular, practice has become the preserve of institutionalised modernist power positions in the art sector (Kester, 2004, p.137) and become worthy and detached from its activist roots (Kelly, 1984). The creative exchange is authored and mediated (Bourdieu, 1986) by the artist in a position of prescribed power, enacting upon a subject that has been defined *a priori* in need of the artists' facilitation (Bishop, 2006, p.184). Whilst Critical Art Ensemble (1998, p.73) [n.6] sees a 'co-mingling' of artist and community in community art, this is still dialectically oppositional, the artist acting as a signifier for the oppressed, speaking on their behalf (Kester, 2004, p.137), and set up as a mirror to the urban positivist structures that created it (Kwon, 2004, p.139) and where the notion of the community is not always transparent to those it assumes comprise it. Community art here is a disempowering process, of sanitised 'modest gestures' rather than 'singular acts' (Bishop, 2012, p.23) and productive of unoriginal, practiced responses from the participants. In this power relationship, community art is not a politically resistant force but 'diluted' and supporting of the political hegemony (Bishop, 2004, in Whybrow, 2011, p.29). Lastly, terminology will be ascribed to participatory practice in a vernacular fashion, perpetuating the confluence of terminology still – if an artwork is site-specific to a community, ergo, it is community art for example. Artist Sophie Hope, of Preston's *In Certain Places* [n.7] describes her participatory work thus:

> To be seen as an open process, filled with democratic possibility, drawing together on an equal basis amateurs and professional artists, people with a range of skills and experience, who congregate and collaborate, shifting their previously fixed positions, becoming alternately producers and spectators, viewers and evaluators. (Hope, 2010, p.69)

What is described above though is a mode of collaborative engagement that is more sophisticated than that nominally ascribed to participatory art and thus calls for a more complex and nuanced understanding of levels and modes of participation, which this text draws upon from social practice arts to the benefit of social practice placemaking, which underpins both practices. This moves discussion here away from participation to collaboration and the dissolving of the artist/non-artist binary – and in the case of social practice placemaking, the dissolving of urban expert/non-expert – positions. The common confluence of the terms participatory, interactive, collaborative or relational art under the participatory art banner has led to distinctions now being made between projects designed by artists and projects designed collaboratively between artist and participants (Finkelpearl, 2013, p.4). The 'binary logic' of participation is being overthrown by a constellated practice (Beech, 2010, p.26). This nuances the understating of participation as being 'taking part' and collaboration as 'working together', with the neglect of this ontological and phenomenological difference being the sacrifice of arts engagement to a one-dimensional terrain (ibid., p.24). Thus, through the prism of Dewey's (1958) understanding of participation as cooperative action for joint problem solving and as a democratic process that creates cooperative and social individual beings (Finkelpearl, 2013, p.345), and through the practice of artists that term themselves participatory, when in fact their practice is based on a deeper process of engagement, the term participatory art is found lacking.

The nature of participation and its critique through social practice placemaking practice is a central concern of the text, as well as the political conscientisation of participant, the dissolving of authorial position of artist and urban expert and the restoration of social bonds, all concerns of the work of the case studies. Notions of participation and collaboration in the arts, their depth and breadth, relative merits and social concerns, reference relational and dialogical aesthetics as modes of interaction and intervention between people and place, and this thinking will be introduced below.

2.3 Relational and dialogical aesthetics

Bourriaud (1998/2006, p.165) defines the *relational aesthetic* as 'an art that takes its theoretical horizon the sphere of human interactions and its social context, rather than the assertion of an autonomous and private symbolic space'; this is a cluster of artistic styles on a 'common trajectory' that, via 'convivial' modes of social exchange with the 'viewer', were concerned with human interactions, social context and collective meaning making. Relational art then is of meetings, encounters, events and various types of collaboration between people that works around intersubjective exchange (Kester, 2011, p.29) of seven forms: cooperation, interaction, participation, sustainability, responsibility, authorship and feedback (Bosch and Theis, 2012, p.14). This is akin to Debord's (1958/2014, pp.96–9) *unitary urbanism*, where the art work must be lived by its constructors and the use of all arts techniques is 'a means of co-operating in an integral composition of the environment' that aimed to 'appropriate social forms as a way to bring art closer to everyday life' (Bishop, 2006, p.10, 2012, p.11).

The relational aesthetic is a 'critical materialism' process, the art not being the outcome of work, but the process of work, and which is forever transitive, never complete and constantly discursive (Whybrow, 2011, p.28). It has a social responsibility at its core and to its integral aesthetic enhancement (Jackson, 2011, p.46), its social and evental outcomes concerned with provoking and sustaining individual and collective encounters, this concern of inter-human relations being the only true common denominator in relational practice (Bourriaud, 1998/2006, pp.163–5; Doherty, 2004; Whybrow, 2011). With the conceit of this text being located in the urban realm and in arts practices informed by relational aesthetics, a link is made between the two. Bourriaud (ibid., p.160) locates relational aesthetics in the 'urbanisation of the artistic experience', facilitating an increase in social exchanges and in individual geographical and mental mobility, the artwork operating as a 'social interstice', a space that is located within an overarching system but that suggests other possibilities for exchanges and operates from the bottom up (ibid., p.161). This relational artwork moves into the realm of the everyday and away from the events, spectacle and monumental of Public Art. Relational arts re-input relationality into the cityscape and interject the artist into the social fabric, rather than being placed as an agent that reacts to it (ibid. p.162), precipitative of an activist 'social aesthetic' (Bang Larsen, 1999/2006). However, Rancière (2004/2006, p.91) – in accord with adverse urban experience – perceives the phenomenological rise of relational aesthetics as responding to a lack of connections in society, acting as a service to 'repair weak social bonds'.

Kester (2004, 2011, p.32) proposes a *dialogical aesthetic*, borne from a critique of the relational, scripted encounter as presupposing an *a priori* participation invitation from the artist to the subject(s), which levels at it the same critique as that above for community and participatory art disciplines. The dialogical art aesthetic has three keystones. Firstly, that the dialogical experience is an embodied one of social knowledge (Sennett, 2012, p.211), art being 'concerned with a somatic social experience' (Willets, in Kester, 2004, p.91) in a connective aesthetic between self and society (Gablik, 1992, p.6). Secondly, and consequent to the first, the inter-relation of self to other where the pre-existing sense of self is transcended through the inter-relation and encountering of difference, an *empathetic identification* (Kester, 2004, pp.77–8) where people become more self-aware and see one's lifeworld anew through exchange with another (Sennett, 2012, p.19) and a process of distanciation from the everyday. This results in a transformative 'performative interaction' of an 'an empathetic feedback loop in which we observe the other's responses to our statements and actions' (ibid., p.77) that in turn results in a shared identity formation. The third keystone of the dialogical aesthetic, and as operative of the second, is that of dialogue itself, as a practice (Finkelpearl, 2013, p.114), the focus being on dialogical, collaborative encounters within communities (Beech, 2010, p.24). This offers a view of the artist as one that listens to others and is receptive to 'a positon of dependence and intersubjective vulnerability relative to the viewer or collaborator' (Kester, 2004, p.110). This is reciprocal and thus demands of the collaborator openness and willingness to change, to 'accept the transformative effects of difference' (ibid., pp.173–4).

Where these theories are pertinent for this text is in providing insight into individual and community encounters with emplaced art. Community in this case is founded on 'a series of relational encounters that require the ongoing negotiation of difference as well as identity' (Kester, 2011, pp.221–2); this is where the relational and dialogic aesthetics move into the realm of Amin's (2008) micropublics, the creation, in this instance through social practice placemaking, of diverse and temporary communities (Froggett et al., 2011, p.100) that are relational and engaged in dialogue, akin to Kester's (2011, p.29) strategies of micro-communities of human interaction. Theis text views a conjoining of relational and dialogical aesthetic in a *social aesthetic* where relational and dialogical understandings are integrated into each other phenomenologically (Bang Larsen, 1999/2006, pp.172–3) and on a spectrum. In the social aesthetic, art activity is a means of symbolic communication and dialogue, as an arts process, is productive of further dialogue by giving participants a voice (Froggett et al., 2011, p.92), precipitative of a comprehension and representation of their lifeworlds (Kester, 2004, p.69).

This section has located art work both in the urban context and out of the gallery into the public realm, and into and then beyond participatory art and new genre public art. It now moves further still into relational and dialogical performative practices with a turn to firstly emplaced art and then social practice arts.

2.4 Emplaced arts

The text now turns to the notion of emplaced arts as a critique of participatory arts, devising this with the use of the dialogical aesthetic. Stepping on from Bourriaud's (1998/2006, p.161) 'urbanisation of the artistic experience' and its function as a social interstice this text is not concerned with the all-encompassing and monumental Public Art (Leeson, 2008) but that art practice which is specifically emergent from contextual issues of place, situated but not static (Lehmann, 2009, p.18), that this text terms *emplaced arts*.

Tonkiss (2013, p.8) states that social actor urban form production is 'routine, unintentional, even accidental': the text conceives that emplaced arts, especially in their social practice form, are in fact intentional, however informally they are located or its practice and process may appear. It is rather an 'instantaneous experience evoked by authentic art places' (Kester, 2004, p.50). Emplaced art further collapses the process/product binary positions of art production through their dialogic aesthetic (Kester, 2004, 2011) and their parasitic (Serres, 1980/2007) role of the third in the people–place–process and liminal, countersited situation. Emplaced arts are associated with a participatory orientated practice, a re-articulation and renegotiation of the aesthetic autonomy of the artist, firstly by the move away from spectacle to participatory or immersive practices and secondly, by the artist working across and through others disciplines such as activism or ethnography. This text takes the position that distinctions need to be drawn between modes of participation as different modes of practice. This in turn will go on to inform the placemaking typology creation [3.7]. Arnstein's (1969) Ladder of Participation has been returned to recently as a model of the hierarchy of forms of citizen participation (Finkelpearl, 2013, p.1), the lowest rung the least participative, manipulation, through therapy, informing, consultation, placation, partnership, delegation and lastly the highest, of citizen control, moving in effect beyond a conceptualisation of participation to a decision-making power located in the citizenry (The Citizens Handbook [n.8]). A critical stance of participatory art would locate it in the 'tokenism' rungs, participation seen as the solution or salver to a host of cultural enragement problems (Beech, 2010, p.25); an exemplified, co-produced stance would be located in the rungs of citizen control. Furthermore, in a participatory art model, the community and its interests are commonly narrowly pre-identified (Kwon, in Finkelpearl, 2013; Lewis, 2013, p.71), this being derived solely from location and not from a social or psychological construction (Philips, 1988, in Miles, 1997, p.97). It is seen as failing to create a 'public' or citizenry by being legitimised by public funding (Miles, 1997, pp.85–7), maintaining formal public/private divisions and the honorific title and traditional role of the artist (Becker, 1974/2003, p.87). Artists that work in the community are often positioned, somewhat romantically, by the authorities as catalysing creative-problem solvers who can help communities solve issues (Fleming, 2007, p.24; Mancilles, 1998, pp.337–9).

Mention is given here of murals as emplaced art, relevant to this text as this was key in much of the Big Car [n.9] practice in Indianapolis and wider-USA commissions. Mural refers to large-scale, largely painted, artworks that may or may not have been sanctioned by the city administration or the local resident community. For the purposes of the research project, the text is focused on legal, sanctioned sites, and in this context how they are used as a point of reference and shared experience in its process to facilitate community connections; a sense of ownership of place; communicate political and/or social dissensus; educate; engage marginalised people; achieve social justice and wellbeing goals (Chonody, 2014, pp.30–1). Murals come to 'symbolise the city' (ibid., p.5) and beautify urban spaces and are a simple and visual marker that the resident community values self and art. In seeking direct creative input from resident community members, mural art is a dialogical mode of production (Kester, 2004, p.173): 'The involvement of local people can be just as important as the mural itself, which is what distinguishes it from commissioned or commercial artwork' (Chonody, 2014, p.30). As Chonody states, though, there is a research gap in how the creation of mural art can be used in practice.

2.4.1 The aesthetic dislocation of emplaced art

The experience of viewing and/or participating in emplaced art is that of an 'aesthetic shock or dislocation' (Kester, 2004, p.84) that functions, just as with the functional site of site-specific art (Meyer, 1995/2009, p.38), to counter the dominant, pejoratively top-down urban cultural narrative or the 'habit-memory' (Connerton, 1989, in Sen and Silverman, 2014, p.4) of everyday place interactions (Merleau-Ponty, 1962). The aesthetic of the 'wrong place' of artworks in unexpected places is disorientating, functioning to 'destabilize our sense of time and place' (from Kwon, 2000, in Doherty et al., 2015, p.127). The role of the artist here is not to act as protagonist of the artwork but to locate the 'wrong place' and open up that space for relational and dialogic creative interrogation (Kwon, ibid.) – the countersite (Holsten, 1998, p.54, in Watson, 2006, p.170). A key component of emplaced art is shock, 'which can be produced by something as simple as seeing a city street from a new perspective' (Kester, 2004, p.83). The *aesthetic dislocation* consequently engenders a 'heightened capacity' in the subject to the hidden political structures of that place through its revealing of the strange (ibid., p.84; Lippard, 1997, p.288). Thus the diversionary message of emplaced art has the capacity to provide a new discursive framework as an embodied, situated experience, and where collaborative, a practice of co-produced narrative formation. It offers 'emancipatory aesthetic knowledge' where the 'aesthetic is defined as an immediate (prediscursive) somatic experience (a shock or epiphany) that is only subsequently "made sense of" in terms of an existing discursive system' (Kester, 2004, p.84).

2.4.2 The aesthetic third

Emplaced arts are located in the countersite and form an *aesthetic third* in the urban realm, with a relational or dialogic aesthetic agency. Both actions constitute a border crossing between art and non-art that in the everyday context creates a 'third way' of a 'micro-politics of art, between the opposed paradigms of art becoming life and art as resistant form' (Rancière, 2004, p.86). The artwork – as object or process – with aesthetic third agency is symbolic and representative but also creates an interactional embodied connection between the person and the world outside (Froggett et al., 2011, p.92). By (re)stimulating a link between an individual and the cultural field, the aesthetic third, in turn, enhances the relational capacity (ibid.): the artist and participant meet through the artwork; the artwork becomes a cultural form for their experience; and in providing a form that shows what could not be said, experience is symbolised and brought into being in a new language. Thus, the aesthetic third takes something that exists in the imagination of those who participate in its production or reception, and by finding a cultural form that can be understood by others, is shared. By enabling experience to be shared the aesthetic third creates a vital link between individual and community; and this new thing is a link which has its own vitality (ibid., pp.93–4). The aesthetic third is de Certeau's (1984) *bricolage* in action, where the 'goal is to inscribe order through seemingly small rituals so that people can get along as harmoniously as possible' (Sennett, 2012, p.204). It is also akin to Serres' (1980/2007) parasitic theory; the art work created operates as the third in the parasitic triad of people and place and the breaker of the signal/noise normative urban condition. This has a ritualistic element of the creative praxis of artist-to-collaborators (Green, in Kester, 2011, p.3), an 'invention of tradition' (Sennett, 2012, p.88) that creates a sense of tradition in collective creative projects, builds group cohesion and elevates the routine to the ritual, guiding the 'process of compare-and-contrast' (ibid., p.94), of self in others (Kester, 2011).

From an understanding of arts in the urban realm and a critique of participatory art, and having focused in on situational and emplaced art forms and the modes of relational, dialogical and social aesthetics, the chapter now moves to the central concern of the art practice of this research project, social practice arts, and closes with a schematic positioning the art forms discussed with that of placemakings by way of a move into the typology of placemaking [Chapter 3].

2.5 Social practice art(s)

The text will now turn to the art form at the core of its endeavour as informing the practice of social practice placemaking as its emplaced form – social practice arts. The consideration of this practice marks the culmination of the text's thinking with regards to emplaced arts practices, the social interstice of space and place and the arts, moving then subsequently to placemaking and the consideration of the arts in this practice.

Social practice arts has 'kinship' with other forms of social practice, 'activist art, social work, protest performance, collaborative art, performance ethnography, community theatre' (Jackson, 2011, p.17), amongst others. A number of interchangeable terms across social practice arts are common within the sector, such as socially engaged, participatory, collaborative, situated, relational, dialogue (Reiss, 2007, p.10). These all share modes of approach where social interaction leads the artistic practice (ibid.); where the locus of control is (contextually) at the lowest possible level (Hoskins, 1999, p.301); and where artist and non-artist co-create objects, events and activism (Reiss, 2007, p.10). Social practice arts is a critical practice (Yoon, 2009, p.76), borne from a change in direction in how art and artists are being asked to define and operate within society. It is concerned with the interpretation and consumption of art, where and by whom; often the relation between art and city space; and the social role of the artist (Lossau, 2006, p.47), reflecting a creative drive for artists to be involved in a more social process and move away from making signature object pieces (McGonagle, 2007, p.6). Social practice arts ontology is 'shared ground' of Bourriaud's (1998/2006, pp.165) relational aesthetic, Bishop's antagonistic social relations (2006) and Kester's dialogic aesthetic and grouped under a 'new social ontology of art' of the performative encounter (Kester, 2011, p.20), bestowing on it counter-cultural agency (Froggett et al., 2011, p.95). This text argues locating social practice placemaking in place is beyond site-specific art as principally *place-led*, and thus, a critical spatial practice (Rendell, 2006).

Social practice art is centrally located in the social aesthetic where meaning is collectively created via inter-subjective encounters (Bishop, 2012, p.257). The artist's role here is to critique the dominant culture from the micro level upwards, the role of art being to draw attention to issues and encourage reflexive reassessment via new thinking and emotions (Murray, 2012, p.256). This is a cultural practice including the 'non-artist' (in quotation here as this term would not be necessarily be recognised as such in the horizontal social practice arts practice) public in creation of the art object and the social action in a pluralist co-authorship (Gablik, 1992, p.6; Kwon, 2004, pp.106–7). The artist deliberately 'bypasses cultural gatekeepers' and in gaining more devising and performance/exhibiting control, 'gives' this back to the audience (Brown, 2012, p.9). When these constituents meet each other, they are affected by each other; the audience is not observer but participant in a collaborative process and 'corporeally and intellectually part of the work, forming an instrument of social change (Madyaningrum and Sonn, 2011, p.358). However, based on an ethos of interconnectedness and intersubjectivity realised through the articulation of community voices (Gablik, 1992, p.4), the artist is not shy or anonymous about their expertism nor their agenda, a rethink of authorship for Bishop (2006) that is informed by Guattari's (2006) 'resingularisation'. The artist's expertism is as creative thinker, disruptor and/or negotiator (Kravagna, 2012, p.243; McGonagle, 2007, p.6; Reiss, 2007, p.11), 'the artist does not pretend to be a facilitator of others but is explicitly self-reflexive about his/her role as motivator and manipulator' (Bishop, in Barok, 2009). The artist works in 'radical relatedness' and is a 'connective, rational self' (Gablik, 1992, p.2), bringing people together via a subjective and differentiated experience from one person and instance to another (Grodach, 2010, p.476).

2.5.1 Social practice arts practice

A *performative* principle is central to social practice arts. Performativity is the production of identity of protagonists through situational encounters (Kester, 2004, p.90), the integration of 'body, actions and audience in relation of exchange and duration' (Jackson, 2011, pp.33–4), forming a 'live body' (Rounthwaite, 2011, p.92). Performativity 'deliberately seeks to "make strange" the everyday, proposing that people look at what they take for granted in new ways' (Froggett et al., 2011, p.97), making way for new forms of participation through the creation of new situational encounters between protagonists and with emplaced art (McCormack, 2013, p.189) – an aesthetic dislocation. As performative, the urban event-space functions as a relational structure (Bryan-Wilson, in Fletcher and July, 2014, p.146) that 'calls into question both the materiality of the built-world and the immateriality of the event revealing the real world as a space of virtuality' (Hannah, 2009, p.117). The urban event-space is a 'spatial acting out', events presented through the urban architectural artists frame, a Deleuzian 'theatre of matter' that renders 'our discursive and material environments more mobile, dynamic and dangerous' (ibid., p.118). This is a political, social and aesthetic process and is a source of socially engaged performance research and learning (Jackson, 2011, p.13; Loftland, in Franck and Stevens, 2007, p.19). The definition of artist is broadened in this practice to include a role of that of citizen; and similarly, the notion of art is extended to that of cultural and material intervention (Hicks, 1998, p.93–4), where the artistic medium changes according to the demands of the intervention; and where the meaning of the work is dependent on the active participation of the audience. The artist position is (self-) identified as that of catalyst and mediator of exchanges that create and maintain an empathetic critical analysis (Barok, 2009; Bishop, 2006; Kester, 2004, p.118; Kravagna, 2012, p.243; McGonagle, 2007, p.6), the 'radical relatedness' (Gablik, 1992, p.2) that relationally connects (Grodach, 2010, p.476).

As catalyst however, the artist will be in a leadership position, as instigator and networker (Lowe, in Lowe and Stern, in Finkelpearl, 2013, p.147). This position is fractured for the artist: the artist sees their role as that of uncovering meanings of place and delegating leadership by then creating opportunities for people to give that meaning a reality (ibid., p.138), desiring a collectivised approach (Hope, 2010, p.69). This is as the 'logic of informality: assist, don't direct' (Sennett, 2012, p.53) that through a collectivised process, such as a workshop or arts lab, forms a performative sociability. This is a social aesthetic, 'this is not Dewey's "art as experience" but socially cooperative experience as art' (Finkelpearl, 2013, p.361).

Artist autonomy and heterogeneous community agendas can come into conflict (Jackson, 2011, p.27) but this is a source of creative impulse that poses questions of the milieu and authorship (Froggett et al., 2011, p.104); the audience–viewer artist–non-artist subject positions are dissolved in performative practice (Beech, 2010, p.21; Bishop, 2006, p.16) to the extent that they are integral to it (Walwin, 2010, p.125). The artist has particular agency to act as a mirror to the community in question (Froggett et al., 2011, p.96) to provoke subsequent reflection and possible action: social practice arts is a means of mobilising social movements through a raised consciousness, to 'decode' the art process within its cultural loci (Larsen, 1999/2006, p.173), challenging the macro, dominant structures and negative social representations of the producing community (Murray, 2012, p.257) and the hope and expectation of change may indeed be a primary motive for engagement in the arts' (Froggett et al., 2011, p.91). Social practice arts participants are empowered in the practice as 'authors of their own lives', thus, 'their voice should be privileged in the process' (Chonody, 2014, p.2). The appeals of the non-artist 'community' protagonists in the encounter take precedence over that of the artists and the practice is to build intra-community dialogue with the end goal of participant-driven empowerment towards change (ibid., pp.32–40).

The notion of *relative expertism* provides a critical framework that 'problematizes the "artist-as-community-helpmate" role' (Jackson, 2011, p.44) and goes some way to collapse notional siloes of co-production, in which the locus of artistic control may shift between any of those involved' (Froggett et al., 2011, p.104). Whilst artist-collaborator categories are not collapsed into each other per se, but resingularised (Guattari, 2006) as a reciprocal practice, each constituent is valued for their knowledge and skills: the artists' expertism is as creative thinker, disruptor and/or negotiator (Reiss, 2007, p.11) and 'the community is expert in being the community' (van Heeswijk, 2012), a mutual aesthetic autonomy that is integral to its criticality (Gablik, 1992, p.2; Grodach, 2010, p.476; Jackson, 2011, p.50; O'Neill, 2014, p.201; Till, 2014, p.168). This is a non-aggressive criticality by virtue of its reciprocal negotiation (Kester, 2011, p.145) and also holds transformative agency: where the expert speaks singularly from their expert position, collectively these partial positions are created in equitability. This leads to a transformation of thinking of the 'expert' being not a pre-existing and fixed individual with finite resources (Bresnihan and Byrne, 2014, p.11) but through 'role dissonance' (Sennett, 2012, p.203), the expert position being de-siloed, extended and socially encountered (Bresnihan and Byrne, 2014, p.11). Collaboration calls for professional partnerships, which can be complex and demanding of capacity, 'involving collisions of organisational culture and frequently gaps in expectations and understanding which require patience and skilled diplomacy' (Froggett et al., 2011, p.103) but which open up the experience to non-fiscal motivations and outcomes and new forms of professional contribution (Shirky, 2008, p.109). In the reciprocal negotiated position, the artist sits both as insider and outsider in the socially engaged terroir, whether they are resident of that community or not; the artist must be knowledgeable of the community as social practice arts practice starts with the understanding of situated context (Chonody, 2014, pp.2–3, 31), located in an embedded spacetime (Munn, 1996, in Low, 2014, p.20) in which they become 'part insider, part outsider', (Froggett et al., 2011, p.96) occupying a liminal, parasitic (Serres, 1980/2007) space which has its own disrupting transformative potential (McCormack, 2013, p.10). Thus, spacetimes can aid the understanding of participation through urban spaces as the aesthetic third, or 'diverted space' (Lilliendahl Larsen, 2014, p.319) and can act as an exemplar to other arts institutions and social institutions in the urban realm of a collaborative process (Froggett et al., 2011, p.99). Tension though is felt in the artist's inherent privileged status of their 'expressive position', 'even when they have a long-standing identification with a given community' (Kester, 2004, p.174). Artists can also act from the 'wrong place', whereby they provocatively insert themselves into a site to uncover the usual forms of knowledge and identity there – which may be uncalled for and from which the artist, having provoked, then leaves (Kwon, 2004).

2.5.2 Social practice arts process

The process that this engenders of one of co-production (O'Neill, 2010, p.208) and will operate across and through empathic axes (Kester, 2004, pp.114–5) based on an ethos of interconnectedness and inter-subjectivity realised through the articulation of community voices (Gablik, 1992, p.4), holding a 'plural and polyphonic understanding of the subject' (Kester, 2011, p.31). This is a transformative process which 'enables the discovery of new forms for feeling which connect selves and communities' (Froggett et al., 2011, p.91). This acts as the aesthetic third, 'providing a point of articulation where the imaginations of individuals meet shared cultural forms', a psychosocial experience 'whereby the social can be creatively internalised by individuals' (ibid., p.94). These inter-subjective relations are that of a Lacanian sublimation, working to explore and critique social and political concerns (Bishop, 2012, p.39) which moves beyond participatory art (ibid., 2006, p.10) to, through its 'imaginative presence' to merge individual and common interests. Social practice art here acts as an agency of social change (Brown, 2012, p.10; Hicks, 1998, p.98; Miles, 1989, p.2; Murray, 2012, pp.256–7). Through cooperation in the social practice art process, a 'process of metamorphosis' is begun that can negotiate community tensions or issues and culminate in the community doing 'their own repair work' (Sennett, 2012, p.215). Thus, the informal logic and aesthetic of social practice art is not 'unconsidered' but intentional and considered.

The process of social practice art is concerned with the creation of connections, intra- and inter-community, through the performative production of social encounters that may be ambiguous and indeterminant (Froggett et al., 2011, p.95). The process itself is of the aesthetic third functioning to illuminate alternative ways of living, through process of symbolisation, dialogue and emplaced criticality that work to 'experience and represent the world from a different point of view' (ibid., p.98). Placing this in the public realm, social practice art becomes a critical site-specific process, incorporating context as critique of the artwork and site, which in turn, facilitates an increase in the contribution of art to wider cultural and social practice (Deutsche, 1991, in Miles, 1997, p.90). Art here is praxis and poiesis on a social scale (Lefebvre, in Whybrow, 2011, p.18) and involves a spatial performativity of the event and time and action (Hannah, 2009, p.114). Site is not a prescribed resource upon which to enact an *a priori* vision, but is generative and contingent (Kester, 2011, p.152). At the same time however, it has constructed situational qualities, a counter movement to the discursive fluidity of site-specific art (Kwon, 2000) and a movement to find a grounded, practical meaning. Site in social practice art is not a by-product of the design of the built environment but a space actively constructed by physical, social and political processes in constant negotiation (Yoon, 2009, p.70). Site is part of an embodied 'situationist impulse' of negation – the loosing of the original sense – and of prelude – the organisation of a new meaning – in relation to art and the city, this process occurring in the street and in response to it, concordantly, disrupting the macro operational level and forming a socially revolutionary 'ludic city' (Whybrow, 2011, p.14–5). The ludic city engenders a participatory citizenship, and drawing on Nancy (1986), an 'un-working' reflexive and legitimating process for those involved (in Whybrow, ibid., p.19).

Duration is part of the polylogic process of social practice art: duration may be ephemeral or longitudinal, each with relative qualities. A capped timeframe of an intervention can prompt urgent identification of specific problems and a focus on working to clearly defined and achievable goals; the 'splash' durational performance may have a particular vitality in the upsetting of the everyday habitus: 'as people go about their everyday life, they click you, they notice you' (Hamblen, 2014, p.83). Working over a longer timeframe 'could be seen as embracing more social and cooperative forms of artistic co-production for specific sites, situations or environments…through various modes of both local and dispersed forms of participation' (O'Neill, 2014, pp.198–9). The more time spent situated allows for the sense of issues to evolve and foster trust between protagonists through more sustained encounters (Jackson, 2011, pp.68–9; Kester, 2004, p.171). Change does not happen quickly either: 'You've got to have slow change' (Stern, in Lowe and Stern, in Finkelpearl, 2013, p.145) as it is the element of time unfolding that allows participants to 'attend to' the process and in doing so 'acquire new ways of producing things' (Froggett et al., 2011, p.95). Both conditions are a 'transitory state of becoming' (O'Neill, 2014, p.198) and part of an 'ethic of participatory ethnography' whereby the inherent challenge is 'to allow duration to have a different kind of aesthetic palpability' (Jackson, 2011, p.69). The durational commitment of the artist is a technique employed as a 'radical experiment' (ibid., p.70) by its de-stabilising of the fixed time and place in which to experience art or means to participate in it (O'Neill, 2014, p.198). The process of being together in a common endeavour, over a period of time itself is constitutive of relational and dialogic process, a 'cohabitational time' (Latour, in O'Neill, ibid.) that locates social practice art in a dispersed 'time-place' beyond the 'moment when the curator-producer or artist are embedded in place' (ibid.).

In the process of being in dialogue with another, in social practice art protagonists are required to systematically articulate themselves and attempt to comprehend the other's response, a reflexive experience whereby one sees the self from the other's perspective, consequently becoming more critical and self-aware of one's own position and coming to understand the other more fully (Chonody and Wang, 2014, p.96). This self-criticality can create the capacity to view the self as 'contingent and subject to creative transformation' (Kester, 2004, p.110) and 'look beyond the art object itself to the open-ended possibility of art and in the process of communication it catalyses' (Finkelpearl, 2013, p.47). In social practice art process, one form of dialogue is storytelling. This is a 'natural way to communicate with others' (Chonody, 2014, p.3); a meaning-making means of self-expression that connects self to others and cuts across intersectionality; and is empowering for the storyteller (Chonody, 2014, pp.3–4; Chonody and Wang, 2014, p.90). It is a process of discursive distanciation from the lifeworld of the storyteller that allows for self-reflexive exploration (Hamblen, 2014, p.83; Kester, 2004, pp.93–4) removing the conversationalist aside from their normative cognitive aesthetic experience, a process of 'collaboratively generated insight' (Kester, 2004, p.95–108). This is part of a micropublic place-attaching process or enactment, where the discussion of ideas publically can reinforce positive community aspects, as well as change community narrative and include marginalised voices, leading to increased intra- and inter-awareness of the community from this and the bridging of cultural differences without the cost of a cultural flattening (Chonody and Wang, 2014, pp.100–1).

Social practice art is gestural in that 'the gestural realm is not simply a "contextual" effect of the art piece but interestingly integral to its interior operation' (Jackson, 2011, p.28). Gestures help expand the understanding of art and art interactions and can have an aspect of performative and repetitive or ritualistic physical labour (Sennett, 2012, p.199). One instance of such is in gardening in social practice art projects. A 'social horticulture', recreational and community gardening represents and enacts a 'return to the simplicity of the natural environment' and 'restores balance and reduces stress' (Anderson and Babcock, 2014, p.141). Analogous to work of social practice art, the natural and social world encounters provide opportunity for sociability, learning and urban form, individual and social transformation (ibid., p.147) and 'can be a simple way to incorporate creativity into practice that does not necessarily require any artistic skills' (Chonody, 2014, p.4).

2.5.3 The re-materialised art object as output

Whereas there was a dissatisfaction with the art object from the 1960s which rejected its autonomy as the medium through which the art experience takes place and whilst this text has talked of the *process* of art, not the *product-as-object*, there is a *re-materialisation* of the art object in social practice art that forms its output. This art object though is not autonomous but collectivised by way of its production and function (Lowe, in Lowe and Stern, in Finkelpearl, 2013, p.149). As processual, the social practice art object poses an 'opticentric challenge' (Jackson, 2011, p.33) of what constitutes the material of the art object: this art object can be of material form, as well as of people and action; this art object is *used* rather than *viewed* (Beech, 2010, p.20), and is interacted with physically as question-forming, explorative and meaning-making and re-making on the occasion of an object's repurposing (Chonody, 2014, p.4; Dunk-West, 2014, p.158). Objects symbolise identity and the creation of self and relation to others and the formation of an emotional state through social interaction around the object (ibid., p.156); thus from interventional agency of the object comes a *dialogic discussion* (Sennett, 2012, p.24) experience where objects operate functionally and symbolically (Dunk-West, 2014, p.156). As co-produced and of collectivised authorship, the social practice art object is endowed subjectively with aspects of the creators' selves and is productive of 'another third, which then exists as a common cultural object' (Froggett et al., 2011, p.95). In this status, the art object creates an 'intermedial' that challenges art-object-relation boundaries: 'It is to make art from, not despite, contingency' (Jackson, 2011, p.28). Blanchette (in Doherty et al., 2015, p.165) gives the example of a bread oven as the social practice art object: the oven 'reflects a technique, a physical environment, a standard of living, a spatial organization, indeed a whole way of life. It reveals a great deal about the perceptual and conceptual schemes of the people using it. The oven may therefore be considered a total cultural fact'.

2.5.4 Outcomes of social practice art

An outcome of social practice art is reflexivity itself, individuals and communities gaining new self and social awareness, 'thereby expanding the possibilities of experiencing and representing the world differently' (Froggett et al., 2011, p.96), and expressive conscientisation (Chonody, 2014, p.2). Social practice art can dissolve 'inherent' subject positions through the social encounter (Dunk-West, 2014, p.157), an outcome experienced equally by artist and non-artist, where artists can uncover previously hidden or ignored issues and the community challenges the artists' ideas of their role and of that of the community (Kester, 2004, p.95). Reflexivity involves a change in behaviour (Dunk-West, 2014, p.157), a second outcome of social practice art resulting from its illusory capacity 'to help protagonists think and act "as if" things were different' (Froggett et al., 2011, p.95), forming 'experimental communities' (Till, 2014, p.169). Chonody (2014, pp.32–9) cites mural creation as a means of integrating local and shared knowledge and beyond being 'spots for beautification', these locations offer an opportunity for the micropublic to 'exert power and control in their immediate environment'. Thus emplaced interaction affects a sense of ownership and identification with place (Kester, 2011, p.204) and a sense of group identification and solidarity, through the breaking down of social barriers through that emplaced interaction, that in turn affects a sense of increased self-efficacy (Chonody, 2014, p.39). This has a further outcome of new, more active forms of citizenship (Kaji-O'Grady, 2009, p.113), which from a starting point of individual transformation can precipitate political change of institutional protocols; broad social values; and 'claims of spatial autonomy which result in the literal physical occupation and control of space' (Kester, 2011, p.204). However, social practice art 'is not a panacea in the sense that all you have to do is put a couple of arts programs in a poor neighbourhood and it'll be transformed' (Stern, in Lowe and Stern, in Finkelpearl, 2013, p.144). This is a process of *exemplification*: 'a mode of presenting a sense of how participation within relation-specific affective spacetimes might be considered to make a difference to the sensibility through which thinking takes place' (McCormack, 2013, p.12). In the urban context, exemplification aids an articulation around change (Till, 2014, p.165; Stern, in Lowe and Stern, in Finkelpearl, 2013, p.146). It complicates the term of encounter by presenting specifics in tandem with a transformative potential; what is exemplified is both what is within and without the spacetime conditions (McCormack, 2013, pp.12–3).

2.5.5 'Social practice arts as social work'

An issue for the social practice art field is its relation to social activism and social work. As an example, social practice art takes a political and anti-institutional stance and is active in working to 'help us imagine sustainable social institutions' (Jackson, 2011, p.14). It does this by drawing on the skills of the artist of interdisciplinary critical thinking; its 'communicative rationality', enabling discursive exchange based on intersubjective communication; and the facilitation of reflection into action (Kester, 2004, pp.90–101) – social practice arts' *critical barometer* (Bishop, in Jackson, 2011, p.48). Its melding of creative practice with 'social intent and pragmatic approach to problem solving' (Finkelpearl, 2013, p.49) is a core component of the social practice art aesthetic and draws on social practice art as the intermedial (ibid., p.28) to dissolve or play with the boundary positions of the art event. In the urban realm and of emplaced arts, this aesthetic, conjoined, can activate interest in urban sites as both spaces of activity and as resistance (Buser et al., 2013, p.624). But whilst some social practice artists and arts organisations work in an ambivalent field or see no tension between the connecting of social intent and art, 'between "excellence" and "usefulness"' (Froggett et al., 2011, p.9) as ameliorative, for others, the instrumentalisation of art renders its critical and resistant position redundant and risks becoming neutralised and homogeneous (Bishop, 2006). The instrumentalisation of art is a process levelled at funders and the state as promulgated through outreach programmes initially which have now become the *raison d'être* of arts administration in its quest for new audiences, particularity from disenfranchised or impoverished demographics (Beech, 2010, p.17). In this respect, artists – in neoliberal economies at least – have been 'asked to pick up the pieces of [US] education, health, and welfare systems that have been increasingly "rolled back"' (Jackson, 2011, p.27). Bishop (2006) though requests that art is removed from this overt social pragmatism function and ethical stance: instead, in the placing of the aesthetic and the social together, social practice art is an exemplar for a different way of thinking and doing.

2.6 Art in the public realm and placemakings schematic

Moving from the urban and arts thinking to placemaking in the following chapter, as a deeper rendering of an understanding of placemaking as well as its critique and extension of theory by way of the placemaking typology, this section closes with a schematic timeline of the development of art in the public realm from its post-Modern move out of the gallery/museum sector to new genre public art and social practice art aligned with a similar transition and sharing of attributes across placemakings. The terms of these placemakings will be introduced and discussed in the following chapter. It is intended for this to serve as illustrative of the relative progression of positioning of social practice arts and social practice placemaking in particular, but also vis-à-vis other identified art genres and placemakings.

ART PRACTICE	PUBLIC	NEW GENRE PUBLIC		SOCIAL PRACTICE
PLACEMAKING PRACTICE	PUBLIC REALM	CREATIVE	PARTICIPATORY	SOCIAL PRACTICE
PROCESS	Top down Public as object	Consultative Participatory Public as subject		Social practice art Public as co-producers Performance
FORM	Monumental Macro Commissioned	Macro to meso Commissioned Monumental to grassroots		Micro/hyperlocal Self-initiating or informally generated

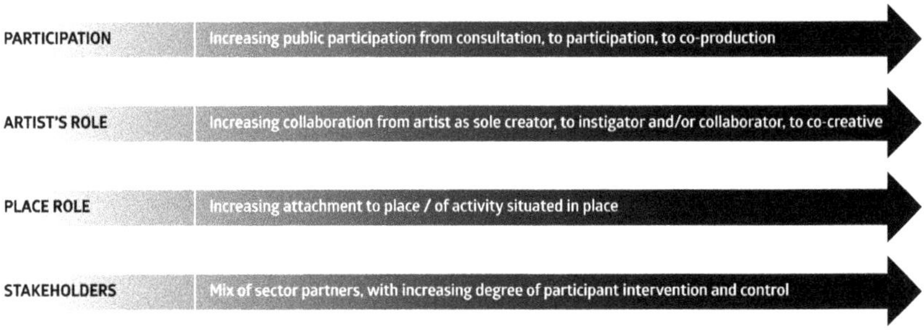

PARTICIPATION	Increasing public participation from consultation, to participation, to co-production
ARTIST'S ROLE	Increasing collaboration from artist as sole creator, to instigator and/or collaborator, to co-creative
PLACE ROLE	Increasing attachment to place / of activity situated in place
STAKEHOLDERS	Mix of sector partners, with increasing degree of participant intervention and control

Figure 2.1 Schematic of art in the public realm and social practice art and placemakings.

2.7 Bibliography

Adamek, M. and Lorenz, K. (2008) 'Be a crossroads: public art practice and the cultural hybrid' in Cartiere, C. and Willis, S. (eds.) *The practice of public art*. New York: Routledge.

Amin, A. (2008) 'Collective culture and urban public life' in *City* 12, 1 (April): 5–24.

Anderson, K. A. and Babcock, J. R. (2014) 'Horticultural therapy: the art of growth' in Chonody, J. M. (ed.) *Community art: creative approaches to practice*. Champaign, IL: Common Ground.

Arnstein, S. R. (1969) 'A ladder of citizen participation' in *AIP Journal*, 216–23 [Online]. Available at: http://geography.sdsu.edu/People/Pages/jankowski/public_html/web780/Arnstein_ladder_1969.pdf. (Accessed: 25th August 2015).

Bang Larsen, L. (1999/2006) 'Social aesthetics' in Bishop, C. (ed.) *Participation*. London: Whitechapel Gallery and The MIT Press.

Barok, D. (2009) 'On participatory art: interview with Claire Bishop', Interview made after workshop Monument to Transformation organised by Tranzit Initiative in Prague, July 2009 [Online]. Available at: http://cz.tranzit.org/en/lecture_discussion/0/2009-07-10/workshop-monument-to-transformation-copy. (Accessed: 28th August 2013).

Becker, H. S. (2003) 'Art as collective action' from American Sociological Review. 39 (1974) 767–76 in Tanner, J. (ed.) *The sociology of art: a reader*. London: Routledge.

Beech, D. (2010) 'Don't look now! Art after the viewer and beyond participation' in Walwin, J. (ed.) *Searching for art's new publics*. Bristol: Intellect.

Bishop, C. (2004) 'Antagonism and relational aesthetics' in *October* 100: 51-80.

Bishop, C. (2006) 'Introduction – viewers as participants' in Bishop, C. (ed.) *Participation*. London: Whitechapel Gallery and The MIT Press.

Bishop, C. (2012) *Artificial hells: participatory art and the politics of spectatorship*. London: Verso.

Bosch, S. and Theis, A. (eds.) (2012) *Connection: artists in communication*. Interface: Centre for Research in Art, Technologies, and Design. Belfast: Dorman and Sons Ltd.

Bourdieu, P. (1986) 'The forms of capital' in Richardson, J. (ed.) *Handbook of theory and research for the sociology of education*. New York: Greenwood.
Bourriaud, N. (1998/2006) 'Relational aesthetics' (1998) in Bishop, C. (ed.) *Participation*. London: Whitechapel Gallery and The MIT Press.
Bresnihan, P. and Byrne, M. (2014) 'Escape into the city: everyday practices of commoning and the production of urban space in Dublin' in *Antipode* 47, 1: 36–54.
Brown, A. (2012) 'All the world's a stage: venues, settings and the role they play in shaping patterns of arts participation' in *Perspectives on non-profit strategies*. [n.p]: Wolfbrown.
Buser, M., Bonura, C., Fannin, M. and Boyer, K. (2013) 'Cultural activism and the politics of place-making' in *City* 17, 5: 606–27.
Chonody, J. M. (2014) 'Approaches to evaluation: How to measure change when utilizing creative approaches' in Chonody, J. M. (ed.) *Community art: creative approaches to practice*. Champaign, IL: Common Ground.
Chonody, J. M. and Wang, D. (2014) 'Realities, facts, and fiction lessons learned through storytelling' in Chonody, J. M. (ed.) *Community art: creative approaches to practice*. Champaign, IL: Common Ground.
Colombo, M., Mosso, C. and de Piccoli, N. (2011) 'Sense of community and participation in urban contexts' in *Journal of Community and Applied Social Psychology* 11: 457–64.
Connerton, P. (1989) *How societies remember*. Cambridge: Cambridge University Press.
Critical Art Ensemble (1998) 'Observations on collective cultural action' in *Art Journal* 57, 2 (Summer 1998): 72–85.
de Certeau, M. (1984) *The practice of everyday life*. Berkeley: University of California Press.
Debord, G. (1958/2014) 'Theory of the dérive and definitions' in Gieseking, J. and Mangold, W. (eds.) *The people, place, and space reader*. New York: Routledge.
Debord, G. (2006) 'Towards a situationist international' in Bishop, C. (ed.) *Participation*. London: Whitechapel Gallery and The MIT Press.
Dewey, J. (1958) *Art as experience*. New York: Putnam.
Doherty, C. (2004) *Contemporary art: from studio to situation*. London: Black Dog.
Doherty, C., Eeg-Tverbakk, P. G., Fite-Wassilak, C., Lucchetti, M., Malm, M. and Zimberg, A. (2015) 'Foreword' in Doherty, C. (ed.) *Out of time, out of place: public art (now)*. London: ART/BOOKS with Situations and Public Art Agency Sweden.
Dunk-West, P. (2014) 'Building objects for identity and biography' in Chonody, J. M. (ed.) *Community art: creative approaches to practice*. Champaign, IL: Common Ground.
Finkelpearl, T. (2013) *What we made: conversations on art and social cooperation*. Durham: Duke University Press.
Fleming, R. L. (2007) *The art of placemaking: interpreting community through public art and urban design*. London: Merrell Publishers Ltd.
Fletcher, H. and July, M. 'Hello' in Fletcher, H. and July, M. (eds.) *Learning to love you more*. Munich: Prestel.
Franck, K. A. and Stevens, Q. (2007) 'Patterns of the unplanned: urban catalyst' in Franck, K. A. and Stevens, Q. (eds.) *Loose space: possibility and diversity in urban life*. Abingdon: Routledge.
Freire, P. (1972) *Cultural action for freedom*. Harmondsworth: Penguin Books.
Froggett, L., Little, R., Roy, A. and Whitaker, L. (2011) New Model of Visual Arts Organisations and Social Engagement, University of Central Lancashire Psychosocial Research Unit [Online]. Available at: http://clok.uclan.ac.uk/3024/1/WzW-NMI_Report%5B1%5D.pdf. (Accessed: 12th August 2015).
Froidevaux, S. (2012) '"60 x 60": From architectural design to artistic intervention in the context of urban environmental change' in *City, Culture and Society* 4 (2013): 187–93.
Gablik, S. (1992) 'Connective aesthetics' in *American Art* 6, 2 (Spring 1992): 2–7.
Gieseking, J. and Mangold, W. (eds.) (2014) *The people, place, and space reader*. New York: Routledge.
Grodach, C. (2010) 'Art spaces, public space and the link to community "development"' in *Community Development Journal* 45, 4 (October): 474–93.

Guattari, F. (2006) 'Chaosmosis: an ethico-aesthetic paradigm' in Bishop, C. (ed.) *Participation*. London: Whitechapel Gallery and The MIT Press.

Hall, T. and Smith, C. (2005) 'Public art in the city: meanings, values, attitudes and roles' in Miles, M. and Hall, T. (eds.) *Interventions: advances in art and urban futures* (Vol 4). Bristol: Intellect Books.

Hamblen, M. (2014) 'The city and the changing economy' in Quick, C., Speight, E. and van Noord, G. (eds.) *Subplots to a city: ten years of In Certain Places*. Preston: In Certain Places.

Hannah, D. (2009) 'Cities event space: defying all calculation' in Lehmann, S. (ed.) *Back to the city: strategies for informal intervention*. Ostfildern: Hatje Cantz.

Hein, H. (1995/2005) 'What is public art? Time, place and meaning' in Neill, A. and Ridley, A. (eds.) *Arguing about art* (2nd edn.). Routledge: London.

Hicks, E. (1998) 'The Artist as citizen: Guillermo Gómez-Peña, Felipe Ehrenberg, David Avalos and Judy Baca' in Frye Burnham, L. and Durland, S. (eds.) *The citizen artist: 20 years of art in the public arena*. New York: Critical Press.

Hope, S. (2010) 'Who speaks? Who listens? Het Reservaat and critical friends' in Walwin, J. (ed.) *Searching for art's new publics*. Bristol: Intellect.

Hoskins, M. W. (1999) 'Opening the door for people's participation' in White, S. (ed.) *The Art of facilitating participation: releasing the power of grassroots communication*. New Delhi: Sage Publications.

Jackson, S. (2011) *Social works: performing art, supporting publics*. Abingdon: Routledge.

Kaji-O'Grady, S. (2009) 'Public art and audience reception: theatricality and fiction' in Lehmann, S. (ed.) *Back to the city: strategies for informal urban interventions*. Ostfildern: Hatje Cantz.

Kastner, J. (1996) 'Mary Jane Jacobs: an interview with Jeffrey Kasther' in Crabtree, A. (guest ed.) *Public art, art and design profile no 46*. London: Academy Group Ltd.

Kelly, J. (ed.) (1993/2003) *Essays on the blurring of art and life: Alan Kaprow* (extended edition). Berkeley: University of California Press.

Kelly, O. (1984) *Community, art and the state*. [n.p]: Comedia.

Kester, G. H. (2004) *Conversation pieces: community and communication in modern art*. Berkeley: University of California Press.

Kester, G. H. (2011) *The one and the many: contemporary collaborative art in a global context*. Durham: Duke University Press.

Kiiti, N. and Nielsen, E. (1999) 'Facilitator or advocate: what's the difference?' in White, S. (ed.) *The art of facilitating participation: releasing the power of grassroots communication*. New Delhi: Sage Publications.

Klanten, R., Ehmann, S., Borges, S, Hübner, M and Feireiss L. (2012) *Going public: public architecture, urbanism and interventions*. Berlin: Gestalten.

Kravagna, C. (2012) 'Working on the community: models of participatory practice' in Deleuze, A. (ed.) *The 'do-it-yourself' artwork: participation from Fluxus to new media*. Manchester: Manchester University Press.

Kwon, M. (1997) 'One place after another: notes on site specificity' in *October* 80 (Spring 1997): 85–110.

Kwon, M. (2000) 'One place after another: notes on site specificity' in Suderberg, E. (ed.) *Space, site, intervention: situating installation art*. Minneapolis: University of Minnesota Press.

Kwon, M. (2004) *One place after another: site-specific art and located identity*. Cambridge, MA: The MIT Press.

Lacy, S. (2008) 'Time in place: new genre public art a decade later' in Cartiere, C. and Willis, S. (eds.) *The practice of public art*. New York: Routledge.

Larsen, L B. (1999/2006) 'Social aesthetics' in Bishop, C. (ed.) *Participation*. London: Whitechapel Gallery and The MIT Press.

Leeson, L. (2008) Art with communities: reflections on a changing landscape. Available at: http://ixia-info.com/new-writing/loraineleeson/ (Accessed: 5th December 2015).

Lefebvre, H. (1984) *The production of space*. Translated, Nicholson-Smith, D. Malden, MA: Blackwell Publishing.
Lehmann, S. (2009) *Back to the city: strategies for informal intervention*. Ostfildern: Hatje Cantz.
Lewis, J. (2013) 'Adapting underutilised urban spaces, facilitating "hygge"', 16th September 2013, Cities for People, Gehl Architects [Online]. Available at: http://gehlcitiesforpeople.dk/2013/09/16/adapting-underutilized-urban-spaces-facilitating-hygge/. (Accessed: 4th December 2013).
Lilliendahl Larsen, J. (2014) 'Lefebvrean vagueness: going beyond diversion in the production of new spaces' in Stanek, Ł., Schmid, C. and Moravánszky, Á. (eds.) *Urban revolution now: Henri Lefebvre in social research and architecture*. Farnham: Ashgate Publishing Ltd.
Lippard, L. (1973) *Six years: the dematerialization of the art object from 1966 to 1972*. Berkeley: University of California Press.
Lippard, L. (1997) *The lure of the local: senses of place in a multi-centered society*. New York: New Press.
Lossau, J. (2006) 'The gatekeeper – urban landscape and public art' in Warwick, R. (ed.) *Arcade: artists and placemaking*. London: Black Dog Publishing.
Low, S. (2014) 'Spatializing culture: an engaged anthropological approach to space and place (2014)' in Gieseking, J. and Mangold, W. (eds.) *The people, place, and space reader*. New York: Routledge.
Madyaningrum, M. E. and Sonn, C. (2011) 'Exploring the meaning of participation in a community art project: a case study on the Seeming Project' in *Journal of Community and Applied Social Psychology* 21: 358–70.
Maksymowicz, V. (1990) 'Through the back door: alternative approaches to public art' in Mitchell, W. T. J. (ed.) *Art and the public sphere*. Chicago: University of Chicago Press.
Mancilles, A. (1998) 'The citizen artist' in Frye Burnham, L. and Durland, S. (eds.) *The artist as citizen: 20 years of art in the public arena*. New York: Critical Press.
McCormack, D. P. (2013) *Refrains for moving bodies: experience and experiment in affective spaces*. Durham: Duke University Press.
McGonagle, D. (2007) 'Foreword' in Butler, D. and Reiss, V. (eds.) *Art of negotiation*. Manchester: Cornerhouse Publications.
Merleau-Ponty, M. (1962) *The phenomenology of perception*. New York: Humanities Press.
Meyer, J. (1995/2009) 'The functional site, or, the transformation of site specificity' in Doherty, C. (ed.) *Situation*. London: Whitechapel Gallery and The MIT Press.
Miles, M. (1989) 'Questions of language' in *Art for public spaces: critical essays*. Winchester: Winchester School of Art Press.
Miles, M. (1997) *Art, space and the city: public art and urban futures*. London: Routledge.
Munn, N. (1996) 'Excluded spaces: the figure in the Australian Aboriginal landscape' in *Critical Inquiry* 22(3): 446-65.
Murray, M. (2012) 'Art, social action and social change' in Walker, C., Johnson, K. and Cunningham, L. (eds.) *Community psychology and the socio-economics of mental distress*. Basingstoke: Palgrave Macmillan.
Nancy, J. L. (1986/2006) 'The inoperative community' in Bishop, C. (ed.) *Participation*. London: Whitechapel Gallery and The MIT Press.
O'Neill, P. (2014) 'The curatorial constellation – durational public art, cohabitattion time and attentiveness' in Quick, C., Speight, E. and van Noord, G. (eds.) *Subplots to a city: ten years of In Certain Places*. Preston: In Certain Places.
Puype, D. (2004) 'Arts and culture as experimental spaces in the city' in *City* 8, 2: 295–301.
Rancière, J. (2004/2006) 'Problems and transformations in critical art' in Bishop, C. (ed.) *Participation*. London: Whitechapel Gallery and The MIT Press.
Reed, S. (2005) 'Art and citizenship' in Charity, R. (eds.) *ReViews: artists and public space*. London: Black Dog Publishing.
Reiss, V. (2007) 'Introduction' in Butler, D. and Reiss, V. (eds.) *Art of negotiation*. Manchester: Cornerhouse Publications.

Rendell, J. (2006) *Art and architecture: a place between*. London: I. B. Tauris.
Rounthwaite, A. (2011) 'Cultural participation by group material between the ontology and the history of the participatory art event' in *Performance Research: A Journal of the Performing Arts* 16, 4: 92–6.
Sen, A. and Silverman, L. (2014) 'Introduction – embodied placemaking: an important category of critical analysis' in Sen, A. and Silverman, L. (eds.) *Making place: space and embodiment in the city*. Bloomington: Indiana University Press.
Sennett, R. (2012) *Together: the rituals, pleasures and politics of cooperation*. London: Allen Lane.
Serres, M. (1980/2007) *The parasite*. Minneapolis: University of Minnesota Press.
Sherlock, M. (1998) 'Postscript – no loitering: art as social practice' in Harper, G. (ed.) *Interventions and provocations: conversations on art, culture and resistance*. Albany: State University of New York Press.
Shirky, C. (2008) *Here comes everybody: how change happens when people come together*. London: Penguin Books.
Suderberg, E. (2000) 'Introduction: on installation and site specificity' in Suderberg, E. (ed.) *Space, site and intervention: situating installation art*. Minneapolis: University of Minnesota Press.
Till, K. E. (2014) '"Art, memory, and the city" in Bogotá: Mapa Teatro's artistic encounters with inhabited places' in Sen, A. and Silverman, L. (eds.) *Making place: space and embodiment in the city*. Bloomington: Indiana University Press.
Tonkiss, F. (2013) *Cities by design: the social life of urban form*. Cambridge: Polity.
van Heeswijk, J. (2012) Public art and self-organisation, London (conference), August 2012. Available at: http://ixia-info.com/events/next-events/public-art-and-self-organisation-london/. (Accessed: 15th January 2016).
Vaughan Williams, K. (2005) 'We need artists' ways of doing things: a critical analysis of the role of the artist in regeneration practice' in Blundell-Jones, P., Petrescu, D. and Till, J. (eds.) *Architecture and participation*. Abingdon: Spon Press.
Walwin, J. (2010) 'Introduction' in Walwin, J. (ed.) *Searching for art's new publics*. Bristol: Intellect.
Watson, S. (2006) *City publics: the (dis)enchantments of urban encounters*. Abingdon: Routledge.
White, S. (1999) 'Participation: walk the talk' in White, S. (ed.) *The art of facilitating participation: releasing the power of grassroots communication*. New Delhi: Sage Publications.
Whybrow, N. (2011) *Art and the city*. London: I. B. Tauris.
Yoon, M. J. (2009) 'Projects at play: public works' in Lehmann, S. (ed.) *Back to the city: strategies for informal urban interventions*. Ostfildern: Hatje Cantz.

2.8 List of URLs

1 Joseph Beuys: www.tate.org.uk/art/artists/joseph-beuys-747
2 Allan Kaprow: www.allankaprow.com/
3 Fluxus: www.fluxus.org/
4 The Diggers: www.diggers.org/default.htm
5 European Expert Meeting on Percent for Art Schemes: www.publicartonline.org.uk/resources/reports/percentforart/percent_schemes.php
6 Critical Art Ensemble: www.critical-art.net/
7 In Certain Places: http://incertainplaces.org/
8 The Citizens Handbook: www.citizenshandbook.org/
9 Big Car: www.bigcar.org/

3   A typology of placemaking

In the current time, urban arts projects can be undertaken as a 're-imagining' of public space and to foster a social connectivity, placemaking and place activation, the 'latest must-haves in the tool kits of city planners' with creative individuals playing an active role in this (Cohen, 2009, p.142). Having approached the subject matter from an arts perspective, the text now turns to placemaking, situated as it is, and the case studies are, in the built environment and with creative approaches to urban function. As a practice, placemaking is a spectrum of urban design interventions that include physical interventions and a concern with the cognitive, social and cultural aspects of place. The text argues that whilst the placemaking sector as a whole shares within it some core similarities regarding aims, objectives and outcomes, as a developed field, it has a breadth and depth of process that goes unrecognised and a diversity of approaches and practices consumed under the meta 'placemaking' category. There are tensions along creative and political lines inherent in the sector, and thus, this chapter will define the two widespread terms of placemaking and creative placemaking and then go on to delineate other forms of placemakings as yet unarticulated in the sector, *Public Realm placemaking* and *participatory placemaking*, and then detail the field of social practice placemaking. Together, these forms of placemakings mutually indicate a similar break from normative practice as *new genre public art* (Lacy, 2008) did from Public Art in the 1980s, as presented schematically at the end of the preceding chapter. The chapter will close with the presentation of a placemaking typology [3.7].

3.1 Understanding placemaking

In its broadest sense, placemaking is the term used by the architectural and planning professions to describe the process of creating the material and social spaces of place so that they are desirable for the public to visit and spend time in (Marshall, 2009; Olin et al., 2008; Uytenhaak, 2008; Wiedenhoeft, 1981). Placemaking recognises that being in a place is an affective and social phenomenon, that sights, sound, environmental factors, ambiance, and imagination all play a role historically and contemporaneously in the making of place and how it is used (Bachelard, in Coglan, 2010; Herbert and Thomas, 1990, p.255) – the sense of place as above. Placemaking holds an assumption of a strong 'mutually constitutive' (Watson, 2006, p.6) and positive affect between the built environment and behaviour: as with positive urban experience, public spaces are positioned in a value framework, viewed as essential for urban civility and for creating rich relations between strangers (Sennett, in Watson, 2006, p.14) where people shape the place around them according to their needs and desires (Gotteliener and Hutchinson, 2006, p.18; Watson, 2006, p.8). Place has been 'rediscovered' in city thinking (Nicholson, 1996, p.117) and is now viewed as an imperative organising principle in the delivery of spatial policy, especially in times of economic austerity, a product of the discourse of social relations and 'the promotion of social justice and cohesion, and the delivery of responsible economic development' (Roberts, 2009, p.441). New forms of cultural placemaking are being devised and produced by and with creative organisations and citizens, using arts as a means of urban revitalisation and acting at the hyperlocal as a point of intervention and delivery; placemaking is a 'critical arena in which people can lay claim to their "right to the city"' (Silberberg, 2013, p.6), taking place as it does, in the public – and civic – realm.

Numerous definitions of placemaking are found across academic and popular placemaking literature which converge and diverge in various ways, illustrative of the contestation around issues of place management and analysis (Roberts, 2009, p.439). Placemaking is concerned with the creation or improvement of (predominantly) urban environments that reflect community values and their socio-economic and environmental context (Legge, 2012, p.34). It is an 'art' and concerned as much with the function of places as their aesthetic and relationality, its focus on the generative and co-dependent relations between people and the physical environment to ensure the 'success' (CABE and DETR 2000, in Sepe, 2013, p.xvi) of places, however success is measured. Global placemaking agency Project for Public Spaces (PPS) [n.1] defines placemaking as an 'overarching idea and a hands-on tool' (PPS [b]) for urban improvement. The BMW Guggenheim Lab [n.2] stresses the professional constituent cross-cutting inclusive design process of placemaking (Nicanor and Antilla, 2012), an approach supported by 2007 report from The Social Impact of the Arts Project (SIAP) [n.3] research group at University of Pennsylvania. This report includes an economic imperative in placemaking, 'to increase economic opportunity, the quality of public spaces, and investment and development activity in distressed places' (SIAP, 2007). As a process, for Schneekloth and Shibley (2000, p.132) placemaking is an active act of both creating and maintaining places through cultural work that 'allows for multiple standpoints and momentary meanings that facilitate or hinder daily life'. For Silberberg (2013, p.12), this process is similarly fluid and should include multivariate entry points for community participation. Thus, placemaking is the 'set of social, political and material processes by which people iteratively create and recreate the experienced geographies in which they live' and is a networked process 'constituted by the socio-spatial relationships that link individuals together through a common place-frame' (Pierce et al., 2011, p.54). It involves participation in both the production of meaning and the means of production of a locale (Lepofsky and Fraser, 2003, p.128) to deliberately shape space to improve a community's quality of life in place (Silberberg, 2013, pp.1–2).

3.2 Understanding creative placemaking

Placemaking that has an explicit arts element to it, in the vernacular, is grouped under the term creative placemaking (PPS [e]). Creative placemaking though has a boundaried definition as proposed in a white paper from the USA's National Endowment for the Arts (NEA) [n.4]:

> In creative placemaking, partners from public, private, non-profit, and community sectors strategically shape the physical and social character of a neighbourhood, town, city, or region around arts and cultural activities. Creative placemaking animates public and private spaces, rejuvenates structures and streetscapes, improves local businesses viability and public safety, and brings diverse people together to celebrate, inspire and be inspired (Markusen and Gadwa, 2010 [b], p.3]).

Thus, creative placemaking utilises arts in public–private partnerships in the proactive making of place, with an economic imperative to precipitate localised economic development, creating jobs; fostering entrepreneurs and cultural industries as well as new products and services; and attracting and retaining businesses and skilled workers (Gallant, 2013; Markusen and Gadwa, 2010 [b], p.3). The artist in the creative placemaking process is 'an entrepreneurial asset ripe for development' and the arts a tool of 'revitalisation of the city to encourage its businesses and citizens to undertake their own making of place' (Markusen and Gadwa, 2010 [a], p.3), a form of 'urban homeopathy' whereby 'the art is a cure in itself, with the intervention symbolizing the curing' (Bishop, in Barok, 2009). The creative placemaking understanding has further been instrumentalised (Lilliendahl Larsen, 2014, p.330) to strive towards a number of key 'liveability goals', of 'public safety, community identity, environmental quality, affordable housing, workplace options for creative workers, collaboration between civic, non-profit and for-profit partners', and aims to instil 'more beautiful and reliable public transport' (ibid., p.5). Arguably, this is the creation of the city for Florida's (2011) creative class that desire fluid 'live–work–learn–play' places that have 'active, participatory forms of participation' and like to live in mixed-use live–work districts where amenities and leisure functions sit side-by-side (ibid., pp.167–8), a culturisation (Zukin, 2010, p.3) that assigns the arts an entrepreneurial role in city-making (Nowak, 2007) and in purporting casual links between cultural activity and neighbourhood improvement (SIAP, 2007) and consequently, market improvement (Knight Soul of the Community, 2010; Mitchell, 2003; Tonkiss, 2013, p.165). The NEA creative placemaking definition is important: as Weger (2013) states, this is setting the benchmark for practice and funding in the USA and, as a sector thought-leader, globally. It has understandably directly informed the NEA's *ArtPlace* [n.5] funding scheme, an Obama administration funding initiative of thirteen foundations, eight federal agencies and six banks. Creative placemaking is a process of 'investing in art and culture at the heart of a portfolio of integrated strategies that can derive vibrancy and diversity is powerful that it transforms communities' (Gage,

2013). It is also driving the placemaking mission of grassroots neighbourhood revitalisation organisations like Community Development Corporations (CDC) in the USA (Hou and Rios, 2003, p.19).

Placemaking credited as creative placemaking though is meeting with critique. Cohen (2009, p.145) attributes the term 'place-faking' to the process whereby artists are placed into a project and where placemaking is *done to* a community, not emergent *from* it. The authors of the NEA White Paper themselves critique creative placemaking and arguably what has been deliberately adopted by city administration and funders is the fiscal imperative of the definition, with the arts instrumentalised and entrepreneurialised and the questions of the paper discourse largely ignored. Markusen states that there is confusion as to what the outcomes of creative placemaking should be, ranging widely from job creation and tourism to an increase in property values and the provision of residential services (Schupback, 2012), a culturised (Zukin, 2010, p.3) flattening of arts value that is ignorant of arts socially transformative values (Stern, in Lowe and Stern, in Finkelpearl, 2013, p.150). The definition also attests to a creative participation approach being a guarantor of participation, problematic as 'the arts' are for many people an exclusive arena, thus including arts in a project is no guarantee of participation (Arts Council England, 2011) the modes of participatory arts problematic too of course. The NEA White Paper makes no differentiation in participation in arts activity between formal arts activity and grassroots cultural activity, nor does it address the demographics of people that attend arts events – those that are time and cash rich (ibid.).

3.3 Aims and benefits of placemaking and creative placemaking

In broad terms, placemaking aims to 'improve the quality of a public place and the lives of its community in tandem' (Silberberg, 2013, p.2). It aims to improve 'social comfort' (Whyte, 1980, p.32) and foster Gehl's 'hygge' (Lewis, 2013), an intimate atmosphere that in the placemaking context equates to a place that one wants to stop and linger and socialise in. Its beneficial outcomes are stated as including an empowered community, cross-sector partnerships, and an ongoing processual approach to the making of place, creating alternative financing and diverse income models (Silberberg, 2013, pp.7–12) and creating and/or enhancing social capital (Putnam, 2001, in Silberberg, 2013, p.6). The benefit of community has facilitated a nuanced and applied understanding of community, power differentials and social capital and marked a turn from '"what makes a good place" to "what – and who – make a good placemaking process?"' (ibid., p.51). Thus, a benefit of community in placemaking has been a sector shift in some practices away from 'place making' to 'place shaping', a '"making-focused" paradigm for the practice' (ibid., p.11) – again, the gap between placemaking theory and practice. This position accepts that places are already existent before a placemaking intervention, and that placemaking is a subsequent, additional mode of intervention. *Community driven placemaking* (Hou and Rios, 2003) is a collaborative process involving community members in the process, as leaders of it (Lepofsky and Fraser, 2003, p.132); it builds social capital through social cohesion (Silberberg, 2013) and signals the community *loci* in placemaking theory. Placemaking here is concerned with the formation and operation of place identity from a location that includes the wider constituents of placemaking as the community in which the activity is taking place and the associated psychosocial attributes they bring to a process. Aligned with new urbanism, there is a notion that 'well-designed public space, centrally located within an urban village, will foster or create community by bringing people closer together' (Ivseon, 1998, in Gieseking and Mangold, 2014, p.188) through horizontal and self-determining (Carmona et al., 2008, p.14) participation models. A 'virtuous cycle of placemaking' is created of mutual and consequent community and place transformation (Silberberg, 2013, p.3).

Placemaking as top-down urban design is non-contextual, produces generalist outcomes and generates further fixed notions of community and public space based on a 'pseudo-participation' (Petrescu, 2006, p.83) model that is organised and manipulated, idealised, uncritical and concerned with reaching consensus, the process effectively silencing the voices it is meant to articulate (ibid.). If creative placemaking plays a 'pseudo' role, then its transformative agency has to be questioned. Artists in Nowak's (2007) creative placemaking process are romanticised and given an assumption of universal competency in their capacity to uncover and re-create place based on their innate civic and entrepreneurial skills. A critical dissatisfaction with notions of creative placemaking was seen at the April 2013 inaugural meeting of the Placemaking Leadership Council (PLC) [n.6], a global group of those considered to be leading placemaking and formed by PPS (PPS [d]). During the three-day meeting, a splinter group formed of placemaking practitioners who felt their approach to placemaking was not being represented by the main conference programme. Self-named 'low-income placemaking',[1] this group felt a sense of alienation from the rhetoric of many speakers on the value of creative placemaking for retail and land values. Their work was located often (though not exclusively) in communities of severe social, economic, cultural and environmental need. This type of creative placemaking is evidence of a 'community turn' (Hou and Rios, 2003) in placemaking and is more aligned to more expansive definitions of creative placemaking of multivariate partnerships (Froggett et al., 2011, p.103), motivations that are not fiscal-centric (Shirky, 2008, p.109) and not culturising (Zukin, 2010, p.3) the place of the arts (Stern, in Lowe and Stern, in Finkelpearl, 2013, p.144). For Weger (2013), creative placemaking is a grassroots design tool used by and for the community in question, the tool kit comprising arts-based field research, social organising, community rituals, public interventions, educational workshops, audio documentation and performance art (Rochielle, [n.d.], p.40–1). For Puype (2004, p.301), creative placemaking should be used as a tool for strengthening local culture. Thompson (2012, p.86) views creative placemaking as 'a mode of work which is people working with culture in the realm of the social' which encompasses urban planning, sociology and pedagogy. For Hermansen (2011, pp.2–3) creative placemaking is part of the democratisation of creativity, of the move from elitist to populist and exclusive to inclusive – not so much 'social acts' as 'social creativity' – and the depth of the work is determined by the degree of meaningful change that work has made in the social group. As will be seen, this is more aligned to social practice placemaking, showing that placemaking theory has not caught up yet with or is not reflective of nor understanding of a diversity of arts-based placemaking practice. This discussion is continued below, with a consideration of the relation between architecture and placemaking.

3.4 Architecture and placemakings

To many urban realm professionals and non-professionals alike, the architecture sector is an exclusive and closed sector, operating as an 'informed' or 'synthetic' disciplinarity, informed by others but with its focus as singular or combing theories but keeping them still distinct (Lattuca et al., 2004, in Lehmann, 2009, p.15). This is a practice concerned solely with the design of space, free of consultative requirements and thus independent of consideration of the users but designed for them (Gehl and Svarre, 2013, p.2; Sime, 1986, pp.49–50) – an architecture of space. The agency lies with the architect whose role is to 'uncover' or 'infuse into' the cultural characteristics of a place (Guest, 2009, p.1; Wortham-Galvin, 2008, p.32), an ageographic perspective of space (Schneekloth and Shibley, 2000, pp.137–8). Architecture of space is thus aligned to a top-down regeneration, itself aligned to a Public Arts–based regeneration strategy where place becomes a mechanism of value perception and where, through a design-led approach, places are 'given' a sense of place and the practice is viewed as an emerging economy in the property sector. This placemaking may have no or low levels of public involvement: 'participation' may only manifest as consultation and a survey of varying degrees of preconceived ideas (Ball and Essex, 2013; Hou and Rios, 2003). An 'architecture of place' however heralds an 'architectural localvore movement' where practices work locally in an embedded way (Klinkenberg, 2013) and not on the behalf of end-users (Sime, 1986, p.49). This practice mode has followed the same move into the urban realm as art, anchored in the ethos of the Situationists and interventionist art, and following from Jacobs (1961) and rising alongside the democratisation of urban design as a social concern.

It is thus a place-led practice. The Office for Metropolitan Architecture [n.7], with Koolhaas [n.8] as its lead, 'emphasised the importance of researching the activities that already exist in urban locations, bringing these to the centre stage of architecture, as well as documenting the design process and publishing this as visual material along with images of final buildings' (Rendell, 2006, p.66). Such architecture of place practices such as Gehl [n.9], Foreign Office Architects [n.10], muf [*sic*] [n.11], public works [sic] [n.12], AOC [n.13] and Turner Prize [n.14] 2015 winners, Assemble [n.15], amongst many others, approach design as a response to contemporary life, leading the way for a wave of architectural practice which questions the materiality of architecture. Daisy Froud, formerly of AOC, for example, bases her approach on a positive agonism (Mouffe, 2005) principle, accepting other groups into the decision-making process, the same premise as that of relative expertism and the urban co-creator [2.5.1, 3.5.2]. muf works according to a design studio model, not an architecture practice one, with decreased compartmentalisation of specialism and a cross-disciplinary process (Lehmann, 2009, p.14), similar to that of public works for example, both taking a creative and people-centred approach.

Where architecture of space has a tendency to privilege the objective expert knowledge over that of the subjective, implaced knowledge, of hard skills over soft ones and the consequent devaluing of the relegated position in the decision-making process (Bosch and Theis, 2012; Miles, 1997; Roberts, 2009, p.446; Schneekloth and Shibley, 2000, p.135), the architecture of place approach for many returns the practice to its roots of looking outside of itself for inspiration (Lehmann, 2009, p.15) and is transdisciplinary and includes a wide and de-siloed knowledge base. Lehmann (ibid.) states that architecture needs to be operational in 'several domains at the same time' and architectural appreciation can only be achieved if there is an active experiential process of interaction between the appreciator and the object of appreciation. A failing of architecture of space for Roberts (2009, pp.438–46) is that it has not acted from a 'common philosophy and approach' of interdisciplinary sharing of skills: the sector instead has been siloed by professional knowledge territories, the knowledge of the community in question often excluded too, forgetting or ignoring its inherent collaborative practice mode. Architects will avoid collaboration as this is a competitive and exclusory sector which views itself as comprised of 'avant-garde specialists' that actively resist social and political pressure to behave differently (Jenkins and Forsyth, 2010, p.167). This has emerged from the relegation of 'common and generic' knowledge to that of 'subject and discipline specific', which has negated transdisciplinary and collaborative knowledge and design production (Roberts, 2009, p.446).

Schneekloth and Shibley (2000, p.130) state that to move architecture beyond its expertism is to deny its privileged status but if one subverts this rigid territorialisation, architecture of place could be seen as an act of recognition of architecture's relative expertism and begin to redress architecture's 'long history of failure to connect itself firmly to the larger concerns confronting…society'. Thus an architecture of place approach could then act as a repositioning of architecture rather than an attack and destruction of it. There has been a deliberate avoidance of using a formal/informal binary to describe these modes of practices; here the architecture of place is no less based on formal professional skills, but these skills are deployed in a different way and to different ends. An architecture of place approach is vital for the architecture sector if it is to claim that it is relevant to urban design and that it can create new ways of living for cities based on based on evidence and the answering of real questions (Provoost and Vanstiphout, 2012, p.105). This is a liminal position for architectural practice between modernist theory of expert knowledge and the post-modernist de-privileging of that expert position (Schneekloth and Shibley, 2000, p.130) – the liminal space of relative expertism. This makes architectural practice more complex and responsible and moves it closer to a social art practice or new genre public art (Lacy, 2008) one. Jenkins and Forsyth (2010, p.168) argue that if the architecture sector is recognised as providing an important societal service, which includes an understanding of social values, then architecture will naturally become embedded as a '"normal, not a nice" role'. To do this though, architecture does needs to recognise the skills of others and work with them rather than exclude them. Douglas (2012, p.43) points to a new practice of informal and spontaneous interventions by architects where there is no fee-paying client but a community group to work with that has a desire to self-determine the planning and design of their place. Art and architecture share elements of practice and process – both work with proportion, material, colour, planning and delivery for example (Lehmann, 2009, p.16) and it is the tension in the 'same but different' that creates creative vigour; the introduction of the Lefevbrean (1984) concept of the everyday brings architecture of place closer to social practice arts, and also placemaking.

This is not as simple as adding art into architecture to 'ameliorate' its complexity or professionalisation (Guest, 2009, p.4), but a holistic revision of practice that changes the architect's levels of responsibility when working collaboratively (Sulzer, 2005, pp.149–52). The role of social participation in architecture is to widen the engagement between architects and society (Jenkins, 2010, p.1) and the inclusion of a diverse body of voices in the design process will 'leave room for unpredictability and bottom-up proposals issued from real claims' (Petrescu, 2006, p.82) and position places for social encounter above that of the materiality of space (Klanten et al., 2012, p.9). In such placemaking projects the architect is required to collaborate and engage with the everyday and this progresses the architecture sector's understanding of placemaking to being an act of 'active, conflicted cultural work' that 'allows for multiple standpoints and momentary meanings' in the creation and maintaining of places (Schneekloth and Shibley, 2000, p.132). Placemaking here requires three levels of knowledge for the architect: place knowledge, the material conditions of place; local knowledge, the knowledge that 'people-in-place' have of their lives and places; and situated knowledge, the expert knowledge, situated in expert roles or practices, which include the architect (and any other designated profession included in the project) and the citizen (ibid., p.134).

3.5 Expanding the understanding of placemaking

Thus far, the two overarching terms for placemaking – placemaking itself and creative placemaking – have been introduced; this text though asserts that both are inadequate, singularly or used in reference to each other, to describe what is a multiscalar practice field. This section expands placemaking terminology by introducing new concepts of practice that reflect this. Fleming (2007, p.13) identifies in the sector common confusions of the term placemaking: it is often conflated to mean a vague sense of place and has become an 'ill-defined buzzword' which implies much but is open to misunderstanding. PPS [f] sees another confusion: the term placemaking is used in many settings, from planners and developers that latch on to the phrase as a branding tool, 'place wash' as Legge (2013) terms this, to grassroots community organisations that use it as a mechanism for making improvements to their place. The definitions of placemaking above are used in many of the same contexts, proving these terms inadequate for a sector that has myriad types of practice and which demands more nuanced and complex understandings of its practice other than a binary duality, and a granular understanding of the arts in placemaking. Legge (ibid.) is correct in seeing a current adverse stratification of two categories of placemaking, of the formal planning sector and the creative and citizen-led approach; as is Wyckoff (2014) to identify a 'standard' placemaking practice and then subset practices of strategic, creative and tactical placemaking. A schematic reading of de Certeau's (1984) strategy and tactics could be mapped – with certain limitations as regards an exact overlay of one theory to another – to types of placemaking. The strategic that is location-driven and is a projection of hegemonic power onto space is analogous to top-down placemaking; the tactical that is of the other, operating independently and on contextual impulse, in isolated spaces through detailed actions and that is opportunistic, is the same for bottom-up placemaking (Mozes, 2011, p.10). This is useful to inform a thinking of different types of placemaking and to consider where these practices could be located and their relative power relations (ibid., p.11). A strategic position defends a position of power and imposes a view of urban form and function, whereas a tactical position is defined in effect by the absence of power and in iterative manoeuvrings within place. These new practices have emerged out of a critique of the dominant placemaking discourses above, informed by new genre public art (Lacy, 2008), participatory and community art and revised considerations on the differentiation between meanings of space and place towards a multivariate 'manifold commons' use of space (Bresnihan and Byrne, 2014, p.10).

To borrow from Lacy, *new genre placemaking* concerns the creating of spaces with an emotional awareness of, and linking to, place by an individual (Sime, 1986, p.50) and to community (Boenau, 2014), a urban porosity (Stavrides, 2007) of placeshaping (Roberts, 2009) and placeshaking (Doyon, 2013). This equates to a relative expert 'user-generated urbanism' or 'collaborative city-making' (Marker, in Kuskins, 2013) that has the personal-as-political (Lacy, 2008) ethos threaded through it and is aligned to community-driven placemaking (Hou and Rios, 2003). Its framework dissolves the binary professional/non-professional demarcation of knowledge and acknowledges the multiple stakeholders and processes involved. This aids the placemaking sector to move beyond a 'top-down versus bottom-up approach' to instead view the placemaking terrain as faceted of multiple stakeholders in degrees of participation, with the foundations of this approach resting in a discourse-building process to frame issues and construct meanings collectively, a process itself that engenders shared understandings of the uses of space by others. This section will present newly termed categories of placemaking that work to make delineations in practice and process in the placemaking sector.

3.5.1 Public Realm placemaking

Public Realm placemaking (capitalisation deliberate) utilises 'culture as a means of framing space' (Zukin, 1995, p.6), and as with Public Art, it is co-opted to meet market and political ends, and is rendered complicit as a means of domination

62 *A typology of placemaking*

(Best, 2014, p.286), both cultural and material. When Fleming (2007) talks of the 'art of placemaking' the projects offered as examples are ones of a Public Realm and masterplanned strategy (ibid., pp.34–207) analogous to the practice of monumental Public Art and regeneration that utilises arts as a consumer product and as a backdrop to space. Public Realm placemaking 'draws on a managerial and visual language of urban development that works against differentiation' and 'is in large part about making urban spaces that are recognisable to, safe for and accommodating of transnational investment flows and that class of economic actors who attempt to ride them' (Tonkiss, 2013, p.11). It has limited to no public engagement (Bosch and Theis, 2012; Keast, 2012) and a legacy of 'neutral impersonal public spaces' and 'concrete monuments' (Sime, 1986, p.61), lacking the 'psychological fit between people and their physical surroundings' (ibid., p.49).

3.5.2 Participatory placemaking

Participatory placemaking is akin to a model of participatory arts and as such, is open to the same critique as such practices. Participatory placemaking is motivated by a commitment to active citizen engagement in its process, over and above a consultative one as found primarily in Public Realm placemaking. This has emerged both in response to an increased demand from developers and governments for all project stakeholders to share responsibility and decision-making (Legge, 2012, p.34) as well as from a body of placemakers that view citizen participation in placemaking as a moral imperative and the start of a process that is based on Lefebvre's (1968, in Harvey, 2008) right to the city of local citizenship. Participatory placemaking operates with an inclusive design approach (Lehmann, 2009; Newman, 2001; Sorensen, 2009) that begins to level professional and non-professional constituents in dialogue (Friedmann, 2010, p.159–62) and holds that the placemaking professional should have a level of personal involvement in a place and recognise a place's diverse range of users and patterns of behaviour and experience which give it meaning (Thompson, 1984, in Sime, 1986, pp.57–60). This approach is exemplified by PPS, which employs a participative and collective process across its numerous global placemaking projects. PPS asserts that the notion of who is creative needs to widen to cultivate the creativity of the people 'in place' rather than those 'professional' creatives that will be brought into place (PPS [b]).

Whilst participatory placemaking may be evidence of a sector turn to develop practice out of criticism of Public Realm placemaking above, grounded as this may be in a consultative approach to public engagement (Provoost and Vanstiphout, 2012, p.104), this practice is not without its own challenges, centring, as with participatory arts, around issues of inclusivity, gentrification and capacity and standardisation. Firstly, participatory placemaking may not be the inclusive process it appears to be on the surface but another pseudo-participation that in turn manifests 'pseudopublic spaces' (Sorkin, in Crawford, 1999, p.22). The professional and non-professional are wary of each other in a participatory process, compounded by the use of professionalised and exclusory language (Friedmann, 2010, p.162). The including of artists in participatory placemaking may also exacerbate this as a further professionalised class that may alienate the community (PPS [c]) and confuse urban professionals (Fleming, 2007, p.14). Secondly, there is tension between the threads of some participatory placemaking practices claiming both economic and community benefit, which may be mutually exclusive. Participatory placemaking can assume a common community and economic driver that results in gentrification (Baker, 2006, p.147; Deutsche, 1991, in Miles, 1997; Kwon, 2004) that focuses attention on the attraction and securing of a creative workforce in an area, rather than the appreciation and use of local resources (Mici, 2013). Thirdly, there are issues of capacity and practice standardisation. Not all projects will follow a linear professional phased project timeline, nor may they be completed in the short term but require a shift from normative practice to one that is more fluid in terms of non-binary roles and timeframes (Hou and Rios, 2003, p.26). So a technique such as PPS's *Lighter, Quicker, Cheaper* (PPS [a]) [n.16] approach, a 'low-cost, high-impact incremental framework', that is 'capitalizing on the creative energy of the community to efficiently generate new uses and revenue for places in transition' (PPS [a]) may fail in some contexts. There is danger when a longitudinal view is not facilitated by funders or commissioners that the *Lighter, Quicker, Cheaper* approach becomes the sector's *go-to one-size-fits-all* solution to placemaking, attractive to city authorities in a time of austerity (PPS [e]): *Lighter, Quicker, Cheaper* appeals to impoverished administrations and is therefore an attractive box ticking and cheap solution that acts as a salver to urban realm problems without any structural and meaningful change. As Cleveland (2001, p.21) warns, 'there are no short cuts to place-making [sic]' and a placemaking process of citizen participants is collaboration intensive in terms of both time and capacity (Froggett et al., 2011, p.95; Stern, in Lowe and Stern, in Finkelpearl, 2013, p.145). Whilst an approach such as *Lighter, Quicker, Cheaper* might 'activate urban capabilities' (Fernandez, 2012) expediently, The Public Art Development Trust [n.17] for example advocates a three- to five-year project timeline that includes periods for research, beta testing and redevelopment (Rendell, 2006, p.62). As will be seen in the placemaking typology [3.7.1], different types of placemaking are thus suited to different types of placemaking endeavours.

This then, brings the text to tactical urbanism[2], similar to, but a precursor of, PPS's *Lighter, Quicker, Cheaper*. Tactical urbanism is a process of urban bricolage, 'an approach to neighborhood building and activation using short-term, low-cost, and scalable interventions and policy' (Lydon and Garcia, 2015, p.2). It is iterative, capacity and material efficient, bespoke, intentional and flexible and utilises the 'creative potential unleashed by social interaction' (ibid.). Tactical urbanism interventions will commonly be fleet-of-foot; informal; hyperlocal in scale; not requiring developed infrastructure or large investment; and spontaneous and participatory, enabling individuals and communities to enact change on the 'otherwise hegemonic urban landscapes' (Hou, 2010, p.15) and redefine 'the boundaries, meanings and instrumentality of the public sphere' (ibid., p.14). Driven by community issues and motivated by grassroots activism (Fernando, 2007), tactical urbanism participates in the city by changing its cultural fabric through its interventions, opening the lived experience up to questioning and challenging the usual operations of art and architecture (Lehmann, 2009, p.14). It is employed by a range of governmental, organisational and individual citizen actors together aiming to create a 'more responsive, efficient, and creative approach to neighbourhood building' (Lydon and Garcia, 2015, p.20) and aims to trounce 'proscribed functional urban design strategies' (Klanten and Hübner, 2010, p.2). Lydon and Garcia (2015, p.10) take the view that governments should work tactically and citizens strategically, that strategies and tactics are of equal value and should be used in concert with one another. It signifies 'a porous – and more productive – relationship between grassroots activists and local government' (SPUR, 2010) as emanating from the community 'uniquely positioned to initiate community policy or programming that has far reaching effects' (Mancilles, 1998, p.339). Tactical urbanism in fact stems from the powerlessness of people vis-à-vis the urban design and planning process that does not serve their interests, 'The Challenge of Getting Things Done' (Lydon and Garcia, 2015, p.79).

Central to both tactical urbanism and *Lighter, Quicker, Cheaper* is the aim to lead by example in urban design and planning by interventions being used to pilot interventions that could lead to longer-term and permanent projects (Lydon and Garcia, 2015; PPS [a]; Silberberg, 2013, p.10). In this regard it operates at three levels:

> For citizens, it allows the immediate reclamation, redesign, or reprogramming of public space. For developers or entrepreneurs, it provides a means of collecting design intelligence from the market they intend to serve. For advocacy organizations, it is a way to show what is possible to garner public and political support. And for government, it's a way to put best practices into… practice (Lydon and Garcia, 2015, p.3).

Done by the grassroots, tactical urbanism can act then as 'placeholder' (Lydon and Garcia, 2015, p.16) projects that have a trickle-up agency (Burnham, 2010, p.139; Silberberg, 2013, p.10). Thus, duration is a mechanism of this practice. As temporary, tactical urbanism interventions capitalise 'paradoxically on the nature of their ephemeral power, namely the affect and effect they exercise on the liquid sociocultural statistics of our cities' (Klanten et al., 2012, p.9). As Rebar [n.18] – the 'interdisciplinary studio operating at the intersection of art, design and activism' that founded Park(ing) Day [n.19] – founder, Marker, asserts 'culture changes faster than infrastructure', (in Kuskins, 2013), thus formal placemaking, as argued above, cannot keep pace with the users of a space so these 'quick adaptations' can work as a pilot to generate responsive data on an intervention that could be used to inform the formal planning process (Lewis, 2013), as with Rancière's (2004, p.86) and Bishop's micro-politic border-crossings (2012, p.258) and Debord's (1958/2014) *La Dérive*. As an artform, tactical urbanism is ephemeral in urban context and display, akin to outsider art, a decentralised and democratised art in space and place. It represents a development in the dialogue between vernacular and formal street design, identifying and filling gaps in formal design. Artist tactical urbanists, as with social practice arts artists, view process and outcome as the 'art installation' (Suderberg, 2000, p.2). As pop-up or of a short and limited duration, tactical urbanism is similar to 'splash' art interventions, interventions that 'catalyse creative thinking in the viewer and encourage us to question our relationship with the city and what it offers, as well as what we can offer back' (Legge, 2012, p.78). This is, taking from placemakings' formative text by Whyte (1980), a triangulated urban encounter, where the unexpected or remarkable causes someone to break their routine and/or speak to a stranger in an act of social barrier-breaking (ibid., p.94; Beyes, 2010; Kaji-O'Grady, 2009; Klanten and Hübner, 2010; Lydon and Garcia, 2015; Rendell, 2006) – the aesthetic dislocation of social practice art in place.

With the introduction of two new terms to the placemaking field, Public Realm placemaking and participatory placemaking, and the discussion of modes of interventions such as tactical urbanism and *Lighter, Quicker, Cheaper* (PPS [c]), the chapter now turns to the central focus of the research project, that of the identification and investigation of social practice placemaking. The following section will define this term and outline its practice, process and aims and outcomes.

3.6 Social practice placemaking

As this text has positioned its thus far, social practice placemaking is a post–new genre public art practice and process. This section will define social practice placemaking as a performative and co-produced art form and advance social practice placemaking as the latest development in social practice art and art in the public realm as joining both an urban arts and placemaking timeline and practice.

Social practice placemaking is not creative placemaking as an 'arts-plus-placemaking' practice; principally its art form intentionality functions around social issues and a questioning of normative urban form and function. As posited in this thesis, creative placemaking has failed or been misaligned as an umbrella term for 'placemaking-plus-art' and where participatory placemaking limits the agency of those 'invited' to participate, social practice placemaking is more closely aligned to that of dialogical aesthetics (Kester, 2011) and the social aesthetic. It also works beyond participative practice of participatory arts and participatory placemaking as a collaborative practice enacted at the hyperlocal level, at the other end of the scale to Public Realm placemaking for example. Whilst Public Realm placemaking, creative placemaking and participatory placemaking are practices with their own integrity, to place social practice placemaking-type arts-based placemaking under these practice headings is misrepresentative and inaccurate. Following Hou and Rios (2003, p.27) social practice placemaking recognises the active social engagement of a new wave of placemaking practice. As such, social practice placemaking represents a progression in practice for the placemaking sector and a deepening of the work of social practice art as a situated emplaced art form. Where social practice placemaking is a term created in this thesis, the following section uses and extends thinking on emplaced social practice arts through social practice placemaking as a deeper observation and understanding of emplaced social practice art; put simply, thinking to date to a large extent has been describing a practice that is a step beyond new genre public art, creative placemaking, participatory placemaking and tactical urbanism and this thesis employs a more nuanced reflection on practice to demarcate the specific practice of social practice placemaking.

This section will first look at social practice placemaking practice and then its process; it should be noted however, just as with social practice arts in the vernacular, practice is often transposable with process and vice versa, where the city is used by artists as a lab for social interaction and invention, the practice is a process and the artworks are evolving and mobile (Graham, 2009, p.124).

3.6.1 Social practice placemaking practice

Social practice placemaking is concerned with the social aesthetic encounter – an 'encounter art' (Rancière, 2004/2006) – found in both relational (Bourriaud, 1999/2006, p.161–5) and dialogic (Kester, 2011) aesthetics and the 'relational specificity' of the interactions between objects, people and spaces (Kwon, in Rendell, 2006, p.33) and of space, site and process (Beech, 2010, pp.17–8), where site has relational and dialogic agency as networks, systems and processes (Pierce et al., 2011, p.59; Rendell, 2006, p.36). As an arts practice, social practice placemaking is an immersed art/artist placemaking process (Guest, 2009, p.5) that involves a deeper level of engagement in the process of art making than found in other placemakings, the artist an 'un-artist' (Kaprow, 1996/2006) that works in the public interest. This is a practice that includes an art object and event situation in the participant-art encounter, Debord's (1958/2014, p.96) 'unitary urbanism', pertaining to the use of the arts practice and process as a methodology in the creation of the urban environment. Just as with aesthetic dislocation [2.4.1], social practice placemaking works to 'jolt cultural assumptions', using the performative everyday to engender at an 'engagement through alienation' (Klanten and Hübner, 2010, p.3; Rendell, 2006) and a 'disruption of the sensible' expected urban norm (from Rancière, Beyes, 2010, p.231) in the urban event space (Hannah, 2009, p.117). Social practice placemaking is performative: it is 'enmeshed unselfconsciously within social life, rather than a conscious decision to participate in capital "C" culture' (Messham-Muir, 2009, pp.120–3) of Public Realm placemaking or creative placemaking for example. It affects by drawing people into the work through its performative disruption of the urban norm and engenders new urban semantics by making the lived experience active, a dissonance of open space. This precipitates a process of reflection, imagination and the uncovering of hidden or silenced vernaculars (Amin, 2008, pp.17–8; Lippard, 1997). Social practice placemaking performs duration and temporality as a concept and as a way of generating projects (Bishop and Williams, 2012, p.121), temporary here pertaining to the duration and intention of a project: interventions may be temporary in their first or whole iteration, either as a deliberate act of meanwhile use, or evoking a longer-term or permanent installation.

Social practice placemaking opposes fixed meanings of spatial use, being a durational practice of 'urban porosity' (Stavrides, 2007), discursive and peripatetic and with a 'loose' unintended use of space (Franck and Stevens, 2007, p.2).

It involves the 'revitalisation of left-over urban spaces' (Lehmann, 2009, p.17), operating from liminal and border positions that is precipitative and symptomatic of a relational practice (Schneekloth and Shibley, 2000, p.131). This is a placemaking of 'spaces of uncertainty' (Cupers and Miessen, 2002, in Petrescu, 2006, p.88) which are 'heterogeneous, fragile, indefinite, fragmented and multiple' (Petrescu, 2006, p.88). The social practice placemaking site is materially localised as site-specific (Klanten and Hübner, 2010, p.63) and concerned with creating an urban built environment that enfranchises urban dwellers through an active appropriation of space (Sherlock, 1998, pp.219–20). Social practice placemaking spaces are 'open-ended' (from Rappaport, 1968, in Fernando, 2007, pp.55–71). Site is a place where people can come together to exert control over a space and readapt it, explaining how space can accommodate multiple discursive and cultural uses and express socio-cultural identity and have agentive potential as countersite (Holsten, 1998, p.54, in Watson, 2006, p.170). In the *performative event space*, interventions can come to define the public space itself as free spaces where rules are suspended, becoming 'spaces of uncertainty' (Cupers and Miessen, 2002, in Petrescu, 2006, p.88), 'recasting' the urban space in question as 'heterogeneous, fragile, indefinite, fragmented and multiple' (Petrescu, 2006, p.88; Ostwald, 2009, p.97).

Artists in this practice are moving beyond participatory art models to co-production and 'bypassing cultural gatekeepers and gaining more creative control over the entirety of the arts experience, if only to relinquish it back to the audience' (Brown, 2012, p.9). The 'trans-aestheticisation of art into place' (Whybrow, 2011, p.6–8) recasts the city as the site of art and the polylogic inclusion of the participant creates a practice of an 'embodied placemaking' (Sen and Silverman, 2014, p.2) with its own *spiritus loci* (Fleming, 2007) of place and the body's agency in it as mutually constitutive. With this fusing of site and the artwork comes an increased participation of art in a wider social and cultural milieu (Deutsche, 1992, p.159). This subverts any lack of public autonomy in participatory practices (Miles, 1997, p.1). Social practice placemaking interventions are contextual creative acts and as self-made urban spaces, are new acts of expression of the urban city (Hou, 2010), tapping into the zeitgeist of people's desire to be part of a community (Boenau, 2014) and place-attached and thus aligning with positive urban experience thinking. This is the dialogic aesthetic of relative expertism and urban co-creators where 'knowledge of the professional, the place, and the local people are shared, disrupted, negotiated and considered' (Schneekloth and Shibley, 2000, p.136). It works from a collaborative arts epistemology in which all participants have legitimate claims to knowledge construction and devising that aligns to a people-in-place-based placemaking and the democratisation of urban design as a social concern, a conscious raising of architecture methodology to a critical spatial practice and a mode of cultural production that is a step removed from economic and functional concerns (Rendell, 2006, p.191). A transdisciplinary working, social practice placemaking references Giroux's cultural codes and border pedagogy (ibid., p.131) where expert knowledge is exchanged in an act of reciprocal learning, that is an actant for power and control to be located in the

68 *A typology of placemaking*

citizenry, creating a 'space in which to discuss the social and cultural construction of meaning' (ibid., p.130). This holds creativity at its 'widest expression, thinking and doing' (Legge, 2012, p.5), urban creativity as such a realm of reflexivity and non-consensus co-production (Richardson and Connelly, 2005, p.80). As a co-produced endeavour, knowledge creation is not singular, but relational (Eisner, 2008), occurring between community members that possess a holistic view and knowledge of their place and create a shared language (Nicholson, 1996, p.115).

3.6.2 Social practice placemaking process

Social practice placemaking interventions are a self-activating process: people recognise the potential of a space; determine to use the space in varying degrees of creativity; and create it to suit their needs and desires (Franck and Stevens, 2007, p.10; Murray, 2012, p.257; Pierce et al., 2011, p.54). This is a common process for interventions and a process of artists engaging in political dialogue that is both a collective process and an interpretive one – the 'new situationism' (Doherty, 2004) again which is a polylogic of situational-relational impulse and performativity in a processual contextual interaction of people, site and object. As with a social aesthetic, we are reminded of Koolhaas' 'specific indeterminacy' (Mozes, 2011, p.8) and Whybrow's (2011, p.28) 'critical materialism' where art does not equal an outcome of labour but is the labour itself. Social practice placemaking process is a critical spatial practice of art being a 'doing' of work and a contextual performance, manifested via an engagement with the multifarious urban lived experience (Rendell, 2006; Whybrow, 2011, pp.25–6) where there is both an interaction of material objects as well as a relation of rhizomatic exchanges, movement and communication (Deleuze and Guattari, 2002). This process re-materialises the art object, creating both a phenomenological event and an object (McGonagle, 2007, p.7) for people to both create and respond to, the critical materialism.

The function of the social practice placemaking art object is to act ideologically (Baker, 2006, p.147), using it to interrogate and contest the common functions of the urban realm (David, 2007, p.250). The artwork – as process and re-materialised object – sits in a cultural and subjective position, both the site and object of reflexive art experience, animated by this interaction of in-situ evocation and its participatory meaning-making productive role (Whybrow, 2011, p.36). The artwork has a parasitic quality where it acts as the aesthetic third in the social aesthetic, which through its artistic integrity 'activates new interpretations' and 'opens up ways of seeing things differently' to in turn 'generate new relational forms or dialogues' (Froggett, 2011, in Doherty et al., 2015, pp.15–6). Such a place-framing approach harnesses socio-spatial relations and networks and 'socio-spatial positionality' (Martin, 2013, pp.85–6) holding a transformative agency to motivate a place-based activism towards an aim of effecting change – collective action frames of *communitas* (Turner [n.d], in Tuan, 2014, p.106) with motivational, diagnostic and prognostic analytic elements (ibid., p.89). In this framework the situation of the process is 'not local or empirical or especially "real", but it is related to a notion of situatedness, a placing of self, conflict, and/or other' (Martin, 2013, p.90). If a residential community, its closeness to the site is to the project's necessitation, not disadvantage, as Jacobs feared (in Kasther, 1996, p.42), mobilised to benefit a 'collaborative urbanism' that is local, informal and about 'changing the "me" into a "we"' (Legge, 2012, p.33).

Integral to social practice placemaking is its co-productive process. In social practice placemaking the term 'participant' is effectively dissolved to that of 'urban co-creator'. This term is an amalgamation of Klanten and Hübner's (2010, p.2) 'urban creative' and Kageyama's (in Lydon and Garcia, 2015, p.90) 'co-creator'. 'Urban creative' is based on the notion that the city is predisposed to be fecund creative ground and that its creative populace demands a wider understanding of who is a creative agent. Urban creatives are part of a design practice and process working for social benefit (Bishop and Williams, 2012; Jenkins and Forsyth, 2010, p.168), the Critical Art Ensemble 'co-mingling' (1998, p.73) of a creative co-produced practice (Bresnihan and Byrne, 2014, p.12; Kearton, 1996; Kravagna, 2012, p.254; Madyaningrum and Sonn, 2011, p.358). 'Co-creator' refers to a mixed professional/non-professional, artist/non-artist cohort. By merging the two terms to form urban co-creator, there is created a constituency of 'professional' and 'non-professional' (schematically, the public community gathered around a project) working in an assemblage in which expertism is de-siloed and rendered progressively more inclusive by omitting the noun of 'creative' to a wider productive framework. Urban co-creating goes beyond a top-down and proscribed 'I manage, you participate' participation model (Saxena, 2011, p.31) – the 'pseudo-participation' that Petrescu talks of – to a horizontal, collaborative process with a deeper level of engagement with who traditionally would have been thought of as the participants. As such, this is the creation of the micro-public in process, working in equitability and a relative expertism.

Whilst the co-produced process may often involve professionals working outside of strict professional skills boundaries (Zeiger, 2011), the social practice placemaking professional is not anonymous in their expertism (Roberts, 2009). The communitys' – however this may be determined in the project, functional, neighbourhood, interest or geographic community – expertism is that is of expert in their lived urban experience (Chonody, 2014, p.2; van Heeswijk, 2012). Relative expert skills will be deployed strategically and tactically at different stages of a project, with the locus of power with the community (Klanten and Hübner, 2010). This collaboration results in diverse outcomes, personal growth and knowledge exchange (Lehmann, 2009, p.18) where residents become co-designers in the process of urban regeneration, acting as a connect between the proscribed urban designers and the users of the city (Klanten et al., 2012, p.209; Miles, 1997). This creates Miles' 'convivial city' of user-centered urban design and planning strategies (ibid., p.2) with artists employed as co-creators of urban design and planning in a 'mutually interrogative dynamic' (ibid., p.188), a new urban vernacular of 'architecture without architects and urban space without planning' (Bresnihan and Byrne, 2014, p.12; Kirshenblatt-Gimblett, 1999, p.19). The social practice placemaking process then is one of dissolved categories of artist and non-artist, participant or audience (Maksymowicz, 1990, p.151), where the artist is moved from the participatory arts stance of 'elevated outsider' to co-producing 'engaged partner' (Adamek and Lorenz, 2008, p.57). This has a consequential deconstruction of relationships and roles to and in urban public space that 'release possibilities for new interactions, functions and meanings' (Hou, 2010, p.15).

3.6.3 Aims and outcomes of social practice placemaking

Social practice placemaking is then a process of co-production by 'collective actors working together to create shared meanings' which is a self-empowering process for communities (Sorensen, 2009, pp.207–8) that engages them in a process that connects with social, environmental, technological and physical issues (Lange et al., 2007, p.101; Yoon, 2009, p.76). Social practice placemaking operates at the psychological interstices of social identity, social representation and power and facilitates individual and social awareness about social identities and realities (Madyaningrum and Sonn, 2011, p.360). Through a performative dialogic encounter, social practice placemaking results in an unsettling of cultural assumptions (Stevens, 2007, p.74; Whybrow, 2011), the 'social friction' (Sennett, 2012, p.6) that is found also in the micropublic (Amin, 2008) between diverse groups of people that would not normally coalesce. This is a networked practice, working with diverse groups as socially multicentred (Lippard, 1997, p.286) and a specific social process where networks themselves are created (Froidevaux, 2013, p.189; Saegert, 2014, in Gieseking and Mangold, 2014, p.401). Community conscientisation via co-production in social practice placemaking is a process of developing a community level critical awareness of lived experience (Campbell and Jovchelovitch, 2000, in Madyaningrum and Sonn, 2011, p.360; Colombo et al., 2011, p.457; McGonagle, 2007, p.7; Reiss, 2007, p.13; Sorenson, 2009, p.207). Communities experience a physical and social differentiation which results in a sense of affiliation and emotional inter-related connections and the creation of their own narrative (Colombo et al., 2011, p.460; Grodach, 2010, p.489; Madyaningrum and Sonn, 2011) – the process of place attachment and which leads to individual and community empowerment (Bishop and Williams, 2012, p.23; Hall and Smith, 2005, p.175).

This conscientisation process leads to a subconscious desire to be involved in culture at a deeper level (Boenau, 2014; Messham-Muir, 2009, p.123) and can form new spatial, cultural and social identities (Franck and Stevens, 2007; Hall and Smith, 2005, p.176). Community conscientisation challenges the notion of citizens as passive to active (Bishop, 2006 [a]; Franck and Stevens, 2007, p.4), developing intra-community social capital (Lydon et al., 2013, p.1) as personal growth and knowledge exchange from interdisciplinary collaboration, creative process and co-production (Lehmann, 2009, p.18). A 'cause and effect of local action' (Chavis and Wandersman, 1990, p.73, in Prezza and Schruijer, 2001, p.401), social practice placemaking precipitates a re-valuing of community and is a means of community self-validation (McGonagle, 2007, p.7). Sense of community is garnered via belonging, influence, fulfilment of needs and emotional connections which emerge via active participation. This cause-and-effect is positively mutually beneficial and self-perpetuating (Prezza and Schruijer, 2001, p.401), resulting in increased community capital: through participation people gain skills, influence and control over the conditions affecting their own lives, *ergo* community responsibility comprehended within the individual gives them a sense of competence and control to respond to new situations during and post-participation (Colombo et al., 2011, p.461) through this conscientisation process which enhances their social capital and civic participation and makes audible otherwise marginalised voices in the intra-urban and extra-urban realm (Doherty et al., 2015, p.15). This exploration is part of the place identity formation process and transforms the understanding of space as processual and of interrelational differentiation (from Massey, 2005, in Beyes, 2010, p.231): space is organised along plural, temporary and inclusive lines (Amin, 2008, p.17) and paradoxically, even though often ephemeral, interventions have a long-lasting affect and effect on the sociocultural aspects of the city (Klanten et al., 2012, p.9; Silberberg, 2013, p.3).

Social practice placemaking can be viewed as symptomatic of a street level activism, a type of politics where people are involved at the hyperlocal in issues that affect them, from a point of disaffection with formal politics (Bishop and Williams, 2012, p.138) and 'The Challenge of Getting Things Done'. Functions and aesthetics of the city are questioned (Burnham, 2010; Lehmann, 2009) as is private ownership of the public realm, its access and use rights (Hou, 2010, p.1; Sherlock, 1998, p.220). Miles (1997, p.164) sees the value of place-located creative practice as being to create an ongoing process of social criticism, in which, when grounded in an understanding of the groups' needs and view of the world, how others view them and the wider macro socio-political context (Madyaningrum and Sonn, 2011, p.360; Murray, 2012), social and material alternatives can be postulated and prototyped. Through the performative dialogic encounter comes new forms of intersubjective experience and a consequent production of a new form of insurgent urban citizenship grounded in social or political activism (Crawford, 1999, p.23–4; Graham, 2009, p.125) that reinvigorates political systems, social structures and local economies (Bishop and Williams, 2012, p.147; Hirsch, n.d., p.21). This challenges traditional and fixed space hierarchies and urban planning and activates a 'citizen energy', 'which brings about urban growth through inventive strategies' (Mozes, 2011, p.13). Intra-urban realm groups act to undermine power through their actions, through their particular practice aiming to find and inhabit 'radical postures and resistance to power' (ibid., p.11), and affect a 'trickle-up' process (Burnham, 2010, p.139; Silberberg, 2013, p.10). This is activist, communitarian, works in mixed traditional and non-traditional media, forges direct intersections with social issues, 'encourages community coalition building' in pursuit of social justice and is an 'attempt to garner increased institutional empowerment for artists to act as social agents' (Kwon, 2004, p.106). The micro-events of social practice placemaking are a challenge to overall city stability (Hannah, 2009, p.116), anchored in community struggles which in the urban context can offer 'prefigurative models of mobilisation' (Tabb, 2012, p.202) being collectively organised by people to make a claim on an urban space where those people express demands (Uitermark et al., 2012, p.2546). When social practice placemaking is used as a framing device for the social condition of the group, projects can lead to sustainable change, creating and employing other social representations or narratives to do this, giving validity to the knowledge of the community and engaging it in dialogue (Jauckelovitch, 2007, in Murray, 2012, p.235), concurring with Freire (1972) that the potential for change lies with 'ordinary people'.

Social practice placemaking represents a new performative aesthetic lens for a certain form of placemaking. It is encountered, relational and dialogic qualities are its aesthetic and its art practice and process is both revealed and inserted into everyday situations and practices (Froggett et al., 2011, p.101) and disrupts the everyday habitus (Saegert, 2014, in Gieseking and Mangold, 2014, p.400). These heterogeneous encounters form an immediate 'urban-aesthetic discourse' (Deutsche, 1996, in Beyes, 2010, p.231), required to emancipate new urban collective experiences (Stavrides, 2007) and strengthen socially connective tissue (Crawford, in Bishop and Williams, 2007, p.89). Artists turn to social practice placemaking from push factors of wanting to work 'under their own enquiry' (Cornford, 2008), with the everyday (Gage, 2013; Miles, 2008) and to be disruptive to macro politics, utilising situationist concepts of negation and prelude (Hannah, 2009, p.115); and pull factors of frustrations with the regeneration process and the desire to materialise a less bureaucratically bound public realm strategy (Cornford, 2008) and to work immersed in the community locale and take artistic queues from the site (Brown, 2012, p.10). Artists and communities work in depth with each other and share an aim across the practice genre to improve the urban lived experience and environment by cultivating the social aesthetic and the material ouput. Seen through the prism of the reworking of Freire's (1972) education theory, social practice placemaking can be seen as the 'problem-posing' (Finkelpearl, 2013, p.30) inversion of the Subject/Object positions to one of mutual teaching and learning through dialogue, in which critical thinking is the goal via a process of transformative conscientisation.

3.7 The placemaking typology

The need for a placemaking typology is manifold. Current explanatory vocabulary is unspecific and inadequate (Schmid, 2014, p.4) and overwhelming (Silberberg, 2013, p.2), which risks the compromise of aesthetic practice (Jackson, 2011, p.15). With generalisations and blurring of subject positions in placemaking the spectrum of activity in placemaking needs to be articulated whilst at the same time not drawing fixed subject boundaries; for placemaking as with social practice arts, there is a current 'interactive moment in public space as an artistic product worthy of analysis. But the language surrounding the practice is still up for grabs' (Finkelpearl, 2013, p.5). The placemaking typology responds to this. The political, economic, social and health impacts being asked of and accredited to placemaking (Silberberg, 2013, p.2) is resulting in a cumulative confusion augmented by the competing demands made and expectations of placemaking (Fleming, 2007;

Markusen and Gadwa, 2010 [a]); an understanding of the scope of each is essential to manage expectations and expedite clearer and more effective outcomes and outputs measuring. Learning from the arts sector, it was community art's failure to construct its own theoretical framework that was the reason for its relative devaluing in the art sector (Kelly, 1984, p.29). If the placemaking sector does not create its own theoretical framework it risks a similar reduction of a 'naïve romanticism' of its claims to outcomes and a side-lining in urban design and planning as a creative, worthy 'welfare arts' (ibid.) adjunct to be deployed for city culturisation (Zukin, 2010, p.3) marketing and regeneration, rather than as a meaningful strategy for urban living (Schneekloth and Shibley, 2000, p.130).

The purpose of the typology is to share knowledge across types of placemaking and to redress exclusory power practices by uncovering the many different types of placemaking undertaken by different ecologies of practice and people to result in the opening up of a continually negotiated border positions (Schneekloth and Shibley, 2000). With this border negotiation, the placemaking sector too can engage those 'outside' of it through a clear articulation of the variety of practice and the value of these practices. Appreciating that no single lens can be adopted in placemaking as it is multiscalar and has a field of varied actors, the typology aims to reflect this whilst at the same time, exhibiting placemaking's coherency and being of developmental, conceptual and pragmatic use to a rapidly evolving sector. It is offered as a classification of observed placemaking practices and as nascent and is designed to be operationally tested against and in practice.

3.7.1 Presenting the placemaking typology

The placemaking typology (Figure 3.1) draws on Legge's (2013) three modal classifications of placemaking: strategic, undertaking and engaging with in-depth research into the local social, political, economic, physical and cultural context to define its placemaking strategy and its implementation; tactical, referring to the collaborative, citizen-led interventions that focus on place improvement, community capacity building and economic development, which can in turn feed into a larger strategy or objective; and opportunistic, ad hoc and unprogrammed interventions on the micro scale enacted by small groups of people or lone citizens in response to an immediate need. The typology further augments and extends this categorisation by including Public Realm placemaking, participatory placemaking, creative placemaking and social practice placemaking as modalities of practice. The typology operates as a matrix of degrees of relationality, the degree to which a placemaking practice is engaged with people-in-place, and of art practice/process: one can define Public Realm, creative, participatory and social practice modes of strategic, tactical and opportunistic placemakings.

Whilst the typology portrays intentional positions of practice, they are not fixed; there is fluidity through the typology that is symptomatic of participatory, creative and co-produced practices where the tactics of placemakings deployed may vary from site to site and temporally within a placemaking project. Across

74 *A typology of placemaking*

placemaking, stakeholders will enter the common field with diverse agendas and knowledge and have inherently diverse and numerous aims and objectives – this is the norm of this activity and cannot be considered any other way (Roberts, 2009, p.442; Tonkiss, 2013, p.6). No single discipline or site of knowledge takes precedent over another in placemaking but each one may be called into primary use at different stages in the process. Any expert appropriation of placemaking for some renders placemaking effectively redundant as it denies the people the opportunity to take control over their lives (quoting from Schneekloth and Shibley, 2000, p.72). Four illustrative examples of how the typology can be used to matrix pattern and name placemaking practices can be found below: strategic Public Realm placemaking; tactical creative placemaking; opportunistic participatory placemaking; and tactical social practice placemaking.

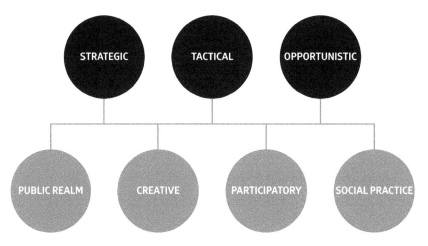

Figure 3.1 Placemaking typology.
Source: Courage, 2014

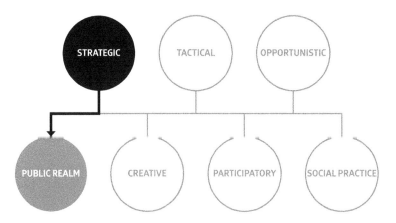

Figure 3.2 Strategic Public Realm placemaking matrix illustration.

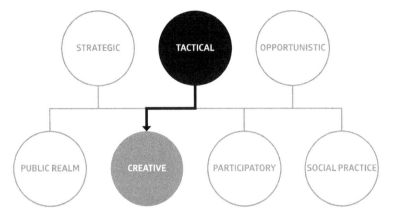

Figure 3.3 Tactical creative placemaking matrix illustration.

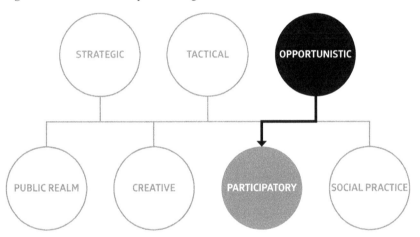

Figure 3.4 Opportunistic participatory placemaking matrix illustration.

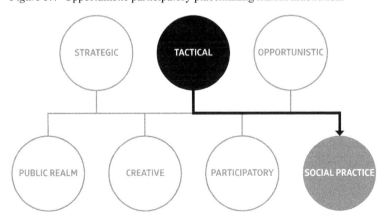

Figure 3.5 Tactical social practice placemaking matrix illustration.

76 *A typology of placemaking*

In these matrices, we see types of placemaking illustrative of the use of the typology matrix: strategic Public Realm, tactical creative, opportunistic participatory and tactical social practice. Equally, one could have a strategic participatory or tactical Public Realm, or any combinations of the modes and modalities of the typology. Thus, the typology takes into account and also celebrates the sector's multiple standpoints and acts as both a promulgation of differentiation of placemaking practices as well as a mode of critique and sector reflexivity. By including the social aesthetic in social practice placemaking it acknowledges the vast span of chronological and contemporary arts practice that placemaking has been consciously influenced by as a multivalent and deep practice. It aims to provide a 'new analytical utility' (Pierce et al., 2011, p.54) for the sector. Roberts (2009) and Julier (2005) both see the collaborative, co-produced aspect of placemaking as its point of difference in practice and the key to its success. The typology also aids Roberts' (2009, p.440) call for 'intra-professional action learning sets' that include a wide network of actors, including the community, by acknowledging the extending of the placemaking sector further towards a practice field where Public Realm placemaking, participatory placemaking, creative placemaking and social practice placemaking can inhabit and display varying strategic, tactical and opportunistic attributes.

It is evident thus far that any placemaking, as being operative in the urban public realm, is subject to, active in and forming of the political milieu. Social practice placemaking projects will have questions of collusion and complicity with neoliberal city policies and gentrification levelled at it specifically. Thus, any study of placemaking of this nature cannot ignore the political context in which the practice operates. This section now turns to this matter, discussing the political implications of social practice placemaking, 'the neoliberal dilemma', right to the city and agonistic perspectives on the subject and citizenship. It closes with a consideration of an operational practice of ambivalent pragmatism, which will prove useful in discussions in the data chapters.

3.8 The political implications of social practice placemaking

With a more detailed and faceted understanding of placemaking from the placemaking typology and sections above, this text can now place social practice placemaking – as a people–place–process polylogic with potential civic and citizenship outcomes through a place attachment process – in a broader political context. This section considers the scope of urban realm politics pertaining to countersites (Holsten, 1998, p.54, in Watson, 2006, p.170) and the re-appropriation of space by a grassroots citizenry, and considers its impact on notions of citizenship and the relation of social practice placemaking projects to a neoliberal city administration. By virtue of being situated in the public realm, forms of emplaced arts and placemaking are implicated in a politics of place of neoliberal planning and policy, citizenship and democracy, and urban regeneration and the marketisation and privatisation of public space (Carmona et al., 2008, p.10; Minton, 2009; Zukin, 2013). This section will frame social practice placemaking through the lens of the place attachment outcomes of social cohesion, civic participation and citizenship, in relation to city administrations and policy in the urban milieu in which they operate. It will broadly position emplaced arts and placemaking, with a leaning specifically to social practice placemaking in this social–cultural political realm, and present the neoliberal dilemma facing emplaced arts – whether it acts contra to neoliberal policy or colludes with it. It will go on to discuss this and politics, citizenship and the introduced notion of ambivalent pragmatism as a response to the neoliberal dilemma, through the lens of social practice placemaking.

3.8.1 'The neoliberal dilemma' and political implications of social practice placemaking

The neoliberal dilemma is thus: social practice placemaking and wider emplaced arts and placemaking practices have been questioned vis-à-vis their acting in support, or not, of neoliberal narratives and polices, from their sometime co-opting by neoliberal administrations as modes of localist politics. Thus, they are made complicit in the politics they may be working to subvert or act outside of and agitate by activist practices. The agency and impact of participation in such arts practices has also been questioned for its effect, or not, on citizen democratic participation, its motivations and forms. Thus, social practice placemaking as an embodied arts practice is both answerable to, and also makes evident, urban realm politics and the agencies of the citizen and the state or city authorities in placemaking, and places placemaking explicitly in a political discourse as a mode of urban revitalisation and design. In the US and the northwest of Europe, a 'communal conservatism' is found that romanticises the virtues of civic life to underpin the privatisation of land and public realm services and the delivery of many former-state run services by volunteers (McAllister, 2014, p.194; Sennett, 2012, p.250). Through the neoliberal lens, it is the role of markets to create 'successful' places (Rybczynski, 2014, p.120), success here equating to the liberal capitalist model of public space signifying a cosmopolitan perception of choice (McClay and McAllister, 2014, p.98). The privatisation and consequent proscription of activities in that space rests cultural creation with the 'private-sector elites' that own the space, these groups determining the public culture permitted in them, and 'marks the erosion of public space in terms of its two basic principles: public stewardship and open access' (Zukin, 1995, p.32), removing the symbolic meaning making of space and foregrounding dominant modes of space production (Lilliendahl Larsen, 2014, p.330). One can see in this process that social practice placemaking interactions may be relegated, consumed and subsumed to the neoliberal enactment of the public realm; this then leads to 'the neoliberal dilemma' and the neoliberal rhetoric of social inclusion and the co-opting of arts-in-place (Kwon, 2004) to culturised ends.

There is a 'dialectical contradiction' (Schmid, 2014, p.7) between top-down and bottom-up urbanisation processes. In the former, art acts as a salver to the neoliberal 'dissolution of community' (Nancy, 1986/2006, p.56) and has social work outcomes and outputs asked of it. Emplaced art is utilised as a spatial diversion

(Lilliendahl Larsen, 2014, p.322) from the structural causes of decreased levels of civic and social participation (Bishop, 2006 [b]) and is an agent in creating 'compliant citizens' that in the face of decreased State intervention will 'look after themselves', while at the same time, the presence of the art makes the State 'look good' (Bishop, 2012, p.14). The art created from this situation is that of a fabricated 'public artopia' (Zebracki et al., 2010, p.786) that produces largely cosmetic artworks (Miles, 1989, p.2). In the hands of developers, art is complicit in an '"urban upgrading" of space as gentrification' (Roy, 2005, p.150); when tactical urbanism for example is enacted by planners it is an 'interpretation from above' (Miles, 2005, p.597, in Lilliendahl Larsen, 2014, p.331), rendering its agency a spatial diversionary strategy, becoming de Certeau's (1984) strategy. The other side of the neoliberal dilemma views emplaced art as a critique of 'capitalist spectacle', the spectacle being that mediated experience which is at once divisive and also pacifying, creating a subjugated and passive subject (Debord, in Bishop, 2006 [a], p.12). In this context, art that does not mirror prescribed, privatised public space or a wider received notion of civic space has the potential to 'disturb the annulment of politics' (Beyes, 2010, pp.242–3). This again is where emplaced arts has a 'parasitic takeover' (Klanten and Hübner, 2010, p.103) agency, emerging from an activist 'grassroots desire to "do something"' (Bishop and Williams, 2012, p.213) by people who been pushed out of conventional economics and politics and are determining their own (ibid., p.138).

As citizen-enacted, social practice placemaking rests in the paradoxical space of the neoliberal dilemma, wherein the rolling back of state provision, including of the arts, has resulted in citizens 'seek[ing] "individual solutions to systemic problems"' (Jackson, 2011, p.27). An increased significance accredited to artists in the cultural production and framing of space, 'involved in challenging previously conceived ideas about a city's identity that they set a new framework for viewing social life' (Zukin, 1995, p.267). The diverging of positions on the neoliberal dilemma is at the belief in the agency of art. For Bishop (2012, p.258), the political concerns of relational art are 'microtopian' and do not change the world; Ostwald (2009, p.94), agreeing with Rancière, sees no established causal correlation between aesthetics and political movements. Just as a public does not automatically constitute an audience (Kaji-O'Grady, 2009, p.109), a public gathering does not inevitably equate to a 'politics of the public realm' and 'accordingly, it is a heroic a leap to assume that making a city's public spaces more vibrant and inclusive will improve urban democracy' (Amin, 2008, p.7). Sorkin (2011) and Petrescu (2006) both view arts in neoliberal pseudopublic spaces as based on normative definitions of public and private, boundaries that are supported by neoliberal public/private division of land ownership and diverting people from public dissent.

However, it is Bishop's microtopian that for some is the (performative) agency of social practice art: 'people make value out of their relationship to the arts in their everyday life, though the relations and processes that happen at a micro-level, in the context of their networks of families and friends, in communities, at home and other private domains, as well as in publically funded institutions'

(Gilmore, 2014, p.17). Social practice art also gives form to inchoate feelings which can precipitate individual transformation (Froggett et al., 2011, p.91). The social practice art process, by acting at the microtopic level, can help the urban co-creators to envision a new urban lived experience (Sennett, 2012, p.53; Till, 2014, p.168), activating people from a passive participation to an active one. Here, social practice art and social practice placemaking hold citizen tactical power as the response to the strategic power of the government (Lydon and Garcia, 2015, pp.9–10), subverting the idea of a citizenry as occupying discursively weak space (Lilliendahl Larsen, 2014, p.319). This is an artform in 'found spots' (ibid., p.38), an appropriation of site as an 'art-place' that has a direct effect on the dominant hegemonic order, making a cleavage in this visible and 'Taking place in a shared public space is tactical and materialist, asserting that the common-sense aesthetic is not an adequate reflection of our collective everyday lives' (ibid., p.43).

3.8.2 Right to the city, agonism and progressive politics

This text potentially places social practice placemaking in the realm of progressive politics. Social practice placemaking, as an applied artform, can extend Lefebvre's (1984) right to the city discourse to show that it can be constructive and solution-finding, of collective articulation of civic needs and desires and a reaction against neoliberal urbanisation (Hirsch, n.d.) that moves beyond emplaced arts as a mediated experience, prescribed by the normative culture as an open-ended site for community autonomy of diverse meaning (Fernando, 2007). Mouffe's (2005) concept of *agonism* is pertinent to introduce here as a compliment to right to the city and also as an adjunct to the arts in the urban context. Agonism is the concept of political conflict having a positive aspect and is counter to Castells' (1983) *antagonism*, of a top-down structuralism that runs counter to bottom-up practices, as the positivity of agonism is centred in the bottom-up (Mouffe, 2005). Agonism is more than an oppositional stance but 'delineate[s] a more fundamental space of epistemological contingency' (Jackson, 2011, p.50). Contingency is a fractious state, required for any genuine democratic or political dialogue (Munthe-Kaas, 2015, p.3); it does not aim to create a 'microtopia' as this is ceded as impossible, as also found in Amin's (2008) micropublic of divergent and dissenting groups aligned in a moment in time.

In the placemaking context, the socially cohesive social practice placemaking is a means to counter neoliberal urbanisation (Finkelpearl, 2013, p.356) as part of progressive politics, the art made agonistic by both facilitating dissensus and revealing the dominant consensus and creating the space for debate across all urban actors (Munthe-Kaas, 2015, p.3). Rather than colluding with localist politics, the 'insurgent' actions of a progressive politic in the public realm in fact 'destabilise[s] the structure and relationships in the official public and release[s] possibilities for new interactions, functions and meanings' (Hou, 2010, pp.15–6). Hou's (2010, p.2) 'insurgent public spaces…challenge the conventional, codified notion of public and the making of space', creating new and alternative identities, meanings and relationships, a 'barometer of the democratic well-being and inclusiveness of our present society' (ibid., pp.15–6). In this light, the emergence of social practice placemaking urban commons from dissatisfaction with neoliberal urbanisation (Boenau, 2014; Bresnihan and Byrne, 2014; Miles, 1997) is a protest response that materialises an alternative city form (Bresnihan and Byrne, 2014, p.13). Re-appropriated and independent spaces in the urban form 'involve an alternative subjective relationship to urban space – one which operates outside the norm of private property and cultivates common forms of belonging' (ibid., pp.9–10). This response has the potential to restructure the underlying power relations of urban space and democracy (Boyer, 2014, p.173), offering by example 'prefigurative models of mobilization that can provide alternative norms of the good society' (Tabb, 2012, p.202).

In the context of planning, a consultative planning system is antagonistic; an agonistic one is 'for making visible and exploring the many different possibilities for the future that exist in the public sphere' (ibid.). Agonism is useful for resolving the paradoxes of the neoliberal dilemma by revealing and acknowledging global power relations that constitute space and place, the acknowledgement being the basis for the creation of a progressive political outlook (Speight, 2014, p.152), a Lefebvrian space for representation where 'public space is a place within which political movements can stake out the territory that allows them to be seen' (Mitchell, 2003, p.129). Social practice placemaking requires of planners the questioning of 'to whom things belong' over where they belong (ibid., p.155), however contested the answer may be, at the heart of right to the city and agonism as a matter of social justice. However, 'informality, and the state of exception that it embodies, is produced by the state' (ibid.) – the informal can only be that contra to the formal, and informal practices, such as that of tactical urbanism, are used by extra-urban realm actors; the issue is the interactions between legitimised and delegitimised informality (Tonkiss, 2013, p.91). Whilst informal planning 'in most cases require only fairly modest government capacities, and will nearly always entail fewer economic and social costs than more frontal approaches to the informal city based on policing, evictions, demolitions and clearances' (ibid., p.98), the attraction of tactical urbanism and *Lighter, Quicker, Cheaper* tactics will appeal in an age of austerity and localism politics as an alternative to strategic and long-term planning.

3.8.3 Citizenship

Citizenship through a social practice placemaking lens is not an extra-urban realm deliberative process (McAllister, 2014, p.198) per se, but discursive and collective and is intentional to engage those in the intra-urban realm. Social practice placemaking civic engagement creates micropublics around problem identification and solution finding and a sense of collective ownership over its process (McAllister, 2014, p.199). This moves beyond a participative democracy (although it is part of this consequential process also) that is 'seen' to be inclusive (Finkelpearl, 2013, p.13) to a non-linear model of democratic power that distributes power evenly, circuitously and horizontally amongst citizens (Chandler, 2014, p.42). This is a transducive process of everyday democratisation that sees citizens as 'differentiated, plural and overlapping social and cognitive communities' (ibid., p.44) imperatively separate from market forces and the formal public sphere, where people are identifying and responding to issues themselves to counter private and public forces (Bresnihan and Byrne, 2014, p.1).

The role of the social practice placemaking art practice/process here is to form the micropublic through urban realm tasks and endeavours and creating space for expression (Chonody, 2014, p.2), and from this, a process of conscientisation and active citizenship begins. Cities and neighbourhoods can thus become the focus of mobilising local identities, a place-framing through place-identity formation (Martin, 2013) that stimulates collective organisation and action (Uitermark et al., 2012, p.2549). The performative social aesthetic exchange can enable the community to go on to participate in the larger political context (Kester, 2004, p.174). As these actors become more active in the urban realm, they also – consecutively and sequentially – become more active citizens (Munthe-Kaas, 2015, p.17). As a lived, yet aesthetic third, space – which is associated with emplaced arts – this is 'the key to revolutionary change' from its affective centre (ibid., pp.285–6). This brings about a condition of flexible citizenship, 'the ability to make claims to space and place at multiple levels' (Lepofsky and Fraser, 2003, p.133), an ideological challenge as to who can be a citizen and how in the agonistic production of space (ibid., p.128). As discursive, flexible citizenship is a performative act of something one *does* rather than something one *has* and within its discursivity are the 'real power relations' amongst participants that contribute to everyday material conditions (ibid., p.127). Flexible citizenship thus makes 'legitimate one's claim to participate in place-making [sic], regardless of traditional markers of identity with a place', nullifying any *a priori* demarcation between community member residents and non-community member non-residents (ibid., p.134). The same transcendence of citizenship boundaries are found in the micropublic (Amin, 2008) where one is not a community member per se, but one participates in community. The research of Carmona et al. (2008, p.19) showed that from a base of public space emotional investments then comes a learning of how to make an impact in it, a civic conscientisation of place where place is not a fixed entity alone, but a 'a grounding or situatedness for some sort of activism' that discursively produces local place as a basis for local politics (Martin, 2013, pp.91–100). This subverts the power/weak dialectic for formal/informal space through active citizenship in changing place with social cohesion and cooperation active in this process (Lydon and Garcia, 2015, p.10), 'the power is derived from the use of direct action to communicate the desire and possibility for change' (ibid., p.12) and materially and culturally change the relations of urban realm production (Bresnihan and Byrne, 2014, p.10).

3.8.4 Ambivalent pragmatism

This text further proposes a pragmatic stance in response to the neoliberal dilemma and citizenship: this rests on the acceptance that a complex public/private interrelation is the lived urban reality, a 'polymorphic spatial politics' which is 'mutually constitutive and relationally intertwined' (Jones, 2013, p.103). *Ambivalent pragmatism* is enacted thorough a 'mobilization structure', 'the informal and formal vehicles through which people mobilize and engage in collective actions' which will be networked, collective and draw on 'external resources, internal innovations, and social capital' (Hou and Rios, 2003, p.20). It enjoins the 'state-driven and top-down functional regionalization and often pre-existing and more bottom-up civic society regionalism' (Jones, 2013, p.103) and the intra- and extra-urban realm interlaced and porous micro- and macro-systems (Sepe, 2013, p.xvi) as a social reality. In Dublin for example, its independent spaces are 'characterised by neither an ideological nor counter-cultural identity' (Bresnihan and Byrne, 2014, p.8) – they have more in common with everyday urbanism in that they 'develop a set of pragmatic practices facilitating access to, and alternative uses of, urban space' and are 'characterised by a more ambiguous and nuanced political significance' (ibid.). Ambivalent pragmatism is also based on the understanding that art and the political are not separate, but also interdependent and that 'social change comes from a complex understanding of cause and effect' (Kester, 2011, p.207). Here, strategy and tactics are not oppositional but are 'tools of equal value' – as seen in the placemaking typology [3.7.1] as used as a matrix of practice – both to be used by city dwellers in a fluid system of co-production and a total independent stance taken by tactics to strategy is futile (Fernández, 2011, p.5).

Social practice placemaking in this context facilitates an understanding that subverts 'simple opposition between (aesthetic) play and (instrumental) work, between a realm of pure "collective desire" and the impure world of bureaucratic compromise and consensus, or between an absolute revolution or overturning and mere reform' (Kester, 2011, p.210). Instead, the urban political realm is 'multivalent and contradictory' (ibid.). This knowledge is produced by the social practice placemaking practice and process which generates an 'increased sensitivity to the complex registers of repression and resistance, agency and instrumentalisation, which structure any given site or context' and also devises and maintains an external relationship with non-governmental organisations, funders and the like 'in order to develop a more formal and coherent understanding of the specific insight generated through practice' (ibid., pp.212–3). Participants may join social practice placemaking projects along a spectrum of partisan to non-partisan artist/protagonist subject positions: 'Each of these actors…can be situated along a continuum of positions, ranging from those who advocate an absolute overturning of existing structures of power, to those who support more gradual or piecemeal change, to those who reject change entirely' (ibid., p.207), and 'the production of these urban commons is not derived from an explicitly political motivation ("we want to make an anti-capitalist society") or ethical stance ("it is good to share"). It is the immediate and practical result of people seeking to escape the enclosure of the city' (Bresnihan and Byrne, 2014).

Thus far, the keystones of urban and arts thinking in relation to the production of space and place and the role of the arts in this as relational and dialogical have been laid, alongside those of the countersite agency of liminal urban spaces – a common ground for social practice placemaking practices – and the practice of social practice art, and its informing of social practice placemaking. The text has paid close attention to processes of place attachment and the function of the arts in place identity formation and causality with regards active citizenship placed in a progressive politics narrative. Social practice placemaking practice has been positioned as an informal aesthetic critical spatial practice with relational, dialogic and social aesthetic agency and as an emplaced artform that acts as an aesthetic third in the urban realm with varying degrees of aesthetic dislocation agency. Social practice art and social practice placemaking have been presented as differentiated – the latter is informed by the former and there is a degree of fluidity between the two; social practice placemaking is a placemaking practice with a rematerialised art object as part of its process and outputs. Common notions of participatory art have been problematised and social practice art and social practice placemaking presented as artforms that operate beyond this conceptualisation and enactment of the artist/non-artist relation to co-production and relative expertism. The role of the arts in place attachment, identity and loss has been discussed and its potential outcomes of individuals and communities that are more civically aware and/or active. Discussion has taken place around the social work claims of social practice art and gone on to situate the research practice in its urban political context and presented it as part of a progressive political movement of active citizenship and ambivalent pragmatism. This chapter has identified more nuanced placemaking practices, beyond the common umbrella practice terms of placemaking and creative placemaking, and presented these as a typology pertaining to degrees of participation and the practices and processes of arts in placemaking.

The text now moves on to discuss the data gathered during the research project and presents these as case study–based chapters. The chapters also though form a broad narrative arc when read sequentially through practice and process and outcomes – moving from arts practice and process though to land use and planning policy – that then are synthesised in the concluding chapter.

3.9 Notes

1 At the time of the conference, though this term has since been debated between members of the group.
2 TU is seen in other guises too: Jugaad urbanism (Mozes, 2011, pp.12–4); hacktivism (Burnham, 2010, p.139); urban acupuncture (Lerner, 2014); citizen-based construction projects (Sustainable Cities Collective, 2012); DIY urbanism (Finn, 2014; Jabareen, 2014; Lydon and Garcia, 2015, p.6); everyday urbanism (Lydon and Garcia, 2015, p.10); and creative counter-urbanism (Klanten and Hübner, 2010, p.2).

3.10 Bibliography

Adamek, M. and Lorenz, K. (2008) 'Be a crossroads: public art practice and the cultural hybrid' in Cartiere, C. and Willis, S. (eds.) *The practice of public art*. New York: Routledge.
Amin, A. (2008) 'Collective culture and urban public life' in *City* 12, 1 (April): 5–24.
Arts Council England (2011) Arts audiences: insight [Online]. Available at: www.artscouncil.org.uk/media/uploads/pdf/arts_audience_insight_2011.pdf (Accessed: 20th December 2013).

Baker, M. (2006) 'Afterword' in Warwick, R. (ed.) *Arcade: artists and placemaking*. London: Black Dog Publishing.

Ball, S. and Essex, R. (2013) A hidden economy: a critical view of MEANWHILE USE in *ixia*. Available at: www.publicartonline.org.uk/downloads/news/FINAL%20VERSION%20A%20hidden%20economy;%20a%20critical%20review%20of%20Meanwhile%20Use.pdf. (Accessed: 2nd July 2013).

Barok, D. (2009) 'On participatory art: interview with Claire Bishop', Interview made after workshop Monument to Transformation organised by Tranzit Initiative in Prague July 2009 [Online]. Available at: http://cz.tranzit.org/en/lecture_discussion/0/2009-07-10/workshop-monument-to-transformation-copy (Accessed: 28th August 2013).

Beech, D. (2010) 'Don't look now! Art after the viewer and beyond participation' in Walwin, J. (ed.) *Searching for art's new publics*. Bristol: Intellect.

Best, U. (2014) 'The debate about Berlin Tempelhof Airport, or: a Lefebvrean critique of recent debates about affect in geography' in Stanek, Ł., Schmid, C. and Moravánszky, Á. (eds.) *Urban revolution now: Henri Lefebvre in social research and architecture*. Farnham: Ashgate Publishing Ltd.

Beyes, T. (2010) 'Uncontained: the art and politics of reconfiguring urban space' in *Culture and Organisation* 16, 3: 229–46.

Bishop, C. (2006) [a] 'Introduction – viewers as participants' in Bishop, C. (ed.) *Participation*. London: Whitechapel Gallery and The MIT Press.

Bishop, C. (2006) [b] 'The social turn: collaboration and its discontents' in Artforum, 178–83 [Online]. Available at: www.gc.cuny.edu/CUNY_GC/media/CUNY-Graduate-Center/PDF/Art%20History/Claire%20Bishop/Social-Turn.pdf (Accessed: 28th November 2015).

Bishop, C. (2012) *Artificial hells: participatory art and the politics of spectatorship*. London: Verso.

Bishop, P. and Williams, L. (2012) *The temporary city*. Abingdon: Routledge.

Boenau, A. (2014) 'Congress for new urbanism 20 report: DIY urbanism', 14th May 2014, in *Urban Times* [Online]. Available at: http://urbantimes.co/2012/05/cnu20-report-diy-urbanism/ (Accessed: 5th March 2014).

Bosch, S. and Theis, A. (eds.) (2012) *Connection: artists in communication*. Interface: Centre for Research in Art, Technologies, and Design. Belfast: Dorman and Sons Ltd.

Bourriaud, N. (1998/2006) 'Relational aesthetics' (1998) in Bishop, C. (ed.) *Participation*. London: Whitechapel Gallery and The MIT Press.

Boyer, M. C. (2014) 'Reconstructing New Orleans and the right to the city' in Stanek, Ł., Schmid, C. and Moravánszky, Á. (eds.) *Urban revolution now: Henri Lefebvre in social research and architecture*. Farnham: Ashgate Publishing Ltd.

Bresnihan, P. and Byrne, M. (2014) 'Escape into the city: everyday practices of commoning and the production of urban space in Dublin' in *Antipode* 47, 1: 36–54.

Brown, A. (2012) 'All the world's a stage: venues, settings and the role they play in shaping patterns of arts participation' in *Perspectives on non-profit strategies*. [n.p.]: Wolfbrown.

Burnham, S. (2010) 'Scenes and sounds: the call and response of street art and the city' in *City* 14, 1–2 (February–April): 137–53.

Carmona, M., de Magalhães, C., and Hammond, L. (2008) *Public space: the management dimension*. London: Routledge.

Castells, M. (1983) *The city and the grassroots: a cross-cultural theory of urban movements*. London: Edward Arnold.

Chandler, D. (2014) 'Democracy unbound? Non-linear politics and the politization of everyday life' in *European Journal of Social Theory* 17, 1: 42–59.

Chonody, J. M. (2014) 'Approaches to evaluation: how to measure change when utilizing creative approaches' in Chonody, J. M. (ed.) *Community art: creative approaches to practice*. Champaign, IL: Common Ground.

Chonody, J. M. and Wang, D. (2014) 'Realities, facts, and fiction lessons learned through storytelling' in Chonody, J. M. (ed.) *Community art: creative approaches to practice*. Champaign, IL: Common Ground.

Cleveland, W. (2001) 'Trials and triumphs: arts-based community development' in *Public Art Review* Fall/Winter 2001: 17–23.

Coglan, N. (2010) 'New architectures of social engagement' in *Aesthetica* 37 (Oct/Nov).

Cohen, M. (2009) 'Place making and place faking: the continuing presence of cultural ephemera' in Lehmann, S. (ed.) *Back to the city: strategies for informal intervention*. Hatje Cantz.

Colombo, M., Mosso, C. and de Piccoli, N. (2011) 'Sense of community and participation in urban contexts' in *Journal of Community and Applied Social Psychology* 11: 457–64.

Cornford, M. (2008) 'Takin' it to the streets', ixia [Online]. Available at: http://ixia-info.com/new-writing/matthewcornford/ (Accessed: 10th March 2014).

Crawford, M. (1999) 'Blurring the boundaries: public space and private life' in Chase, J., Crawford, M. and Kalishi, J. (eds.) *Everyday urbanism*. New York: The Monacelli Press.

Critical Art Ensemble (1998) 'Observations on collective cultural action' in *Art Journal* 57, 2 (Summer): 72–85.

David, E. A. (2007) 'Signs of resistance: marking public space through renewed cultural activism' in Stanczak, G. C. (ed.) *Visual research methods: image, society and representation*. Thousand Oaks, CA: Sage Publications.

de Certeau, M. (1984) *The practice of everyday life*. Berkeley: University of California Press.

Debord, G. (1958/2014) 'Theory of the Dérive and definitions' in Gieseking, J. and Mangold, W. (eds.) *The people, place, and space reader*. New York: Routledge.

Debord, G. (2006) 'Towards a situationist international' in Bishop, C. (ed.) *Participation*. London: Whitechapel Gallery and The MIT Press.

Deleuze, G. and Guattari, F. (2002) *A thousand plateaus: capitalism and schizophrenia*. London: Continuum.

Deutsche, R. (1992) 'Public art and its uses' in Seine, H. F. and Webster, S. (eds.) *Public art: content, context and controversy*. New York: Iconeditions.

Doherty, C. (2004) *Contemporary art: from studio to situation*. London: Black Dog.

Doherty, C., Eeg-Tverbakk, P. G., Fite-Wassilak, C., Lucchetti, M., Malm, M. and Zimberg, A. (2015) 'Foreword' in Doherty, C. (ed.) *Out of time, out of place: public art (now)*. London: ART/BOOKS with Situations and Public Art Agency Sweden.

Douglas, G. (2012) 'Do-it-yourself urban design in the help-yourself city' in *Architect*, August, 43–50.

Doyon, S. (2013) Placemaking vs placeshaking, 25th February 2013 [Online], Place Partners. Available at www.placemakers.com/2013/02/25/placemaking-vs-placeshaking/ (Accessed: 8th July 2013).

Eisner, E. (2008) 'Art and knowledge' in Knowles, J. G. and Cole, A. L. (eds.) *Handbook of the arts in creative research*. Los Angeles: Sage Publications.

Fernandez, M. (2011) 'The time of the temporary city', 17th November 2011 Human Scale City [Online]. Available at: www.ciudadesaescalahumana.org/2012/03/the-time-of-temporary-city.html (Accessed: 5th March 2014).

Fernando, N. A. (2007) 'Open-ended space: urban streets in different cultural contexts' in Franck, K. A. and Stevens, Q. (eds.) *Loose space: possibility and diversity in urban life*. Abingdon: Routledge.

Finkelpearl, T. (2013) *What we made: conversations on art and social cooperation*. Durham: Duke University Press.

Finn, D. (2014) 'DIY urbanism: implications for cities' in *Journal of Urbanism: International Research on Placemaking and Urban Sustainability* 7, 4: 381–98.

Fleming, R. L. (2007) *The art of placemaking: interpreting community through public art and urban design*. London: Merrell Publishers Ltd.

Florida, R. (2011) *The rise of the creative class revisited*. New York: Basic Books.

Franck, K. A. and Stevens, Q. (2007) 'Patterns of the unplanned: urban catalyst' in Franck, K. A. and Stevens, Q. (eds.) *Loose space: possibility and diversity in urban life*. Abingdon: Routledge.

Freire, P. (1972) *Cultural action for freedom*. Harmondsworth: Penguin Books.

Friedmann, J. (2010) 'Place and place-making in cities: a global perspective' in *Planning Theory and Practice* 11, 2 (June 2010): 149–65.

Froggett, L., Little, R., Roy, A. and Whitaker, L. (2011) New Model of Visual Arts Organisations and Social Engagement, University of Central Lancashire Psychosocial Research Unit [Online]. Available at: http://clok.uclan.ac.uk/3024/1/WzW-NMI_Report%5B1%5D.pdf (Accessed: 12th August 2015).

Froidevaux, S. (2012) '"60 x 60": From architectural design to artistic intervention in the context of urban environmental change' in *City, Culture and Society* 4: 187–93.

Gage, C. (2013) 'How artists strengthen communities: the rise of creative placemaking raises interesting questions', 21st May 2013, On the Commons [Online]. Available at: http://onthecommons.org/magazine/how-artists-strengthen-communities (Accessed: 2nd July 2013).

Gallant, M. (2013) 'A vibrant transformation: cities and states take creative placemaking to new heights', 3rd November 2013, NEA Arts 'Arts and Culture at the Core' magazine [Online]. Available at: http://arts.gov/NEARTS/2012v3-arts-and-culture-core/vibrant-transformation (Accessed: 4th December 2013).

Gehl, J. and Svarre, B. (2013) *How to study public life*. Washington: Island Press.

Gieseking, J. and Mangold, W. (eds.) (2014) *The people, place, and space reader*. New York: Routledge.

Gilmore, A. (2014) 'Raising our quality of life: the importance of investment in arts and culture'. *Centre for Labour and Social Studies* [Online]. Available at: http://classonline.org.uk/docs/2014_Policy_Paper_-_investment_in_the_arts_-_Abi_Gilmore.pdf (Accessed: 12th August 2015).

Gottelierer, M. and Hutchison, R. (2006) *The new urban sociology*. Boulder, CO: Westview Press.

Graham, A. (2009) 'The Living City' in Lehmann, S. (ed.) *Back to the city: strategies for informal urban interventions*. Ostfildern: Hatje Cantz.

Grodach, C. (2010) 'Art spaces, public space and the link to community "development"' in *Community Development Journal* 45, 4: 474–93.

Guest, A. (2009) 'Artists and places: the time for a new relationship and a new agenda'. September 2009, Public Art Scotland [Online]. Available at: www.publicartonline.org.uk/resources/reports/repregeneration/artistsplacesAG.php. (Accessed: 4th January 2014).

Hall, T. and Smith, C. (2005) 'Public art in the city: meanings, values, attitudes and roles' in Miles, M. and Hall, T. (eds.) *Interventions: advances in art and urban futures* (Vol 4). Bristol: Intellect Books.

Hannah, D. (2009) 'Cities event space: defying all calculation' in Lehmann, S. (ed.) *Back to the city: strategies for informal intervention*. Ostfildern: Hatje Cantz.

Harvey, D. (2008) 'The right to the city' in *New Left Review* 53: 23–40.

Herbert, D. T. and Thomas, C. J. (1990) *Cities in space, city as place*. London: David Fulton Publishers.

Hermansen, C. (2011) 'Social creativity'. January 2009. SCIBE: Scarcity and Creativity in the Built Environment working paper no 3 [Online]. Available at: www.scibe.eu/wp-content/uploads/2010/11/03-CH.pdf (Accessed: 2nd July 2013).

Hirsch, T. [n.d.] 'Taking space: expressing in the city of fear' in *Public Art Review* 21, 1, Issue 41.

Hou, J. (2010) *Insurgent public space: guerrilla urbanism and the remaking of contemporary cities*. London: Routledge.

Hou, J. and Rios, M. (2003) 'Community-driven place making: the social practice of participatory design in the making of Union Point Park' in *Journal of Architectural Education*: 19–27.

Jabareen, Y. (2014) '"Do it yourself" an as informal mode of space production: conceptualizing informality' in *Journal of Urbanism: International Research on Placemaking and Urban Sustainability* 7, 4: 414–28.
Jackson, S. (2011) *Social works: performing art, supporting publics*. Abingdon: Routledge.
Jacobs, J. (1961) *The death and life of great American cities: the failure of town planning*. Harmansworth: Penguin Books.
Jenkins, P. (2010) 'Introduction' in Jenkins, P. and Forsyth, L. (eds.) *Architecture, particpation and society*. Abingdon: Routledge.
Jenkins, P. and Forsyth, L. (2010) 'Current challenges and recommendations for the UK' in Jenkins, P. and Forsyth, L. (eds.) *Architecture, participation and society*. Abingdon: Routledge.
Jones, M. (2013) '"Polymorphic spatial politics": tales from a grassroots regional movement' in Nicholls, W., Miller, B. and Beaumont, J. (eds.) *Spaces of contention: spatialities and social movements*. Farnham: Ashgate.
Julier, G. (2005) 'Urban designscapes and the production of aesthetic consent' in *Urban Studies* 42, 5/6: 869–87.
Kaji-O'Grady, S. (2009) 'Public art and audience reception: theatricality and fiction' in Lehmann, S. (ed.) *Back to the city: strategies for informal urban interventions*. Ostfildern: Hatje Cantz.
Kaprow, A. (1996/2006) 'Notes on the elimination of the audience' (1966) in Bishop, C. (ed.) *Participation*. London: Whitechapel Gallery and The MIT Press.
Kastner, J. (1996) 'Mary Jane Jacobs: an interview with Jeffrey Kasther' in Crabtree, A. (guest ed.) *Public art, art and design profile no 46*. London: Academy Group Ltd.
Kearton, N. (1996) 'Nicola Kearton interview with Stephen Willets' in Kearton, N. (ed.) *Public art, art and design profile no 46*. London: Academy Group Ltd.
Keast, M. (2012) '10 ways to improve your city through public space' in *Urban Times*. Available at: http://urbantimes.co/magazine/2012/09/10-ways-to-improve-your-city-through-public-space/ (Accessed: 4th December 2013).
Kelly, O. (1984) *Community, art and the state*. [n.p.]: Comedia.
Kester, G. H. (2004) *Conversation pieces: community and communication in modern art*. Berkeley: University of California Press.
Kester, G. H. (2011) *The one and the many: contemporary collaborative art in a global context*. Durham: Duke University Press.
Kirshenblatt-Gimblett, B. (1999) 'Performing the city: reflections on the urban vernacular' in Chase, J., Crawford, M. and Kalishi, J. (eds.) *Everyday urbanism*. New York: The Monacelli Press.
Klanten, R., Ehmann, S., Borges, S, Hübner, M and Feireiss L. (2012) *Going public: public architecture, urbanism and interventions*. Berlin: Gestalten.
Klanten, R. and Hübner, M. (2010) *Urban interventions: personal projects in public spaces*. Berlin: Gestalten.
Klinkenberg, K. (2013) 'Architects: get local' in *New urbanism*. Available at: http://newurbanismblog.com/architects-local/ (Accessed: 2nd July 2013).
Knight Communities (2010) *Overall Knight Soul of the Community 2010: Where People Love Where They Live and Why It Matters: A National Perspective* [Online]. Available at: www.neindiana.com/docs/national-research/soul-of-the-community---overall.pdf?sfvrsn=4 (Accessed: 13th August 2015).
Kravagna, C. (2012) 'Working on the community: models of participatory practice' in Deleuze, A. (ed.) *The 'do-it-yourself' artwork: participation from Fluxus to new media*. Manchester: Manchester University Press.
Kuskins, J. (2013) 'Love or hate it, user-generated urbanism may be the future of cities', 23rd September 2013, *Urbanism*. Available at: http://gizmodo.com/love-it-or-hate-it-user-generated-urbanism-may-be-the-1344794381 (Accessed: 4th December 2013).
Kwon, M. (2004) *One place after another: site-specific art and located identity*. Cambridge, MA: The MIT Press.

Lacy, S. (2008) 'Time in place: new genre public art a decade later' in Cartiere, C. and Willis, S. (eds.) *The practice of public art*. New York: Routledge.

Lange, B., Misselwitz, P., Oswalk, P., Overmeyer, K., Rudolph, I-U., Miller Stevens, J. and Voight, S. (2007) *Urban pioneers: temporary use and urban development in Berlin*. Berlin: Senatsverwaltung für Stadtentwichlung Berlin and jovis Verlag GmBH.

Lefebvre, H. (1984) *The production of space*. Translated, Nicholson-Smith, D. Malden, MA: Blackwell Publishing.

Legge, K. (2012) *Doing it differently*. Sydney: Place Partners.

Legge, K. (2013) *Future city solutions*. Sydney: Place Partners.

Lehmann, S. (2009) *Back to the city: strategies for informal intervention*. Ostfildern: Hatje Cantz.

Lepofsky, J. and Fraser. J. C. (2003) 'Building community citizens: claiming the right to place-making in the city' in *Urban Studies* 40, 1: 127–42.

Lerner, J. (2014) *Urban acupuncture*. Washington: Island Press.

Lewis, J. (2013) 'Adapting underutilised urban spaces, facilitating "hygge"', 16th September 2013, *Cities for People, Gehl Architects* [Online]. Available at: http://gehlcitiesforpeople.dk/2013/09/16/adapting-underutilized-urban-spaces-facilitating-hygge/ (Accessed: 4th December 2013).

Lilliendahl Larsen, J. (2014) 'Lefebvrean vagueness: going beyond diversion in the production of new spaces' in Stanek, Ł., Schmid, C. and Moravánszky, Á. (eds.) *Urban revolution now: Henri Lefebvre in social research and architecture*. Farnham: Ashgate Publishing Ltd.

Lippard, L. (1997) *The lure of the local: senses of place in a multi-centered society*. New York: New Press.

Lydon, M. (ed.) (2011) *Tactical urbanism volume 1*. The Street Plans Collective and NextGen (Next Generation of New Urbanists) [Online]. Available at: http://issuu.com/streetplanscollaborative/docs/tactical_urbanism_vol.1 (Accessed: 28th August 2013).

Lydon, M. and Garcia, A. (2015) *Tactical urbanism: short-term action for long-term change*. Washington: Island Press.

Madyaningrum, M. E. and Sonn, C. (2011) 'Exploring the meaning of participation in a community art project: a case study on the Seeming Project' in *Journal of Community and Applied Social Psychology* 21: 358–70.

Maksymowicz, V. (1990) 'Through the back door: alternative approaches to public art' in Mitchell, W. T. J. (ed.) *Art and the public sphere*. Chicago: University of Chicago Press.

Mancilles, A. (1998) 'The citizen artist' in Frye Burnham, L. and Durland, S. (eds.) *The artist as citizen: 20 years of art in the public arena*. New York: Critical Press.

Markusen, A. and Gadwa, A. (2010) [a] Creative placemaking white paper [Online]. Available at: www.nea.gov/pub/CreativePlacemaking-Paper.pdf (Accessed: 5th October 2013).

Markusen, A. and Gadwa, A. (2010) [b] Creative placemaking white paper executive summary [Online]. Available at: www.nea.gov/pub/CreativePlacemaking-Paper.pdf (Accessed: 5th October 2013).

Marshall, S. (2009) *Cities design and evolution*. London: Routledge.

Martin, D. G. (2013) 'Place frames: analysing practice and production of place in contentious politics' in Nicholls, W., Miller, B. and Beaumont, J. (eds.) *Spaces of contention: spatialities and social movements*. Farnham: Ashgate.

McAllister, T. V. (2014) 'Making American places: civic engagement rightly understood' in McClay, W. M. and McAllister, T. V. (eds.) *Why place matters: geography, identity, and civic life in modern America*. New York: New Atlantis Books.

McClay, W. M. and McAllister, T. V. (2014) 'Preface' in McClay, W. M. and McAllister, T. V. (eds.) *Why place matters: geography, identity, and civic life in modern America*. New York: New Atlantis Books.

McGonagle, D. (2007) 'Foreword' in Butler, D. and Reiss, V. (eds.) *Art of negotiation*. Manchester: Cornerhouse Publications.

Messham-Muir, K. (2009) 'Beneath the pavement' in Lehmann, S. (ed.) *Back to the city: strategies for informal urban interventions*. Ostfildern: Hatje Cantz.
Mici, B. (2013) 'Participation makes for successful placemaking', 8th March 2013. *Planetizen* [Online]. Available at: www.planetizen.com/node/61068 (Accessed: 4th December 2013).
Miles, M. (1989) 'Questions of language' in *Art for public spaces: critical essays*. Winchester: Winchester School of Art Press.
Miles, M. (1997) *Art, space and the city: public art and urban futures*. London: Routledge.
Miles, M. (2008) 'Critical spaces: monuments and changes' in Cartiere, C. and Willis, S. (eds.) *The practice of public art*. New York: Routledge.
Minton, A. (2009) *Ground control: fear and happiness in the twenty-first century city*. London: Penguin Books.
Mitchell, D. (2003) *The right to the city*. New York: The Guildford Press.
Mouffe, C. (2005) *The return of the political*. London: Verso.
Mozes, J. (2011) 'Public space as battlefield' in Fernández Per, A. and Mozes, J. (eds.) *A+T Strategy and Tactics in Public Space, Independent Magazine of Architecture and Technology* Autumn 2011, Issue 38.
Munthe-Kaas, P. (2015) 'Agonism and co-design of urban spaces' in *Urban Research and Practice* 8, 2: 218–37.
Murray, M. (2012) 'Art, social action and social change' in Walker, C., Johnson, K. and Cunningham, L. (eds.) *Community psychology and the socio-economics of mental distress*. Basingstoke: Palgrave Macmillan.
Nancy, J. L. (1986/2006) 'The inoperative community' in Bishop, C. (ed.) *Participation*. London: Whitechapel Gallery and The MIT Press.
Newman, J. (2001) *Modernising governance: new labour, policy and society*. London: Sage.
Nicanor, M. and Antilla, A. (2012) 100 urban trends: a glossary of ideas from the BMW Guggenheim Lab Berlin [Online], New York: Soloman R Guggenheim Foundation. Available at: http://cdn.guggenheim.org/BMW/100_Urban_Trends_1106_3MB.pdf (Accessed: 4th July 2013).
Nicholson, G. (1996) 'Place and local identity' in Kraener, S. and Roberts, J. (eds.) *The politics of place attachment: towards a secure society*. London: Free Association Press.
Nowak, J. (2007) A summary of creativity and neighbourhood development: strategies for community investment TRFund [Online]. Available at: www.sp2.upenn.edu/siap/docs/cultural_and_community_revitalization/creativity_and_neighborhood_development.pdf (Accessed: 4th December 2013).
Olin, L. D. et al. (eds.) (2008) *Placemaking*. New York: Monacelli Press.
Ostwald, M. J. (2009) 'Public space and public art: the metapolitics of aesthetics' in Lehmann, S. (ed.) *Back to the city: strategies for informal intervention*. Ostfildern: Hatje Cantz.
Petrescu, D. (2006) 'Working with uncertainty towards a real public space' in *If you can't find it, give us a ring: public works*. [n.p.]: Article Press/ixia.
Pierce, K. Martin, D. G., and Murphy, J, T. (2011) 'Relational place-making: the networked politics of place' in *Transaction* NS 36: 54–70.
Prezza, M. and Schruijer, S. (2001) 'The modern city as community' in *Journal of Community and Applied Social Psychology* 11: 401–6.
Project for Public Spaces [a] The lighter, quicker, cheaper transformation of public spaces. Available at: www.pps.org/reference/lighter-quicker-cheaper/ (Accessed: 13th February 2016).
Project for Public Spaces [b] What is placemaking? www.pps.org/reference/what_is_placemaking/ (Accessed: 5th October 2013).
Project for Public Spaces [c] How to be a citizen placemaker: think lighter, quicker, cheaper. Available at: www.pps.org/blog/how-to-be-a-citizen-placemaker-think-lighter-quicker-cheaper/ (Accessed: 6th July 2013).

Project for Public Spaces [d] Placemaking leadership council. Available at: www.pps.org/leadership-council/ (Accessed: 7th September 2013).

Project for Public Spaces [e] All placemaking is creative: how a shared focus on place builds vibrant destinations. Available at: www.pps.org/blog/placemaking-as-community-creativity-how-a-shared-focus-on-place-builds-vibrant-destinations/ (Accessed: 4th January 2014).

Project for Public Spaces [f] Times Square. Available at: www.pps.org/projects/timessquare/ (Accessed: 21st March 2014).

Provoost, M. and Vanstiphout, W. (2012) *Make no big plans*. Crimson Architectural Historians: John Wiley and Sons Ltd [Online]. Available at: http://onlinelibrary.wiley.com/doi/10.1002/ad.1469/pdf (Accessed: 2nd July 2013).

Puype, D. (2004) 'Arts and culture as experimental spaces in the city' in *City* 8, 2: 295–301.

Rancière, J. (2004/2006) 'Problems and transformations in critical art' in Bishop, C. (ed.) *Participation*. London: Whitechapel Gallery and The MIT Press.

Reiss, V. (2007) 'Introduction' in Butler, D. and Reiss, V. (eds.) *Art of negotiation*. Manchester: Cornerhouse Publications.

Rendell, J. (2006) *Art and architecture: a place between*. London: I. B. Tauris.

Richardson, T. and Connelly, S. (2005) 'Reinventing public participation in the age of consensus' in Till, J., Blundell-Jones, P. and Petrescu, D. (eds.) *Architecture and participation*. London: Spon.

Roberts, P. (2009) 'Shaping, making and managing places: creating and maintaining sustainable communities through the delivery of enhanced skills and knowledge' in *Town Planning Review* 80, 4–5: 437–54.

Rochielle, J. [n.d.] 'Taking it to the streets; artists hit the road using creativity, communication and food to address social issues' in *Public Art Review* 13, 2, Issue 48: 40–2.

Roy, A. (2005) 'Urban informality: toward an epistemology of planning' in *Journal of the American Planning Association* 71, 2.

Rybczynski, W. (2014) 'The demand side of urbanism' in McClay, W. M. and McAllister, T. V. (eds.) *Why place matters: geography, identity, and civic life in modern America*. New York: New Atlantis Books.

Saxena, N. C. (2011) 'What is meant by people's participation?' in Cornwall, A. (ed.) *The participation reader*. London: Zed Books.

Schmid, C. (2014) 'The trouble with Henri: urban research and the theory of the production of space' in Stanek, Ł., Schmid, C. and Moravánszky, Á. (eds.) *Urban revolution now: Henri Lefebvre in social research and architecture*. Farnham: Ashgate.

Schneekloth, L. H. and Shibley, R. G. (2000) 'Implacing architecture into the practice of placemaking' in *Journal of Architectural Education* 53, 3: 130–40.

Schupback, J. (2012) 'Defining creative placemaking: a talk with Ann Markusen and Anne Gadwa Nicodemus' in *NEA Arts Magazine* [Online] 2012 no. 3. Available at: www.arts.gov/about/NEARTS/storyNew.php?id=01_defining&issue=2012_v3 (Accessed: 3rd July 2013).

Sen, A. and Silverman, L. (2014) 'Introduction - embodied placemaking: an important category of critical analysis' in Sen, A. and Silverman, L. (eds.) *Making place: space and embodiment in the city*. Bloomington: Indiana University Press.

Sennett, R. (2012) *Together: the rituals, pleasures and politics of cooperation*. London: Allen Lane.

Sepe, M. (2013) *Planning and place in the city: mapping place identity*. Abingdon: Routledge.

Sherlock, M. (1998) 'Postscript – no loitering: art as social practice' in Harper, G. (ed.) *Interventions and provocations: conversations on art, culture and resistance*. Albany: State University of New York Press.

Shirky, C. (2008) *Here comes everybody: how change happens when people come together*. London: Penguin Books.

Silberberg, S. (2013) *Places in the making: how placemaking builds places and communities*. MIT Department of Urban Studies and Planning [Online]. Available at: http://dusp.mit.edu/cdd/project/placemaking (Accessed: 13th August 2015).

Sime, J. D. (1986) 'Creating places or designing spaces?' in *Journal of Environmental Psychology* 6: 49–63.

Social Impact of the Arts Project (SIAP) of University of Pennsylvania (2007) *The power of place-making: a summary of creativity and neighborhood development: strategies for community investment* [Online]. Available at: www.sp2.upenn.edu/siap/docs/cultural_and_community_revitalization/power_of_placemaking.pdf (Accessed: 2nd July 2013).

Sorenson, A. (2009) 'Neighbourhood streets as meaningful spaces: claiming rights to shared spaces in Tokyo' in *City and Society* 21, 2: 207–29.

Sorkin, M. (2011) *All over the map: writing on buildings and cities*. London: Verso.

Speight, E. (2014) 'Senses of a City' in Quick, C., Speight, E. and van Noord, G. (eds.) *Subplots to a city: ten years of In Certain Places*. Preston: In Certain Places.

SPUR (2014) DIY Urbanism: testing the grounds for social change. Available at: www.spur.org/publications/article/2010-09-01/diy-urbanism (Accessed: 5th March 2014).

Stavrides, S. (2007) 'Heterotopias and the experience of porous urban space' in Franck, K. A. and Stevens, Q. (eds.) *Loose space: possibility and diversity in urban life*. Abingdon: Routledge.

Stevens, Q. (2007) 'Betwixt and between: building thresholds, liminality and public space' in Franck, K. A. and Stevens, Q. (eds.) *Loose space: possibility and diversity in urban life*. Abingdon: Routledge.

Suderberg, E. (2000) 'Introduction: on installation and site specificity' in Suderberg, E. (ed.) *Space, site and intervention: situating installation art*. Minneapolis: University of Minnesota Press.

Sulzer, P. (2005) 'Notes on participation' in Blunell-Jones, P., Petrescu, D. and Till, J. (eds.), *Architecture and particpation*. Abingdon: Spon Press.

Sustainable Cities Collective (2012) Tactical urbanism: cities built by people for people. Available at http://sustainablecitiescollective.com/embarq/79646/tactical-urbanism-cities-built-people-people (Accessed: 5th March 2014).

Tabb, W. (2012) 'Beyond "cities for people, not for profit": cities for people and people for systemic change' in *City* 16, 1–2 (February–April).

Thompson, N. (2012) 'Socially engaged art is a mess worth making' in *Architect*, August, 86–7.

Till, K. E. (2014) '"Art, memory, and the city" in Bogotá: Mapa Teatro's artistic encounters with inhabited places' in Sen, A. and Silverman, L. (eds.) *Making place: space and embodiment in the city*. Bloomington: Indiana University Press.

Tonkiss, F. (2013) *Cities by design: the social life of urban form*. Cambridge: Polity.

Tuan, Y. (2014) 'Place/space, ethnicity/cosmos: how to be more fully human' in McClay, W. M. and McAllister, T. V. (eds.) *Why place matters: geography, identity, and civic life in modern America*. New York: New Atlantis Books.

Uitermark, J., Nichools, W. and Loopmans, M. (2012) 'Cities and social movements: theorizing beyond the right to the city' in *Environment and Planning A* 44: 2546–54.

Uytenhaak, R. (2008) *Cities full of space: qualities of density*. Rotterdam: 010 Publishers.

van Heeswijk, J. (2012) *Public art and self-organisation*, London (conference), August 2012. Available at: http://ixia-info.com/events/next-events/public-art-and-self-organisation-london/ (Accessed: 15th January 2016).

Watson, S. (2006) *City publics: the (dis)enchantments of urban encounters*. Abingdon, Oxon: Routledge.

Weger, A. (2013) 'Creative what?' in *Changing Media*, 6th June 2013. Available at: http://changingmediagroup.com/iso-a-creative-placemaking-definition/ (Accessed: 3rd July 2013).

Whybrow, N. (2011) *Art and the city*. London: I. B. Tauris.

Whyte, H. W. (1980) *The social life of small urban spaces*. Washington D.C.: The Conservation Foundation.

Wiedenhoeft, R. (1981) *Cities for people: practical measures for improving urban environments*. New York: Van Nostrand Reinhold.

Wortham-Galvin, B. D. (2008) 'Mythologies of placemaking' in *Places* 20, 1: 32–9.

Wyckoff, M. A. (2014), 'Definition of placemaking: four different types' in MSU Land Policy Institute. Available at: http://pznews.net/media/13f25a9fff4cf18ffff8419ffaf2815.pdf (Accessed: 18 June 2016).

Yoon, M. J. (2009) 'Projects at play: public works' in Lehmann, S. (ed.) *Back to the city: strategies for informal urban interventions*. Ostfildern: Hatje Cantz.

Zebracki, M. Van Der Vaart, R. and Van Aalst, I. (2010) 'Deconstructing public artopia: situating public-art claims within practice' in *Geofourm* 41: 786–95.

Zeiger, M. (2011) The Interventionists Toolkit, 31st January 2011, *Design Observer* [Online]. Available at: http://places.designobserver.com/feature/the-interventionists-toolkit/24308/ (Accessed: 5th March 2014).

Zukin, S. (1995) *The cultures of cities*. Malden, MA: Blackwell Publishers.

Zukin, S. (2010) *Naked city: the death and life of authentic urban place*. Oxford: Oxford University Press.

Zukin, S. (2013) 'Whose culture? Whose city?' in Lin, J. and Mele, C. (eds.) *The urban sociology reader*. Abingdon: Routledge.

3.11 List of URLs

1 Project for Public Spaces: www.pps.org/
2 BMW Guggenheim Lab: www.bmwguggenheimlab.org/
3 The Social Impact of the Arts Project (SIAP): http://impact.sp2.upenn.edu/siap/
4 National Endowment for the Arts (NEA): www.arts.gov
5 Art Place: www.artplaceamerica.org/
6 Placemaking Leadership Council: www.pps.org/about/leadership-council/
7 Office for Metropolitan Architecture: http://oma.eu
8 Rem Koolhaas: http://oma.eu/partners/rem-koolhaas
9 Gehl: http://gehlpeople.com
10 Foreign Office Architects: https://eng.archinform.net/arch/1250.htm
11 muf: www.muf.co.uk
12 public works: www.publicworksgroup.net
13 AOC: www.theaoc.co.uk/index.html
14 Turner Prize: www.tate.org.uk/whats-on/tate-britain/exhibition/turner-prize
15 Assemble: http://assemblestudio.co.uk
16 The Lighter, Quicker, Cheaper Transformation of Public Spaces: www.pps.org/reference/lighter-quicker-cheaper/
17 The Public Art Development Trust: http://discovery.ucl.ac.uk/31656/
18 Rebar: http://rebargroup.org/
19 Park(ing) Day: http://parkingday.org/

4 The Drawing Shed, London

> *'Is it art? Brilliant. What a great idea. Look, there's more art over there!'*
> *(The Drawing Shed audience member, in reaction to Ed Woodham (2014)*, Danger Deep Water *[performance], and Site Space (2014) [processional performance], during The Drawing Shed (2014)*, IdeasFromElse[W]here *[residency], Lloyd Park, Walthamstow, London)*

The presentation of the case studies begins with The Drawing Shed [n.1] as this is the prime case study to interrogate social practice art practice and process in place, that which underpins the research project and lays foundations from which to understand the following case studies. It also establishes some common concerns that run through all case studies, including the artist position and the role of the artist in a regeneration and gentrifying context. This chapter will focus on the role and agency of art in a social practice placemaking context, focusing on the co-produced art practice and process and the art experience in and affective spacetime (Munn, 1996, in Low, 2014, p.20); the role of the artist working with a collective body of urban co-creators [3.6.2]; and its arts practice working as reinterpreting the space and cultural activities of the estate, enabling the residents to critically distance themselves from the estates' life-world (Kester, 2004).

4.1 Introducing The Drawing Shed

The Drawing Shed is a contemporary arts organisation founded by artists Sally Labern and Bobby Lloyd, operating mainly in London. The research with The Drawing Shed took place with two projects in Walthamstow, *LiveElse[W]here* [n.2] and *IdeasFromElse[W]here* [n.3], and *Some[w]Here Research* [n.4], in the south London borough of Wandsworth. *LiveElse[W]here* took place on the Atlee Estate, and *IdeasFromElse[W]here* at Winns Gallery [n.5], housed in the Aveling Centre (forming a café, the gallery, public toilets and some disused previously purposed artists' studios), itself placed in Lloyd Park [n.6]. It was established in 2009 as an artist-led project and collaborative platform for the artist's collaborative creative practice. At the time of writing it instigated the operation of three mobile art studios, homed at The Drawing Shed's three lock-ups on the Atlee Estate, the base for socially engaged projects on the estate, and across London.

Research with The Drawing Shed took place during June–July 2015 and on an ad hoc basis until May 2015, the days here totalling just over two months.

4.2 Facets of social practice placemaking practice and process

The following section will turn to aspects of social practice arts found in place in this case study, introducing the notion of the informal aesthetic first, and then including in the discussion relational and dialogical aesthetics, performativity and duration.

4.2.1 The informal aesthetic

> I think these are important, these informal ways. These are how people glue relationships. (The Drawing Shed artist)

Artists from The Drawing Shed remarked that their practice was 'spontaneity' (The Drawing Shed artist) – something echoed verbatim at Art Tunnel Smithfield [n.7] and Big Car [n.8]. This was part of an *informal aesthetic* operated to 'set up the conditions' to subsequently 'open it [the practice] up to synchronicity, let[ting] there be serendipity' (The Drawing Shed artist), an unforced constellation praxis (O'Neill, 2014, pp.195–6) integral to a social practice arts practice. The intention of *IdeasFromElse[W]here* was an 'emphasis to be an open ended play and exploration process' (The Drawing Shed artist). The processual nature of The Drawing Shed's art lab model in place at *IdeasFromElse[W]here* brought a disjuncture to the usual gallery set up. For the artists, this allowed for 'stealing extra days for thought, process, try out' and suited their temperament as an 'experimenter-through-process' (The Drawing Shed artist). For the most part, artists worked from an inspiration point from the gallery and park experience as a whole, with collaborative conversation with others in the space, artists and the public, driving the devising and delivery of the work. This practice aesthetic was a deliberate act to foster the creative conditions of the arts lab: 'those are very critical words, unplanned, not thought out, but that was an aesthetic choice to have this swirling of energy' (The Drawing Shed artist). For some this posed a challenge at the start of the residency, 'when you go in and say "you can do anything", they [the artists] do nothing, they get frozen, freaked out' (The Drawing Shed artist). The intention however became the ideologically driven *raison d'être* of the residency, moving away from 'hardcore, highly intellectualised work' (The Drawing Shed artist) to an informal aesthetic purposed to 'open up spaces to explore, to make, to make mistakes' (The Drawing Shed artist). It was perceived by The Drawing Shed artist cohort that the practice integrity and intent of the informal aesthetic was missed by many outside of the social practice arts and social practice placemaking fields and there was an observed friction between the informality and the need for a professional base level for the projects to deliver.

4.2.2 The relational and the dialogical

The use of objects in art practice as *relational objects* was key to The Drawing Shed's practice: these included the eponymous *Drawing Shed* (2014), *Print Bike* (2014) and *Typing Pool* (2014) used in *IdeasFromElse[W]here*, and the go-karts and soap boxes in *Some[w]Here Research* (Figure 4.1). The objects were made with a parasitic (Serres, 1980/2007) and an intermedial function (Jackson, 2011, p.28) as the relational aesthetic third. These facilitated an engagement with people in their complex lived environment, taking art to them – rather than expecting people to come to the art – and created spaces that were different to their everyday (Rancière, 2004, p.86), a moment of aesthetic dislocation (Kester, 2004, p.84). These objects were named relational objects by the lead artists, directly informed by relational aesthetics (Bourriaud, 1998/2006, p.160) and in recognition of the durational ritual of making as the embodied third in the triad between people and place, based on Sennett's (2012, p.88) 'invention of tradition' that aided intra-group co-operation, creating a sense of tradition and ritual in ephemeral interventions. Parallel to Dewey's (1958) 'art as experience' and 'doing is knowing', The Drawing Shed worked with an ethos of 'knowing through making' of the relational objects. The 'seemingly worthless "non-objects"' (The Drawing Shed artist) of the go-karts made at Nine Elms were re-valued and re-purposed in their making, 'grappl[ed] with the powerful relationship between imagination, survivability and resistance' (The Drawing Shed, 2015 [a]). A 'fluid and ambiguous nature to the mobile "non-objects"' (ibid.) was recognised by the resident collaborators: one resident, after making his 'boat-cart' [sic], stated, 'Ok, I've just realised, this isn't about making go-carts [sic] is it?' (The Drawing Shed participant). Thus, the art object re-materialises in social practice placemaking as a relational object with functional, symbolic and critical functions (Beech, 2010, p.20; Dunk-West, 2014, p.156; Finkelpearl, 2013, p.47; Kester, 2004, p.110; Sennett, 2012, p.24).

It was evident however that in practice the operative modality was closer to that of Kester's (2011) *dialogical aesthetic*. For instance, 'Projects have become extended conversations between ideological positons within defined programme themes' (The Drawing Shed artist). Similarly, in the *IdeasFromElse[W]here*

Figure 4.1 The Drawing Shed (2009–2015) TL, CW, BR, *The Drawing Shed, Printbike, Go-Kart, Typing Pool*, London, 2014.

gallery setting, artwork was 'arranged so as to demonstrate a coherent intertextual relationship between one another' (The Drawing Shed artist), the dialogical aesthetic an evolving curatorial continuum of overlapping artwork and protagonists within the space becoming part of a dialogical structure. Live art performance was used often in The Drawing Shed projects, which were also part of the dialogical aesthetic structure. Jordan Mckenzie's [n.9] (2014) *Sink Estate* during *Some[w] Here Research* used the act of walking – around the estate dragging a stainless steel sink behind him, tethered across the artist's body with rope – as a visually unambiguous performance method to act as a 'possible provocation to thought, possible link to question with what's going on, on the estate' (The Drawing Shed artist). The accompanying cards – flyered through letterboxes and placed in shop windows and surreptitiously amongst products on shelves – were of the same intent, not including a call to action, but a question that could turn interest to conversation, either with the artist or with others, of lived experience gentrification and social cleansing. Artists though did not explicitly articulate a difference in practice between the relational and the dialogic – it was a 'blended continuum' (The Drawing Shed artist) between and through social practice arts and community art, of practice along the informal aesthetic as a process of social exchange (Schneekloth and Shibley, 2000, p.138). The relational object collapsed the binary of the art object as art experience mediation (Chonody, 2014, p.4; Dunk-West, 2014, p.158; Jackson, 2011, p.28; Kester, 2004, p.110) as embedded in the art process. The agency of the object was both of performativity and also of political resistance (Buser et al., 2013, p.624; Rancière, 2004, p.86). It operated on a discursive (Dixon and Durrheim, 2000, p.32) praxis of the interactionist past and potential in relation to place sociability (Altman and Low, 1992; Rowles, 1983; Sixsmith, 1986).

4.2.3 Performativity

The Drawing Shed stated that 'Being there is the work' and, gesturing to the public outside the gallery, 'The art is in the park' (The Drawing Shed artist). Its model of social practice placemaking was articulated as ideologically motivated to 'throw up fault lines' (The Drawing Shed artist), using a social practice arts questioning process to unlock issues. *Some[w]Here Research* was active in creating a space for residents critical discourse: 'There's a critical edge. Even in passing people are talking about their life here … [we're] creating gaps in which this can pass through, creating space for it' (The Drawing Shed artist), Kwon's (2000, in Doherty et al., 2015, p.127) aesthetic of the 'wrong place' of artworks in unexpected places as functioning to undercut habituated notion of place. Artists appreciated the value of the spectacle in their work. This was seen as a means to break the 'seriousness' that social practice arts could sometimes become trapped in and also its own agency to affect reflection in the observers: 'I don't mind if sometimes it's about spectacle because I think that, sometimes it just cheers people up a bit and makes them think about this space in a different way' (The Drawing Shed artist) – the aesthetic dislocation of social practice placemaking and contra to a de-valuing of the spectacle found in some social practice arts thinking (Bourriaud, 1998/2006, pp.163–5; Debord, in Bishop, 2006 [b], p.12; Doherty, 2004; Whybrow, 2011). In the setting of the housing estate also, this regard was heightened as a means to both 'enchant' the space and engender residents to act to 're-enchant' that space, to soften the 'really hardcore functionalism of the council estate' where a space is seen anew as 'not a shitty garage' (The Drawing Shed artist) for example, but as a social and creative space made unusual through the arts, but still imbricated in the quotidian of the locale.

Performativity was also situated in the relational objects and their associated rituals. In *Some[w]Here Research*, the relational objects were heavy and 'clunkily [*sic*] built' necessitating a group manoeuvring of them, a performative act of mobility and burden-carrying that brought focus to social and physical 'mobility on and off the estates' and 'the things we carry with us as human beings in order to be in the world' (The Drawing Shed artist). Cooking featured in much of The Drawing Shed's activity; this was seen as a relational activity and its utensils as relational objects, the act of preparing, cooking and eating a meal being stages in a relational process and eliciting stories from 'being in the doing of the moment' (The Drawing Shed artist). The Drawing Shed interacted with performative practice through acts of 'small motion' (Labern, 2015). *Some[w]Here Research* framed dialogues throughout the residency 'establishing and reaffirming the critical context of discussion' (The Drawing Shed, 2015 [b]). It was the repeated articulation of street play as an act of resilience to abject poverty that was brought into the artist's preoccupations and the creation of the go-karts for example. This play had a subversive function. A procession of the relational objects made at Nine Elms through the estate had social and political implications in that, through play, 'a series of (un)negotiated incidents began to literally play out on the streets' and 'PLAY [*sic*] engaged with in this way is of course subversive' (The Drawing Shed, 2015 [c]). The artists observed that some residents 'understood deeply' the import of the procession and that 'There was nothing "absurdist"' about it, the value and necessity of the 'built on hard work' of the procession 'gave huge joy to the participants' and consequently 'This became the work' (ibid.). In a durational performance, Sally Labern's *On The Rack* (2014), a piece concerned with homelessness, during *IdeasFromElse[W]here*, it was not enough to take on the issues of the streets around the park; they wanted to 'bring in complex ideas to the place' (The Drawing Shed artist): 'it's the *doing* that leads, not the language … that *doing language* of the performative is fundamental … will affect how people approach it materially and aesthetically and experimentally' (The Drawing Shed artist). This informed Labern's later naming of 'doing-through-making' in her *Manual for Possible Projects On The Horizon: That Things Fall Apart … And Thinking Through Making* publication for The Drawing Shed (The Drawing Shed [a]).

A prime intent and outcome across processes in all case studies was that of storytelling and it is in the example of *Live Lunch* (2014) during *LiveElse[W]here* that the central conceit of storytelling as a performative process of the dialogical aesthetic in social practice placemaking was most acutely illustrated. The lunch took place in the gravel and lawned area outside of The Drawing Shed's Atlee Estate lock-up base over an afternoon. Residents were invited to bring dishes that represented them or their culture and one artist, Pablo Perezzarate, decided to tell the story of his and his family's history and journey to the UK from Mexico, whilst cooking Mexican food with the assembled group (Figure 4.2). Throughout the day, food acted as a conversation starter – 'What is going on?', 'Can I join?', 'What is this food?' – which then opened up to conversation on the lock-ups and the work of The Drawing Shed, and then into deeper conversations of identity, from a transitoriality of not knowing ones neighbours to defending and protecting neighbours from immigration officials and people's own life journeys, geographical, political and cultural. Perezzarate's lunch storytelling took place around a table with the internationally and intergenerationally mixed group seated and all given a preparation task, as more still looked on. For Perezzarate the lunch was a collaborative performance – his act of storytelling and of instructing the food preparation and the spontaneous talk amongst the group enabled people to tell their own stories, recalling memories and talking about the social, cultural, economic and political lifeworld of the estate, and the wider global politics of identity, immigration and place. Once the food had been prepared, cooked and shared, the group stayed seated for some time and continued conversation amongst themselves and with others that joined, intrigued by the subject matter. Perezzarate decided upon this collaborative performance as a 'thank you' to the residents of the estate for hosting him and to those that participated in his audio storytelling project, *Sound Map* [n.10–12], that formed his *LiveElse[W]here* residency. Food here was an informal and social aesthetic activity that engendered storytelling that in turn engendered social capital, wellbeing and resilience, and, just as with acts of painting with Big Car or gardening with Art Tunnel Smithfield, was a process ritual encountered in the performative act. The discursive interactions dissolved subject positions (Chonody, 2014, p.3; Dunk-West, 2014, p.157) for a protagonist micro-community (Kester, 2011, p.29) to form and conversation begun that uncovered community issues (Kester, 2004, p.95).

Figure 4.2 The Drawing Shed (2014), *LiveElse[W]Here Live Lunch* [event], Attlee Estate, Walthamstow, London, 2014.

4.2.4 Duration

The Drawing Shed's month-long residency at Winns Gallery was seen as a minimum amount of time to form a meaningful relationship with the place and people and to create work. At the start of the residency it was not known how the public would engage with the residency, either being one-off or regular visitors; the short duration was thought to favour the former, acting with a 'drop-in' function rather than a 'residential' one. At the end of the residency, it was felt there was a successful level of engagement and volume of artworks created but that it took time for the intra-group relationships here to 'bed in'. Over time and through tacit 'co-mingling' (Critical Art Ensemble, 1998, p.73) and overt negotiation, a communal ownership of the space by all artists formed, manifested in a communitarian attitude to work: 'We had got to a point within the gallery that things [co-sharing and co-production issues] had been resolved. It was a really good space, inside and outside, there was a shift in the furniture' (The Drawing Shed artist), metaphorically and literally. By the time of its closure, it had the feeling amongst the artist cohort of being the start of something that could grow but that had not reached its full potential, 'The issue with use of space like this, as durational, is that it takes time … feeling that project is just gaining momentum, just bedding in, and now [it's] time to stop' (The Drawing Shed artist). The *LiveElse[W]here* residencies on the Attlee Estate were felt to have needed a longer duration still for The Drawing Shed–resident relationship to develop and achieve its aims. The short-term, one-off, nature of this residency did not sit well with some; it felt 'weird sharing personal things with people and them to you and then not seeing them again' (The Drawing Shed artist). The end of the *IdeasFromElse[W]here* residency, whilst at once feeling like the start of something longitudinal, the artists expressed that they wished the project to continue as it was felt that it was on the tipping point of 'bedding in' (The Drawing Shed artist). One artist reflected that 'It couldn't achieve any more, it had done what it set out to do, and I wanted to go out on a high as well, I didn't want it to be superseded by another event' (The Drawing Shed artist). The *Live Lunch* also marked an apposite moment of closure for an artist in residence: 'That essence of closure with people, to acknowledge that I am proud of the work and grateful to people for sharing … this seemed like the ideal moment to share my story with them as they had with me' (The Drawing Shed artist).

Thus, the performative and durational art practice here can be seen as affecting the emplaced arts experience of the reinterpretation of the urban realm, through arts process, and of shaping social cohesion. The Drawing Shed's social practice arts process was seen to be of intra- and inter-community social exchange and connection-forming, through the performative intervention (Froggett et al., 2011, p.95). The micropublic (Amin, 2008) coalesced on the day of the *Live Lunch* (2014) formed from an embodied social dialogue wherein those present were able to reflect on their own life experience through learning of others, in dialogue and enacted in the ritual of food preparation. The siting of this in the public realm incorporated this context in the criticality of the work – of the politic of the housing estate – which magnified and focused the contribution of social practice arts to the social milieu (Deutsche, 1991, in Miles, 1997, p.90). This situational-relational

(Whybrow, 2011, p.5) process is Lefebvre's (in Whybrow, 2011, p.18) praxis and poiesis on a social scale, exemplifying Doherty's (2004) 'new situationism' of the city event-space (Hannah, 2009, p.117) polylogic of people, site, object and process (Kwon, 2000).

4.3 The position of the artist

The following section goes on to consider the position of the artist – one that is granted a special dispensation and one that may or may not be embedded in the project locale – in social practice placemaking. Aspects of this same theme will also be seen in the subsequent case-study chapters.

4.3.1 The special dispensation of the artist

> A common response of many is, when realising [or] being told it's art [or] performance, is to accept it as odd, as 'other', [they're] not so worried then of what is happening or why, but let it happen. (The Drawing Shed artist)

In the declared position of the artist, it was witnessed that artists were given a special dispensation by the audience and protagonists to enact things that were out of the usual. This affected an eventual normalisation of the art, which was not necessarily to its detriment. The young people cycling through the Site Space performance (a situationist and processional improvisation with found objects) during the *IdeasFromElse[W]here* residency are an example of how the art had become normalised in the site instances, with both the artist and the public in the same space 'getting on with their day to day' (The Drawing Shed artist). Once the children swarming Mckenzie's *Fence 2014* (2014) (a durational performance where the artist placed himself on the side of one of the pathways into the park from the main road, and his head through a metal fixed leg barrier fence section, found surplus in the gallery courtyard) realised he was not in danger, they understood this as a deliberate act and an act of art, 'they took a different interest, got a taste of the issue of permission of the intent of the piece, that it was "art", got it as performance art, when they knew that, it was ok [to them]' (The Drawing Shed artist). At the end of the *IdeasFromElse[W]here* residency at Winns Gallery, the park café staff and rangers and the artists together remarked they felt like one and the same team and that they would be sad to say goodbye to the other. One park attendee commented, 'But you're *our* artists...' when they found out that it was the last day of the residency and two boys in the café at this moment also commented that they wanted the artists to remain, thinking that The Drawing Shed had in fact become a permanent fixture in the gallery, and expressed a similar disappointment at their leaving. A park warden came to find interventions such as Woodham's *Danger Deep Water* (2014) [4.5] as 'part of park life' (The Drawing Shed participant). For one of the artists, this special dispensation afforded them the space to 'relinquish the responsibility of having to engage with people' and be 'in' their work as a solo performance, '...and somehow people respect that in a way that they would never otherwise' (The Drawing Shed artist) and it was similarly observed that this dispensation removed art from an othered position and in doing so, this act subsequently gave the public permission to behave differently themselves and join in to varying degrees of participation. One artist likened the freeing from gallery formal process to the normalisation of art in the public realm, using the analogy of the urban gym adjacent in the park, 'At one point we'd've all thought that odd, working out in public like that. Now we see it every day, totally normal' (The Drawing Shed artist).

Extending from this positionality, the artists thought it 'vital to place ourselves in the same frame of making as everyone else' both recognising their own relative value as artists but also to show '"we are asking of others what we are also doing ourselves"' (Labern, 2015) and this engagement going on to fuel yet more creative work. It was recognised in The Drawing Shed projects that the artist is a creator of networks through their work, 'the creative action of being the creation of new contacts and relationships', 'bringing 'people that wouldn't have been in that space together, together' and that this is a unique proposition of the artist, 'Its artists that know all those people' (The Drawing Shed artist). The Drawing Shed lead artists saw their role as being to 'hold things open' (The Drawing Shed artist) and to create that holding process. This was how they 'led' a project but other than that, their authorship was redundant, 'This is a project that The Drawing Shed has led. This is not The Drawing Shed' (The Drawing Shed artist).

4.3.2 The embedded artist position

Two artists commissioned across *IdeasFromElse[W]here* and *Some[w]Here Research* both lived and worked on housing estates; Mckenzie co-running *LUPA* (Lock Up Performance Art, 2011–2013) [n.13], a live art and events-based residency working from a lock-up on his resident east London estate and subsequent projects on his estate in East London; and Valz Gen as part of group+work run Latymer Projects [n.14] in West London, working from a former community day-care centre. It was observed that in the reciprocal negotiated position, these artists sat both as insider and outsider in the socially engaged terroir: whether they were resident of that community or not, the artist must be knowledgeable of the community as social practice arts practice starts with the understanding of situated context (Chonody, 2014, pp.2–3 and p.31), located in an embedded spacetime (Munn, 1996, in Low, 2014, p.20) in which they become 'part insider, part outsider', (Froggett et al., 2011, p.96) occupying a liminal, parasitic space which has its own disrupting transformative potential (McCormack, 2013, p.10). This afforded them a degree of embedded prior knowledge of the Atlee and Carey Estates:

> Both live on social housing estates. Both make work and 'show' work on their own estates. Both respond to, intervene in their own [contested] sites of community. Both are interested in questioning through making, all that this provokes in them as citizens, as artists. (Labern, 2015)

For The Drawing Shed lead artists, their time on the Atlee Estate over a number of years had also afforded them a special 'insider' position, 'We have been there longer than some residents, so we are kinda resident non-residents' (The Drawing Shed artist). The lead artists at The Drawing Shed acted as a mediating conduit between the artists in residence at the lock-ups and the community, their known faces and known work reducing the feeling of risk or trepidation in the estate residents to join in, '…and apart from curiosity, I felt people felt "I can trust this person as The Drawing Shed has done so many great things here before, so I want to do this one as well"' (The Drawing Shed artist). A *Some[w]Here Research* artist, as a non-estate or E17 artist, found they relied on the embedded knowledge of The Drawing Shed lead artists to give them access to the estate residents and from this that their work could begin, 'it really cemented in me that to do socially engaged work you need to have someone from the community to introduce you to everyone else, or you need a long time somewhere' (The Drawing Shed artist). They saw their time situated in a community setting also being essential as one 'need[s] to participate in the community to know who the organiser is' (The Drawing Shed artist), this 'organiser' acting as a social gatekeeper that can then begin to open up the community to the artist so that they are considered less of an outsider, 'it gives it a validation' (The Drawing Shed artist). This was noted even though 'we are very much outsiders, we have come into very different estates, in a very different part of the world' (The Drawing Shed artist); the artists were invited 'specifically to carry on with an inquiry that runs through their work' (The Drawing Shed artist). With this position declared, the artist role can open up and be relinquished altogether, 'once they [the estate residents] get inside a project, it doesn't have to be us [doing it] as it's ideologically driven' (The Drawing Shed artist). A social practice arts practice leans towards the embedding of the artist in the site of their work, with a fetishisation of the need for the artist to be of that place for them to have an artistic integrity, or to employ a participative ethnographic (Jackson, 2011, p.69) methodology, Debord's (1957/2006, pp.96–9) 'unitary urbanism', where the art work must be lived by its constructors. However, the case studies give nuance to this position with relative merits seen as empathiser and provocateur of the insider/outsider aspect respectively. Thus, the cross-cutting milieu of the housing estate and its larger-frame politic acted to both uncover social–spatial intersectionality and engender 'empathetic identification' from that sense of otherness (Kester, 2004, pp.113–8), counter to the 'fetishisation of authenticity' (ibid.) of the artists residential position whereby the artists and non-artists questioned the artist position through the lens of working contextually in the neighbourhood community setting.

4.3.3 Problematising the artist position

A number of issues and criticisms were manifest emerging from the positionality of the artist, firstly in relation to their duration in place. Where longitudinal duration was commonly preferred, in *Some[w]Here Research*, the artists also saw benefit to the capped duration of the residencies, its parameters facilitating a useful mode of practice, that of the pilot and the test-bed. The arts lab model of *IdeasFromElse[W]here* was also viewed in this spirit of experimentation and learning, but as an experience that begged a longer duration, one artist stating that the learning from this pilot might be the acknowledgement of needing more time in the space 'so that the relationship we could have with it would be less intense' (The Drawing Shed artist) for the artists and the volunteer team supporting them. Another artist critiqued the practice of splash projects in which 'people helicopter in, "do social engagement" and then clear off again' and went on to comment on their own practice as an embedded artist, 'But I've kinda got a responsibility to where I live and also I live here, so I don't have the luxury of doing something provocative and then running off' (The Drawing Shed artist). However still, across the case studies was a consensus that no matter the duration of a project in terms of day, months or years, all work was done in the moment and this being part of the artist's role to recognise that moment:

> It's about creating relationships in the moment, small, discrete, the more the better for people. [There is] something really interesting in what people will hold up in the moment, as long as it's kept within the moment. (The Drawing Shed artist)

A second issue was in respect to the artist in a place of gentrification, the subject matter of The Drawing Shed's *Some[w]Here Research* project in Wandsworth in particluar, the area described by the artists as 'the largest single site of gentrification in Europe at the moment' and a 'site of social cleansing' (The Drawing Shed artist). The Drawing Shed felt a large degree of responsibility to discharge their skills appropriately in the position of a relative expert in a team of urban co-creators, to 'recognize your own fingerprints' as one The Drawing Shed artist termed it. The *Some[w]Here Research* project was funded by the site developers, Arts Council England (ACE) [n.15], and Wandsworth Borough Council [n.16], part-administered by the borough council's arts team, which was undergoing a process of privatisation at the time of research. Whereas some 'artists are lured and seduced into being part of regeneration, this project is not one of those' (The Drawing Shed artist). Instead, the artists were keenly aware of complicity in the gentrification process by being funded in this way but saw themselves as in a subversive position with an active agency: 'I think there is a way to do an art project and not be complicit. There are other ways of doing stuff' (The Drawing Shed artist). In this agentive position, the artists too were acutely aware of the political ramifications of their affective political agency in this context as artists:

> The money that this is funded by is from developers and this has caused a huge amount of anxiety, creates a discomfort … and part of this project has been to unsettle them [the developers], to take on some of those uncomfortable conversations. And we have done that and it's caused some uncomfortable ripples. (The Drawing Shed artist)

It was observed that The Drawing Shed artists were in a constant debate amongst themselves and with estate residents about what the role of the artist should be – if at all – in areas of gentrification. The Drawing Shed also perceived their funders as viewing the project as 'a feel good, a sticking plaster' (The Drawing Shed artist) to a tense situation of gentrification. The Drawing Shed ideologically and practically worked against a neoliberal communicative planning ethos of consultative participation that aimed to use its practice as 'arts deployed as [a] tool' (The Drawing Shed artist). It positioned itself 'within and against capitalism' (The Drawing Shed artist). It had 'no choice but to operate within the rules', but 'it's up to us which rules we use or ignore' and by taking the funders – the developer and council – monies to do this, The Drawing Shed was 'not taking from the hand that feeds me, but [to the funder], the hand that feeds you' (The Drawing Shed artist) – i.e. the tax-payers' monies that funded the borough council and ACE roles; it was to this group that The Drawing Shed felt its responsibility lay.

The process between place-*making* and place-*shaping* as place-based practitioners was a conflicted one that was on occasion thrown back to the artists as a challenge by residents:

> I find it really difficult to negotiate. Part of me thinks that a lot of the responses, which are valid, are like, "Why are you trying to come in here and do a project with us?" It [placemaking] assumes that there aren't cultures or really important relationships already there and that somehow the artist is going to forge all this. It's bollocks. Of course they're already there. (The Drawing Shed artist)

Agency was seen to be found, located, activated and galvanised through the relational activities and objects employed and devised by the case studies. This was especially pertinent in the work of The Drawing Shed – not only by its self-aware naming of its art objects as relational objects, but its deeply performative practice that worked to uncover and nurture a place-based socio-political conscientisation where people became firstly more aware of their place, and then more active in it. The agency of place was keenly observed through the lens of situated – often ephemeral – work and as with the crossing of physical and conceptual thresholds into levels of participation and how the form of place engendered this. Issues of formality/informality, thresholds and permission to participate were seen across all case studies and the place of the *loci* of interventions was seen to have an impact on the work created and its affectual outcomes.

The go-kart making was an initial activity that gradually drew people into conversation, The Drawing Shed's *navigation* into place and its particular context on the Nine Elms estate. These conversations informed the work made and the politically and socially interrogative – and navigational – approach to it, 'working with things that are hidden on the estates, things that we found as [we] went through' (The Drawing Shed artist). The work here emerged as that of binaries, informed by the macro politics of gentrification and social cleansing, and the micro experiences created from the architectural disconnect. In one regard, 'The world inside and outside, all the walls, cutting off the connection, them and us, others get the priority, a class difference, them a higher standard, us a lower one' (The Drawing Shed artist). In another, at the street level, this was no longer oppositional: 'Here [at street level] we don't feel the difference between our neighbours, it's good, we're all together, we feel at home' (The Drawing Shed participant). The soapboxes and go-karts acted as a metaphor for a culture of resistance through the dialogic and informal aesthetic of the art practice and in the dialectical and material developmental process of their creation, their critical context was apparent. Critical conversations took place with residents – of their lifeworld and the gentrification process on the literal near horizon – and 'leaps of imagination' occurred in these conversations between the function of the art and the lifeworld of the estate: 'It's not just about making go-karts, it's about making conversations' (The Drawing Shed participant). Conversation – the dialogic aesthetic (Kester, 2004, 2011, p.32) – was here the agency in arts participation in place: conversation was a means of conceptual translation of one context to another, creating a critical distance from the estate by being 'out in the world' (The Drawing Shed participant). Conversation alone though was contested as a means to invoke change – it was felt lacking without an action outcome and felt ameliorative when proscribed by a funder. 'Conversation' itself was problematised by one of the artists; whilst its performative value was recognised – 'All problems come down to people not talking to each other' – its ameliorative proscribed function by the project funders was criticised – '"Let's get artists in to get them to talk to each other then all will be happy" … but why do they have to talk to each other? That won't make them not see the problems here' (The Drawing Shed artist).

4.4 The place of the art

Across all case studies a variety of locations were used to site art and to operate in – locations demanded and deliberated by the neighbourhood and community worked with and in. The following section uses The Drawing Shed case study to interrogate a spectrum of scales of locale, starting at the intimate site of the gallery and moving then to the larger site of the housing estate and then the public space of the park.

4.4.1 The gallery

The Art Tunnel Smithfield gallery in Dublin acted to subvert the conventional gallery space by its location in vacant and public space, and its modes of art-creation; so too, the arts lab model of *IdeasFromElse[W]here* at Winns Gallery was deliberately formed to subvert the expected normative gallery space etiquette of exhibition and viewing. It is focused on here as the sole example across the case studies of a conventional gallery space being used in a non-conventional way, knowingly informed by social practice arts and social practice placemaking artists.

At *IdeasFromElse[W]here*, the gallery threshold posed a query in firstly crossing the gallery threshold and then a questioning of 'art', being met with walls and the floor of the gallery space filled in an ad hoc way with found objects and ostensibly 'unfinished' looking art work. A friction was felt and unpicked around the nature of the gallery walls, an ever-present reminder in the first weeks before the space was populated with a critical mass of artworks, that this was a white cube gallery space (the art sector vernacular for a contemporary white-walled art gallery). Artists' questions included, but by no means were limited to: 'Is this a formal space still?', 'Does the notion of process work, work in such a space?', 'Are we still constrained by parameters, etiquette, form, of the white walls?', 'Are we trapped by conventions of a gallery still?' and 'How will we/can we break this habit?' (The Drawing Shed artists). What was to be displayed posed a quandary – was this to be a display of work-in-progress or of works completed? Through a negotiation of and navigation between practice and space – just as the housing estate setting was navigated (below) – this question abated and the walls, windows, floor and doors of the gallery were used to both archive work and to form work in the moment. The gallery was used as a hub to facilitate from and in, and once inside, there were literal pathways to navigate through the space, including at one point, a wall made of foil emergency blankets. These predicaments were representative too of the unease felt by some artists in the freedom they were given in the arts laboratory *as a space*, the premise of 'do whatever you like' set against the perceptual constraints of the space as a white cube gallery space, and its regulations, pragmatics, logistics and health and safety – a barrier to arts creation and collaboration for the artists. Through the process of conversation, physical proximity, space-sharing and the plan of the space itself and of making, questions were continuously asked of the gallery space, its formal confines and of the artists own practice in this space. The residency was a period of time and a physical space used by the artists to dissect their own and group social practice arts, to 'pick and unpick, a critical space to frame issues' (The Drawing Shed artist). Some event-led activities during the residency created 'flash points' (The Drawing Shed artist) that were unexpected, moments of spontaneous activity that were felt to crystallise or represent the intra-group dynamic of art creation and collaboration with the public, and to signal the open-plan nature of the gallery space. It was commented on for example that in the open plan space anxiety and stress, as much as a creative flow, would travel easily from one person to another; and that the art lab methodological use of the space challenged some into action, but furthermore, with a fluctuating constituency of artists over time, challenged too notions of responsibility and authority.

Through the gallery was an outdoor courtyard that could also be viewed from the park; park-goers would often stand at the gates to the courtyard, park-side, and watch rehearsal, making, exhibition or performance; one park-goer commented that it was 'nice to see the gallery going to be used for what it should be' (The Drawing Shed viewer). The gallery was punctured along one side by a glass sliding door; this opened to a forecourt area along the length of the Aveling Centre to the café and play area in front of it, and to the park to the side of it. Artists at the start of the residency asked 'Will this facilitate the actual and conceptual threshold crossing?' (The Drawing Shed artist) and over time the door and forecourt assumed use as a site of creation and performance and an increasingly significant role of a literal and metaphorical threshold crossing to and from the gallery setting, into, through and out of creative work for the artists and arts audience and participants (Figure 4.3). The glass door began to assume an active role following

Figure 4.3 The Drawing Shed (2014), *IdeasFromElse[W]Here* [residency], Winns Gallery, Lloyd Park, London, 2014.

its observational and abstracting one. One function was as a creative material itself, one artist for example using the glass as a surface on which to draw the people outside. A second was in activity on its actual threshold, with the door opened, bridging activities of what was happening in the gallery, the interior–exterior threshold permeated. This facilitated much walk-up to, and then into, the gallery from passers-by, and then a possible participation in the interior space, the large opening of the glass wall seemingly more welcoming and less intimidating than the gallery door to the far-end of this aspect. On sunny afternoons and on weekend event days, an interior–exterior threshold dissolved completely, which formed the space as an interactional whole:

> This was most powerful on the last day, an extraordinary day, with the performance artists outside and as a result of that, the inside was completely activated in relation to the outside. Somehow an interconnection within the park was fundamental. (The Drawing Shed artist)

With both Art Tunnel Smithfield and The Drawing Shed at Winns Gallery is seen the re-creation of the gallery as a countersite (Holsten, 1998, p.54, in Watson, 2006, p.170), playing with the somatic protocols and regulations of white-cube and exhibition norms. The production of space was generative (Lefebvre, 1984, p.170; Lilliendahl Larsen, 2014, pp.329–30; Low, 2014, in Gieseking and Mangold, 2014, p.35; McClay and McAllister, 2014, p.8) and space contingent (Rendell, 2006, p.17; Soja, 1997), though both acting with and against its parameters at the same time. The agency of the gallery space was produced through fluid communication where the exposing of the art-making process affected the artists' intra-group relations, as well as precipitating a participative to collaborative making with the protagonists and an interiority affinity (Friedmann, 2010, pp.154–6) that was constitutive of the gallery space as a symbolic cultural entity (Magnaghi, 2005, p.37, in Sepe, 2013, p.6). The artists collectively uncovered a new meaning (Lowe, in Lowe and Stern, in Finkelpearl, 2013, p.138) to the gallery space over the course of the residency and relinquished sole authorship by creating further opportunities for public participants to join as protagonists (Beech, 2010, p.21; Bishop, 2006 [a], p.16; Chonody, 2014, p.2; Finkelpearl, 2013, p.361; Hope, 2010, p.69; Walwin, 2010, p.125), a performative sociability engendered by the informal – yet rigorous – logic (Sennett, 2012, p.53) of the arts lab. Where there was conflict, this was used to further question the arts lab practice (Froggett et al., 2011, p.104; Jackson, 2011, p.27). The social practice placemaking activity here lay in the bringing into the park and gallery the social and political issues of the borough and street, the arts lab acting as a cultural loci (Larsen, 1999/2006, p.173) discursive mirror to the functional park community (Nicholson, 1996, p.116), simultaneously challenging and re-producing community over the course of the residency (Froggett et al., 2011, p.96; Murray, 2012, p.257).

4.4.2 The housing estate

The housing estate space was the central concern of The Drawing Shed at the time of research and is given special mention here as a location for site-specific ideologically driven social practice placemaking activity. Situated in the physical, social and political context of the housing estate, the arts practice and processes of The Drawing Shed engaged with its urban form and its cultural, social and psychological ramification at the intersect with the political. Walthamstow – site of the *IdeasFromElse[W]here* activity – was referred to by many as both 'the most diverse borough in London' and 'the new Dalston', meaning its gentrification and 'hipsterisation' was imminent, if not occurring already. The Nine Elms – of *Some[w]Here Research* – site was repeatedly given the conversational strapline of 'the largest site of gentrification in London' (The Drawing Shed artist) and the same issues travelled from Walthamstow to Wandsworth through the arts activity and the artists, though more sharply seen in the latter through the prism of an urgent politic of social cleansing, market forces, privatisation and gentrification and the sense of socio-spatial isolation of the closed off estate entrances and pathways as an effective gated community (Seamon, 2014, p.15), both potentially sites of Nancy's (1986/2006, p.56) 'dissolution of community'. The Attlee Estate is a comparatively green space compared to that of Nine Elms, the latter likened to Fort Knox by one artist due to its closed routing into the estate by road, having to circumnavigate the estate in its entirety to drive into the estate – and the need to navigate physically into the estate informed the approach to practice: '[We] navigate our way across three estates, getting to know people bit by bit' (The Drawing Shed artist). Once within the Nine Elms estate, houses and areas of public space are gated and walled off, many direct pedestrian routes and neighbouring access blocked by story-high metal grilles as a security measure, another artist articulating a Foucauldian Panopticon (Foucault, 1977) analogy of the space, with houses circling each other and impermeable to neighbourly interaction but with a feeling of being watched, 'stacked on top of each other, [there is] no horizontal aspect for communication' and 'It's become antisocial housing' (The Drawing Shed artist).

The navigational approach into place and people was taken by The Drawing Shed across the two housing estates it worked on, but over vastly different durations. The Drawing Shed had been resident on the Attlee Estate since 2009, so had grown to know the area well and had become 'resident non-residents' (The Drawing Shed artist) as mentioned above. This helped it embed into the Winns Gallery residency as the area and its gatekeepers were known, and some participants crossed from one space to the other from knowing The Drawing Shed. In Nine Elms, The Drawing Shed had a period of months, over winter 2014 to early summer 2015, to work across the three estates, from a base of its senior's community centre. Arts interventions here acted as analogy through the relational objects created (go-karts, soapboxes) to issues of migration, relocation, residence, transitoriness, temporality and identity, acting as a mirror to the issues in place. The arts project was seen to have its own customs and language, as developed during Woodham's *Danger Deep Water* (2014) [4.5] and just as with those cultures it was encountering and bringing together in the arts space, and The Drawing Shed was concerned with how to open up its community of practice to others, just as other cultural, religious and civic communities in the borough were also asking of themselves and others.

The Drawing Shed residency at the Attlee Estate lock-ups was described as a 'haptic accident' (The Drawing Shed artist) from which projects took place and formed an agentive space from, and in, which residents and invited artists alike both worked, singularly and in co-production. The place of the lock-ups was recognised in this regard as being a blank canvas which any resident could use to their own creative ends, if they so chose. As a place, the lock-ups were perceived to inhabit their own creative and social latent agency, separate to any creative activity placed into it by The Drawing Shed, created and activated by all those that used the space, by virtue of its liminality. The work on Nine Elms was that of making soapboxes and go-karts, objects inspired by the areas' social, cultural and political past and the experiences of play spoken of to The Drawing Shed by some of the estate's older residents. These were made on the street, on a paved area close to a parade of shops. This staging of the work was deliberate, 'not obvious or attention seeking' meaning that 'people had to see it of their own accord' (The Drawing Shed artist), a gradual process of familiarisation working *in regard of* both people's fear of such creative or undefinable activity and working *against* the architectural, topographical and spatial constraints of the housing estate and the politics of gentrification:

> People are threatened by something new. This is a ghost town. People [are] blocked in. [Its] part of what we're discovering about how people live ... I've never had a feeling of architecture stopping people from doing things as here. [It is] endemic of the blocking off of social housing from policies of fear.

Older people saying that the culture here before was much more communal...
(The Drawing Shed artist)

As 'resident non-residents', artists appeared to experience a process of home-making on the estates. The long-term residency at the Attlee Estate engendered this, as stated above, and this was perhaps to be expected over such a length of time in place. However, the frequency of visits, the length of time of visits and the impact of the street-level interventions fostered a rapid, but not as longitudinally deep, implanting of artists into this space at Nine Elms. This was in part facilitated by a cross-cutting understanding of the macro issues of social housing in London that the Nine Elms residents identified and responded to, 'to carry on with an inquiry that runs through their work and to work almost in parallel with us with moments of overlap' (The Drawing Shed artist). The inquiry was the lived experience of the housing estate in its past and present socio-political form and to pose questions of its future, the artists invited to 'intervene or to make provocations, but without any kind of set brief' (The Drawing Shed artist) of how this might be done. Mckenzie brought the experience of *LUPA* (2011–13) and was 'interested in how difference is accepted or understood on the estate' and interrogating what it means to be 'an artist-in-residence and an artist-in-residence-on-an-estate' (The Drawing Shed artist). His performances were processional around the estate and the adjoining streets, enabling a social site-specific interaction with the people and issues of the estate.

Thus, while the sense of place of the Walthamstow and Wandsworth estates differed, both were encountering social, cultural, economic and political issues of gentrification; The Drawing Shed acted to join this concern through the ideological thrust of their work and through commissioning artists who had an understanding of these issues from their own lived experience, a 'same but different' (The Drawing Shed artist) approach to both engender an empathetic response between artists and estate residents and a degree of objectivity in the artist and the work produced. The artists therefore acted from a liminal position where they were working from a limited, implanted, position, at once familiar with the site and also stranger to it, but given some credence as 'artist of residence' of housing estate lived experience. Thus, the relation to space and place of the project site here informed the relative merits of an embedded artist position. In this instance, whilst knowledge of the hyperlocal site and context was not felt required of the artists, knowledge of its issues in a broader scope was thought required to engender an empathetic response and relationship building with the residents. Where the aims of the projects funders were to use this practice as a mediated (Debord, in Bishop, 2006 [b], p.12) spatial diversion (Lilliendahl Larsen, 2014, p.322) to ameliorate the significant physical and social changes to the site and give credence to claims of public consultation and participation (Bishop, 2012, p.14), the aims of The Drawing Shed were subversive to this. Its work acted as a critique of the neighbourhood's gentrification – to 'disturb the annulment of politics' (Beyes, 2010, pp.242–3), acting from a parasitic activist positon to work with residents to help them uncover their own positions in this process (Bishop and Williams, 2012, p.213; Klanten and Hübner, 2010, p.103), a voice that had been denied them thus far.

4.4.3 Public space

The Drawing Shed's *IdeasFromElse[W]here* also took place in the site of Lloyd Park. The park was a well-used space through the day, with different constituencies of people – pre-school children and their parent(s) or guardian(s), young people and adults from across the local demographic. Activities in the park included: the urban gym; the play areas for children and young people; the café and its seating for socialising and use as a meeting or activity space for organised groups; and the generous pathways through and around the park used for dog-walking, cycling, skateboarding and rollerblading. Whilst there was an overspill of activities from one space to another and a tacit co-mingling here too, people tended to remain socialising within their family or friendship groups. The park was used as a creative space for the artists – a literal out-of-the-gallery move – and a range of procession, dance, critical movement, installation, intervention and making activities took place, with the intention to affect people's experience of the park as place and to from there, engage in conversation about borough issues on a wider level. The park was viewed as an 'intensive public space' (The Drawing Shed artist) and many times referred to, understood and used as a 'breathing space' for work-creation and the art experience, a place where issues from the art sector, in terms of the gallery space and the residency, and from the local streets and housing, could be explored: 'There's something about parks and testing public space … it's a breathing boundary it feels like, it becomes a liminal space…things you do in such a spacehas impact…[we can] trigger off these' (The Drawing Shed artist). The siting of work in the park stimulated interplay with the audience that was immediately more interactional than that observed in the Winns Gallery setting. In Mckenzie's (2014) *Fence 2014*, people would gather in groups around the artist, Mckenzie remaining mute throughout a test to limits of responsibility in public places and of public audience/artist interplay:

> Some people were very interested and others who would've then realised it was a piece of performance art, then became part of the…not voyeurs… but they became active participants, active audience. There was a shift in the mode of spectatorship 'cos they then became witnesses to the art, collectively witnesses of what was going to happen, with other people, what was their reaction, and they were very fascinated by that. (The Drawing Shed artist)

In accordance with ambiguous city experience thinking, public space was experienced as an intensive, serendipitous interactive space but one that also, through the art intervention, facilitated – perhaps indeed required – a 'breathing space'. The density of people in public space facilitated this, as did the liminal qualities of spaces as countersites (Holsten, 1998, p.54, in Watson, 2006, p.170).

However permeable the threshold was rendered at Winns Gallery, predominantly there was a distinct spatial experience between the space of the gallery and the place of the park, the park referred to as 'the out there' (The Drawing Shed artist). The gallery and the park operated too in a contested spatial ecology of 'revolving universes' (The Drawing Shed artist) of art and municipality, and the park-goers embodied ownership of the space. In these differentiated spaces, there was a visceral shift in the interplay with the artwork, seeing here again faceted participation. When in the gallery the hung work was observed as objects of visual art, and the activities, such as badge-making, variously participated in. When outside but close to the gallery, artwork was observed from one step back. As soon as the artwork crossed an invisible but tangible threshold deeper into the park area, it was in the territory of the park-goers and became an object of play and low-level derision. The artists were now in the park domain where the park audience was empowered, out of the art domain of the gallery and forecourt, where the artists were perceptually – and relatively as this was not experienced as uniform with all members of the public as many participated and collaborated – empowered over the public. In the instance of some work, this was a dangerous occurrence for the physical safety of the performer, in which other artists had to intervene and bring the performance to a stop. For one of the artists, this particular threshold-crossing:

> …comes back to the breaking of rules energy, how you don't control it so that you are censoring what the artist is doing, but still protect people, how you stop others from engaging in a way that was not safe. (The Drawing Shed artist)

Artists articulated that it was too easy to say the inter-relation was a 'give and take' of territoriality between one group and other. It was instead a complex and self-aware mutuality:

> That territory thing needs to be unpicked more. It's too easy to talk about holding territory, giving up territory, in those contested spaces, as if these boys don't know what contested spaces are. Well of course they know. They spend their whole lives finding ways around that. But I think there is also an energy that comes from these things that young people pick up on. (The Drawing Shed artist)

The section above viewed the practice and process of The Drawing Shed through the prism of space and place thinking; the following section addresses the same though the prism of issues of participation.

4.5 Degrees of participation

The following section focuses on issues pertaining to participation – that of non-artists and artists in social practice placemaking practice and process, degrees and types of participation and non-participation, and barriers to participation.

Over a linear participative progression, instead facets of active participation and consequent engagement were seen in the case studies.

4.5.1 Non-artist participation

This section will focus on The Drawing Shed as an exemplar of faceted degrees of participation through its projects, though all levels were seen with Art Tunnel Smithfield and Big Car also. One-degree participation was *abstracted participation*. This ranged from people who had observed the *Live Lunch*, shouting, humorously, from their balconies on the estate at its close, 'When are you making food again?'; to boys resident on the estate who had not entered Winns Gallery during *IdeasFromElse[W]here*, offering to help de-install and asking 'When is the next thing you're doing?'; to children circling the gallery and cycling through the exterior artworks, both an act of territory marking but often the precursor to entering the gallery at a later stage. People who later returned to the gallery and participated at a deeper level would often begin their journey into the experience through a similar one-level-removed audience gaze or abstracted participation. A further degree of participation was that of a *mediated participation*, or that of an encouragement to speak. For example, a textural invitation to speak was observed at the typewriters, people typing autobiographically and encouraged to continue to do so by the artists when a moment of embarrassment was experienced: 'You may not feel that your typewriting story is interesting, but it is, they're all fascinating' (The Drawing Shed artist comment to participant).

Moving deeper by degree of participation, next there was *engaged participation*. This was seen in the gestural participation between an artist and the audience member: a language of signs was developed between the artist and the audience in Woodham's *Danger Deep Water* (2014) (Figure 4.4) in Lloyd Park as part of *IdeasFromElse[W]here*, during which a community of spectators turned

Figure 4.4 Ed Woodham (2014), *Danger Deep Water* [performance], during The Drawing Shed (2014) *IdeasFromElse[W]Here* [residency], Winns Gallery, Lloyd Park, London, 2014.

protagonists co-created the performance. In this piece, the artist placed himself in a cordoned off dried-up pond area of the park, dressed in white clothing and wrapped in white fabric strips, with a found polystyrene tube on his head with cut-out eye holes, standing on a stool, for a duration of three hours. The community that formed around the piece were from the park-going constituency and of a varied demographic; some stood in silence and watched, others then began to converse with each other, and a number of children began to shout questions to the artist – 'Are you ok?' 'What are you doing?' 'Why?' 'What is your job?' being common. On one boy asking 'What's your favourite shape?' the artist responded with a shape-forming movement which began the formation of a simple gestural language between him and the questioners. Whilst some did not see this as a 'performance' – 'people think that he is in his own thing' (The Drawing Shed artist) – others identified this as an art piece and saw themselves as 'joining in' by joining the conversation with the artist and with others. Engaged participation was experienced in particular during the last weekend with a high level of public walk-up to the gallery, encouraged by the sunny weather, and a critical mass of art activity and performance in and around the space, which combined, led to many people crossing the threshold of observer to participant: 'at one point when there was so much dynamism going on ... there were people engaged in different ways rather than passing through, in very real ways' (The Drawing Shed artist). In this situation it was remarked that the artists had to 'make decision to a certain extent and then open up' (The Drawing Shed artist) their activity to the public and an experiential turn occurred between all then involved.

The deepest level of participation was that of *co-produced participation*, as seen in an initially artist-led workshop, by Woodham, with a writers group formed during *LiveElse[W]here*. This started with exercise in the gallery around a long table, a familiar setup for the group as that experienced during workshops in the lock-up; it then moved to a tour of the park where protagonists were then asked to find an 'unexpected' place to write in, a level of disjuncture from the usual setting of their writing; the workshop then took on an art making aspect, the group asked to make their own art pieces or performances. This fifth level led on for some to a sixth level, *participant-as-sole-artist*. The last stage of the writing group culminated in the group sharing these artworks or performing in the park space to the group and to nearby members of the public. One participant commented that in this process they felt that they 'had become the artwork' (The Drawing Shed participant). With The Drawing Shed, the creative autonomy of protagonists was sacrosanct and in its co-productive model, 'the non-artist is implicitly and explicitly told, "You are an artist too". Really, they're never thought of as not, right from the beginning' (The Drawing Shed artist). The activity of the writing workshop, as dialogic, was one of storytelling also, observed as socially connective, both empowering for the storyteller and remedial for the listener (Chonody and Wang, 2014, p.90) and facilitating a self-reflexive exploration (Kester, 2004, pp.93–4) for the protagonists, individually and as a group (Chonody and Wang, 2014, p.91–2).

4.5.2 *Problematising participation*

Issues to participation were observed across all case studies: at The Drawing Shed sites issues around participation focused on a preconceived notion of the arts held by (at first) the audience from the abstracted or gaze positions, and by nature of the work being sited in contentious areas of gentrification, the role of the artist.

There was an evident pre-conceptualisation of the arts that for some members of the public audience acted as a barrier to participation. As projects sited in areas of poverty where a fear of having to pay for the arts is a disincentive to participation, artists were aware of the concept of arts as being a monied activity as being an issue for potential participants. One parent at the badge making machine at The Drawing Shed's *IdeasFromElse[W]here*, where their child had made a number of badges before the parent realised, became visibly panicked at the anticipation of what this might cost. For The Drawing Shed, the provision of arts for free was part of the ethos:

> The area around the café is the working class area during such events … free things, you can just hang out, bring [your] own food … I felt, I know this [financial] place, I've been here before, it becomes difficult to afford, if you're living on benefits then every single penny is accounted for, there's no slack. (The Drawing Shed artist)

There was a manifestation of active resistance to not just participation but the art activities, *as art*. For example, during Mckenzie's *Sink Estate* (2014), one woman, on passing the artist and remaining at some distance from him, loudly commented 'Is he for real? Why doesn't he just pick it up. Silly man' (Carey Estate passer-by). Others walked past and pretended not to notice; a young person waiting at a bus top, on being handed a question card, angrily retreated against the wall of the bus shelter. It was only when the artist walked the High Street that the reaction from passers-by was more relaxed, people laughing kindly at the artist, turning to watch him walk down the street, or coming out of their shops to watch, asking to each other such open-ended questions as 'Is it April Fools?'; 'Is it penance?'; 'Is it a dare?' (Wandsworth High Street audience members); a pizza delivery moped rider followed the artist down the street at a crawl and someone waiting in a bus stop on the High Street shouted across the road, again, humorously, to the artist, 'Oi mate, is that art?' (Wandsworth High Street audience member). During the go-kart making workshop on the Carey Estate, artist Daniella Valz Gen was met with open hostility and suspicion when performing her walking creative exploration, *Light Trap* (2014), with a mirrored eye hung around her neck; a resident who had been watching the workshop from his balcony for some hours left his residence and walked to the workshop area and aggressively shouted 'You're wankers!' to the assembled group and then walked home again. This was respected by the artists as a choice but also one based on a culture of non-engagement:

People don't have time, they don't want to get involved. They don't want to think it's for them. But then, that comes with the culture of the estate, [the Council] come in to do the gardens, so people might say they don't need to take part. But the sense of satisfaction you can get from that, all the emotions you can get from that, you can feel proud, you can feel you've created something. That takes people to be involved. (The Drawing Shed artist)

At the smallest formation, the mobile structures designed and employed by The Drawing Shed, *The Drawing Shed* (2009) and the *Printbike* (2010), were created to encounter a psychological trepidation of arts participation but also from crossing one estate boundary to another, 'We learned from experience that people don't cross between estates but there's no reason to, no resources to encourage [that]. Mobile interventions to encourage this' (The Drawing Shed artist). At Nine Elms, the go-karts had the same function. This estate was noted as quiet and often devoid of street-level human presence. Artists commented that they felt vulnerable from being watched from invisible and distant eyes and from feeling alone, and thus unsafe, on the empty streets with locked-off escape routes. It was appreciated by the artists that similar psycho-spatial feelings would be informing the lived experience of the residents, let alone giving them any inclination to join in a publically sited arts workshop. However, the repeated go-kart workshops and the further presence of the artists on the streets of the estates when working on their own creative interventions from found objects over time enacted permission-granting for others to join in, the intensity of this programming aiding the artists acceptance and embedding on the estate:

The other thing that's up against us is that nothing happens on this estate [Nine Elms]. Nothing goes on. On Attlee, there is a walk up, people coming and going. Here we have to do things over and over. [It] takes time to become a culture, part of what goes on. (The Drawing Shed artist)

As noted above, The Drawing Shed's mobile structures acted as relational objects and were purposed to work with people in the location of their complex estate-based habitation, crossing the boundary-marker of The Drive road from one side of the Atlee Estate to the other, travelling around to where people were and 'creating spaces that were very different for drawing' (The Drawing Shed artist). In this way, the mobile objects produced splash interventions and an element of spectacle (Hamblen, 2014, p.83; Kester, 2004, p.83; Kwon, 2004) that punctured the life-world of the estate; the relational objects created at Nine Elms functioned in a similar fashion, through street-based workshops and processions. The *push* of the mobile structures into people's estate lived experience created in some princi pally those from the writing group and young people – a *pull* to Lloyd Park and the *IdeasFromElse[W]here* residency at the Aveling Centre, a walk of just under 1km but a psychogeographic barrier for many nonetheless. The art activity across the Walthamstow and Wandsworth sites were linked by commissioned artists working across both; the extending of similar hyperlocal socio-political themes across both; and the mutual knowledge's produced in a process of a *navigational* 'bridging' between the two places:

> We bridge on, we bridge off, 'cos that is what life is like. We want to have new experiences and bring them back, share them, share the experiences that are grounded in the local, and that's why we are here in Wandsworth working across three housing estates. (The Drawing Shed artist)

Artists also problematised their own participation, as crystallised in a conversation held in the planning of The Drawing Shed's *IdeasFromElse[W]here*: the reference of Kaprow, although generally acknowledged and accepted, was met with some conflict with the artist group, articulating a friction in being 'given' Kaprow as a reference point, which seemingly was contra to Kaprow's ethos of a collaborative formation and creation of work, the artists going on to question who would be involved in the creative decision-making process and how. This was somewhat resolved in the unfolding nature of the art lab, through a tacit process of collaboration and move away from a strict adherence to the daily instructions. The ethos of The Drawing Shed was to create relationships around conversation and dialogue, a means of art participation that is not based on 'managing people' and thus 'contra to model of participation where engagement is proscribed' (The Drawing Shed artist). Participation in the former regard was seen as imbricated in a power relation:

> There's this power construct, 'I have this experience that I want you to participate in'. So it's framed straightway, there are boundaries around it. With participatory arts in the arts sector, having a generic use of 'participation', there's a lot of murky water, not a lot of what I would call truly participatory. It may have its own value, but it's not participation, the artist is still the author, directing the experience. (The Drawing Shed artist)

For others though, the participation model as proscribed from what they perceived as 'the arts sector' was thought to be patronising. In The Drawing Shed's project at Nine Elms, friction with the funder was surfaced in this participatory arts model regard. The funder's aim of the project was felt directed towards a 'community creation' or 'cohesion', but 'if we say this is a goal, this is an enforced participation model' (The Drawing Shed artist). Thus the artists role was to 'protect participants against these pressures, of funders seeing them as a vehicle through which to enact things, their autonomy lost in this and also having power enacted on them' (The Drawing Shed artist). The artist's political complicity, agency and impact were problematised:

> You, this artist, are coming in as pawns for the regeneration. You say you're doing stuff for the community but you're just pawns. But where are we

supposed to go? We're just the artists. Who should do the work? (The Drawing Shed artist)

Another saw their agency in being able to use their position, and the funding, to be contentious against those market forces:

> If you're doing an art project, you'll be accused of being complicit in gentrification, which I disagree with, I think there is a way to do an art project and not be complicit, there are other ways of doing stuff. But people will be like, 'don't even start', and it's disheartening. (The Drawing Shed artist)

Others also saw themselves in a 'position where we can critique some of their [the developers] dubious practices' through an 'ability to critique through our practice' (The Drawing Shed artist). A contradiction was seen though in the position of the artist as someone that 'sets the gears in motion' of the participatory model, the illusion being that:

> They're [the artist] sort of in the background giving the impression to everyone that this is made from material from the estate, but actually what they are doing is orchestrating and organising that material ... I have a problem with that ... so I have decided that I will come in and declare my position as an artist and as a fine art lecturer ... (The Drawing Shed artist)

Thus far, aspects of the social practice placemaking practice and process have been discussed, with nuances of participation presented and notions of participation problematised. The chapter closes with look at the outcomes observed here, primarily, that of a reflexive and transformational outcome.

4.6 Reflexive and transformative outcomes

For protagonists, transformation could take the form of small, personal experiential and cognitive shifts to those that worked directly with individual creative agency. The storytelling activity with The Drawing Shed was directly attributed to being empowering, 'creating ways for people to tell their stories in a myriad of ways, sometimes it is very literal and other times more abstract but the method of making a story tangible is very liberating' (The Drawing Shed artist). Protagonists were said to find 'a voice through this work' that allowed 'in a very visceral language' the protagonist to express 'things that perhaps they do not or cannot speak about so freely' in the day-to-day, and go on to produce 'work that raises awareness of situations' (The Drawing Shed artist). Artists commented also that the art lab experience engendered a reflexive experience in them relating to their practice, learning to enjoy the moment and the process itself; questioning what type of artist they were and their habitual approaches to making work and how this could change; and informing them to work more strategically in other community settings, principally asking who the project is for and to what ends, and becoming more aware of nuance in neighbourhood community settings. For others, their reaction to The Drawing Shed was one of a feeling of a shared humanity not encountered in the everyday. These participants commented that the art experience 'made me feel human' (The Drawing Shed participant). A Ukrainian immigrant kept returning to the residency, and in particular Woodham's *Danger Deep Water* (2014) pool intervention performance, as whilst they had 'never seen anything like it' (The Drawing Shed participant), the language of the intervention being liminal and in the public realm engendered in them a reminiscence of the protests they had been involved in their home city. A member of the Atlee Estate writing group became a leader of one of its workshops, a first-in-a-lifetime experience they found emboldening, commenting at the time, 'Now we've got [The Drawing Shed], it's the most amazing thing that has happened!' (The Drawing Shed participant). For another, the art experience engendered thinking towards the community and their place in it: 'All of us have a responsibility to take part in community. We're all here to look out for each other, we can be one happy family and being part of our individual families, a bigger sense of family' (The Drawing Shed participant).

The impact and degree of transformation and reflection engendered by this work and the blanket acceptance of the notion that participation engenders positively beneficial transformation has to be questioned however. So too, a familiarisation with emplaced arts and its reducing or negation of aesthetic dislocation impact (Kaji-O'Grady, 2009) or a wilful ignoring (Hannah, 2009, p.109) of or unobvious communication of political intention. For example, one artist commented on a participant's perception of The Drawing Shed on the estate: 'You ask him if this has made an impact, and he's like "No, not really"' (The Drawing Shed artist). With the council estate setting of The Drawing Shed's work, a criticism of a default 'transformative position' was articulated, in a synergy with the problematics of a placemaking discourse (Cohen, 2009, p.145; Doyon, 2013; Roberts, 2009) as the 'making' of existing places. One effect of the progression of performative participation, as above, to co-production and then authorship of the 'non-artists' in the work for a resident artist was to again stimulate a deeper reflection on the nature of their work with the community:

> I started to think about different ways of engaging the public and also how the public might project something of themselves on to a performance or art work. I began to consider the role of the public beyond spectator. (The Drawing Shed artist)

Friction was recognised as necessary to achieving a common goal, a necessary part of the process being to work through differences of opinion and accepting that not all views could be represented as the process continued nor that a consensus could or would necessarily be reached – as found in the creation and maintenance of the micropublic. Similarly, restrictions of a transformative potential were also felt by the artists by the limitations of the process medium: 'it varies as a shared experience … there isn't a shared experience as such' (The Drawing Shed artist). As an embodied and permeable space, children and young people would weave in and out of (by foot, on rollerblades or on bicycles) the space and the artworks and performances, the space becoming an agentive shared and embodied and creative terrain. This terrain was one The Drawing Shed created to ease the public into the space, and consequently, into the art process, 'that becoming comfortable with things that aren't necessarily things that you want to connect to yourself, has a sense of an energy that doesn't require your participation' (The Drawing Shed artist). The terrain formed a territory also, a boundary position from which children and young people acted around to test permissions and rules, taking up this position also habitually and also reacting to seeing the artists themselves rule-breaking:

> The people that are breaking rules, children love that. They love challenging all the things that they can, when they realise the boundaries are gone. They have no built in mechanism to know how far they've gone and if they've gone too far … on that boundary, those performers trying to engage with those boys, you can do this but you can't do that. (The Drawing Shed artist)

Some children and young people in the space became over-excited, 'not an ebb and flow so much as a push and pull on what do you do in a space like that where there are no parents around' (The Drawing Shed artist), it being recognised at the time that 'there is a wide space between how young people navigate towards a more complex cultural experience' (The Drawing Shed artist) from one of play.

In London with The Drawing Shed and *IdeasFromElse[W]here*, curation was used to programme art in the community that was a 'gentle magnet' (The Drawing Shed artist) for the public, to draw them into place-based activity and subsequent emotional affective place-bonding (Lewicka, 2014, p.351). Interventions that included dance, performance and play that were 'not flexing the performance art muscle, but still quite challenging' and avoiding 'announcing that it was "Performance Art" [*sic*]' but were instead playful, informal and fluid to join in or out of created a 'breathing space' of self-reflexivity, which was 'much more successful' (The Drawing Shed artist) than the outcomes perceived for more formal emplaced arts interventions experienced by the artists. Thus, social practice placemaking and The Drawing Shed's wider social practice arts approach generated spaces in which people could journey through a process of self-actualisation, interpersonal discourse and place awareness and attachment. During *Some[w] Here Research*, The Drawing Shed consciously created a place-based 'Thinking through Making' (The Drawing Shed [a]) critical approach to placemaking. The relational objects created, fragile as they were from their makeshift fabrication, were metaphors 'essential to how we re-think a world and put it back together again, going off route, off streets, on streets, connecting communities' (ibid.). For co-lead artist Labern, work is 'triggered by the time spent in a place … as I have phenomenologically got to know this place' (The Drawing Shed [b]).

4.7 Bibliography

Altman, I. and Low, S. M. (1992) *Place attachment*. New York: Plenum Press.
Amin, A. (2008) 'Collective culture and urban public life' in *City*, 12, 1 (April): 5–24.
Beech, D. (2010) 'Don't look now! Art after the viewer and beyond participation' in Walwin, J. (ed.) *Searching for art's new publics*. Bristol: Intellect.
Beyes, T. (2010) 'Uncontained: the art and politics of reconfiguring urban space' in *Culture and Organisation* 16, 3: 229–46.
Bishop, C. (2006) [a] 'Introduction – viewers as participants' in Bishop, C. (ed.) *Participation*. London: Whitechapel Gallery and The MIT Press.
Bishop, C. (2006) [b] 'The social turn: collaboration and its discontents' in *Artforum*, 178–83 [Online]. Available at: www.gc.cuny.edu/CUNY_GC/media/CUNY-Graduate-Center/PDF/Art%20History/Claire%20Bishop/Social-Turn.pdf (Accessed: 28th November 2015).
Bishop, C. (2012) *Artificial hells: participatory art and the politics of spectatorship*. London: Verso.
Bishop, P. and Williams, L. (2012) *The temporary city*. Abingdon: Routledge.
Bourriaud, N. (1998/2006) 'Relational aesthetics' (1998) in Bishop, C. (ed.) *Participation*. London: Whitechapel Gallery and The MIT Press.
Buser, M., Bonura, C., Fannin, M. and Boyer, K. (2013) 'Cultural activism and the politics of place-making' in *City* 17, 5: 606–27.
Chonody, J. M. (2014) 'Approaches to evaluation: How to measure change when utilizing creative approaches' in Chonody, J. M. (ed.) *Community Art: Creative Approaches to Practice*. Champaign, IL: Common Ground.
Chonody, J. M. and Wang, D. (2014) 'Realities, facts, and fiction lessons learned through storytelling' in Chonody, J. M. (ed.) *Community art: creative approaches to practice*. Champaign, IL: Common Ground.
Cohen, M. (2009) 'Place making and place faking: the continuing presence of cultural ephemera' in Lehmann, S. (ed.) *Back to the city: strategies for informal intervention*. Hatje Cantz.
Critical Art Ensemble (1998) 'Observations on collective cultural action' in *Art Journal* 57, 2 (Summer 1998): 72–85.
Debord, G. (2006) 'Towards a situationist international' in Bishop, C. (ed.) *Participation*. London: Whitechapel Gallery and The MIT Press.
Dewey, J. (1958) *Art as experience*. New York: Putnam.
Dixon, J. and Durrheim, K. (2000) 'Displacing place-identity: a discursive approach to locating self and other' in *British Journal of Social Phycology* 39, 27–44.
Doherty, C. (2004) *Contemporary art: from studio to situation*. London: Black Dog.
Doherty, C., Eeg-Tverbakk, P. G., Fite-Wassilak, C., Lucchetti, M., Malm, M. and Zimberg, A. (2015) 'Foreword' in Doherty, C. (ed.) *Out of time, out of place: public art (now)*. London: ART/BOOKS with Situations and Public Art Agency Sweden.
Doyon, S. (2013) Placemaking vs placeshaking, 25th February 2013 [Online], *Place Partners*. Available at www.placemakers.com/2013/02/25/placemaking-vs-placeshaking/ [Accessed: 8th July 2013].
Dunk-West, P. (2014) 'Building objects for identity and biography' in Chonody, J. M. (ed.) *Community art: creative approaches to practice*. Champaign, IL: Common Ground.
Finkelpearl, T. (2013) *What we made: conversations on art and social cooperation*. Durham: Duke University Press.
Foucault, M. (1977) *Discipline and punish: the birth of the prison*. London: Allen Lane.
Friedmann, J. (2010) 'Place and place-making in cities: a global perspective' in *Planning Theory and Practice* 11, 2: 149–65.
Froggett, L., Little, R., Roy, A. and Whitaker, L. (2011) New model of visual arts organisations and social engagement, *University of Central Lancashire Psychosocial Research Unit* [Online]. Available at: http://clok.uclan.ac.uk/3024/1/WzW-NMI_Report%5B1%5D.pdf (Accessed: 12th August 2015).

Gieseking, J. and Mangold, W. (eds.) (2014) *The people, place, and space reader*. New York: Routledge.

Hamblen, M. (2014) 'The city and the changing economy' in Quick, C., Speight, E. and van Noord, G. (eds.) *subplots to a city: ten years of In Certain Places*. Preston: In Certain Places.

Hannah, D. (2009) 'Cities event space: defying all calculation' in Lehmann, S. (ed.) *Back to the city: strategies for informal intervention*. Ostfildern: Hatje Cantz.

Hope, S. (2010) 'Who speaks? Who listens? Het Reservaat and critical friends' in Walwin, J. (ed.) *Searching for art's new publics*. Bristol: Intellect.

Jackson, S. (2011) *Social works: performing art, supporting publics*. Abingdon: Routledge.

Kaji-O'Grady, S. (2009) 'Public art and audience reception: theatricality and fiction' in Lehmann, S. (ed.) *Back to the city: strategies for informal urban interventions*. Ostfildern: Hatje Cantz.

Kester, G. H. (2004) *Conversation pieces: community and communication in modern art*. Berkeley: University of California Press.

Kester, G. H. (2011) *The one and the many: contemporary collaborative art in a global context*. Durham: Duke University Press.

Klanten, R. and Hübner, M. (2010) *Urban interventions: personal projects in public spaces*. Berlin: Gestalten.

Kwon, M. (2000) 'One place after another: notes on site specificity' in Suderberg, E. (ed.) *Space, site, intervention: situating installation art*. Minneapolis: University of Minnesota Press.

Kwon, M. (2004) *One place after another: site-specific art and located identity*. Cambridge, MA: The MIT Press.

Labern, S. (2015) *Manual for possible projects on the horizon*. [n.p.]: The Drawing Shed.

Larsen, L. B. (1999/2006) 'Social aesthetics' in Bishop, C. (ed.) *Participation*. London: Whitechapel Gallery and The MIT Press.

Lefebvre, H. (1984) *The production of space*. Translated, Nicholson-Smith, D. Malden, MA: Blackwell Publishing.

Lewicka, M. (2014) 'In search of roots: memory as enabler of place attachment' in Manzo, L. C. and Devine-Wright, P. (eds.) *Place attachment: advances in theory, methods and applications*. Abingdon: Routledge.

Lilliendahl Larsen, J. (2014) 'Lefebvrean vagueness: going beyond diversion in the production of new spaces' in Stanek, Ł., Schmid, C. and Moravánszky, Á. (eds.) *Urban revolution now: Henri Lefebvre in social research and architecture*. Farnham: Ashgate Publishing Ltd.

Low, S. (2014) 'Spatializing culture: an engaged anthropological approach to space and place (2014)' in Gieseking, J. and Mangold, W. (eds.) *The people, place, and space reader*. New York: Routledge.

McClay, W. M. and McAllister, T. V. (2014) 'Preface' in McClay, W. M. and McAllister, T. V. (eds.) *Why place matters: geography, identity, and civic life in modern America*. New York: New Atlantis Books.

McCormack, D. P. (2013) *Refrains for moving bodies: experience and experiment in affective spaces*. Durham: Duke University Press.

Miles, M. (1997) *Art, space and the city: public art and urban futures*. London: Routledge.

Murray, M. (2012) 'Art, social action and social change' in Walker, C., Johnson, K. and Cunningham, L. (eds.) *Community psychology and the socio-economics of mental distress*. Basingstoke: Palgrave Macmillan.

Nancy, J. L. (1986/2006) 'The inoperative community' in Bishop, C. (ed.) *Participation*. London: Whitechapel Gallery and The MIT Press.

Nicholson, G. (1996) 'Place and local identity' in Kraener, S. and Roberts, J. (eds.) *The Politics of place attachment: towards a secure society*. London: Free Association Press.

O'Neill, P. (2014) 'The curatorial constellation – durational public art, cohabitattion time and attentiveness' in Quick, C., Speight, E. and van Noord, G. (eds.) *Subplots to a city: ten years of In Certain Places*. Preston: In Certain Places.

Rancière, J. (2004/2006) 'Problems and transformations in critical art' in Bishop, C. (ed.) *Participation*. London: Whitechapel Gallery and The MIT Press.

Rendell, J. (2006) *Art and architecture: a place between*. London: I. B. Tauris.

Roberts, P. (2009) 'Shaping, making and managing places: creating and maintaining sustainable communities through the delivery of enhanced skills and knowledge' in *Town Planning Review* 80, 4–5.

Rowles, G. D. (1983) 'Place and personal identity in old age: observations from Appalachia' in *Journal of Environmental Psychology* 3, 4: 299–313.

Schneekloth, L. H. and Shibley, R. G. (2000) 'Implacing architecture into the practice of placemaking' in *Journal of Architectural Education* 53, 3 (2000): 130–40.

Seamon, D. (2014) 'Place attachment and phenomenology: the synergistic dynamism of place' in Manzo, L. C. and Devine-Wright, P. (eds.) *Place attachment: advances in theory, methods and applications*. Abingdon: Routledge.

Sennett, R. (2012) *Together: the rituals, pleasures and politics of cooperation*. London: Allen Lane.

Sepe, M. (2013) *Planning and place in the city: mapping place identity*. Abingdon: Routledge.

Serres, M. (1980/2007) *The parasite*. Minneapolis: University of Minnesota Press.

Sixsmith, J. (1986) 'The meaning of home: an exploratory study of environmental experience' in *Journal of Psychology* 6, 4: 281–98.

Soja, E. W. (1997/2005/2011) 'Six discourses on the postmetropolis' in Westwood, S. and Williams, J. (eds.) *Imagining cities: scripts, signs, memory*. London: Routledge.

The Drawing Shed (2015) [a] *Manual for possible projects on the horizon: that things fall apart ... and ... thinking through making*. The Drawing Shed: Some[w]Here. [n.p.]

The Drawing Shed (2015) [b] *Manual for possible projects on the horizon: Some[w]Here*. The Drawing Shed: Some[w]Here. [n.p.]

The Drawing Shed (2015) [c] *Manual for possible projects on the horizon work, survival, play, imagination*. The Drawing Shed: Some[w]Here. [n.p.]

Walwin, J. (2010) 'Introduction' in Walwin, J. (ed.) *Searching for art's new publics*. Bristol: Intellect.

Watson, S. (2006) *City publics: the (dis)enchantments of urban encounters*. Abingdon, Oxon: Routledge.

Whybrow, N. (2011) *Art and the city*. London: I. B. Tauris.

4.8 List of URLs

1. The Drawing Shed: www.thedrawingshed.org/
2. *LiveElse[W]here*: www.thedrawingshed.org/live-elsewhere
3. *IdeasFromElse[W]here*: www.thedrawingshed.org/ideasfromelsewhere
4. *Some[w]HereResearch*: www.thedrawingshed.org/projects-2014/somewhere-research
5. Winns Gallery: www.waltham.ac.uk/plan-your-visit.html
6. Lloyd Park: www.walthamforest.gov.uk/content/lloyd-park
7. Art Tunnel Smithfield: http://arttunnelsmithfield.com/
8. Big Car: www.bigcar.org/
9. Jordan Mckenzie: www.jordanmckenzie.co.uk/
10. The Drawing Shed Soundwalks: Home Sounds: http://perezzarate.bandcamp.com/album/the-drawing-shed-soundwalks-home-sounds
11. The Drawing Shed Soundwalks: Living Here: http://perezzarate.bandcamp.com/album/the-drawing-shed-soundwalks-living-here

12 The Drawing Shed Soundwalks: The Past: http://perezzarate.bandcamp.com/album/the-drawing-shed-soundwalks-the-past
13 *LUPA* (2011–13): www.facebook.com/LUPA.E2/info
14 Latymer Projects: http://forumarts.org.uk/members/group-work
15 Arts Council England: www.artscouncil.org.uk/
16 Wandsworth Borough Council: www.wandsworth.gov.uk/

5   Art Tunnel Smithfield, Dublin

'To change the cachet of an area, the people have to manifest it' (Dublin City Council/Comhairle Cathrach Bhaile Átha Cliath (DCC) [n.1] *interviewee)*

It was evident that the Dublin case study, Art Tunnel Smithfield [n.2], had a central concern around land use, re-appropriation and ownership, which from the grassroots up, was seen to be effecting policy. This case study will be used to exemplify, critique and expand thinking of urban arts projects and their material place in the city; it has raised issues of aesthetics, land use, ownership, the community's Lefebvrian (1984) right to the city and the space/place theories of de Certeau (1984) and Massey (2005); of a creative and cultural meanwhile use of vacant urban space; and of the relation of social practice placemaking to city authorities in terms of funding, planning and administration. This chapter starts by introducing the art practice and process of Art Tunnel Smithfield and issues of participation, place attachment and crossing of thresholds and barriers to participation. It goes on to focus on what has been termed 'the vacant land issue' seen in Dublin and the role that Art Tunnel Smithfield and the wider Dublin arts and community sector has to play in this.

5.1   Introducing Art Tunnel Smithfield

Art Tunnel Smithfield was an urban art gallery and garden-based community space in the Smithfield area of Dublin, begun in 2012 by landscape architect, gardener and artist Sophie Graefin von Maltzan of Fieldwork & Strategies [n.3]. The site was initially identified by DCC City Architect Ali Grehan, with its proposal of use developed by von Maltzan with the owner of the site, local artists' studios, local residents and businesses. It was closed in February 2014 after the site owner gave notice to leave. The site was a strip of 'left-over' land from the development of the Luas [n.4] tram line corridor and was planning exempt. It was financed through micro-grants and by local businesses and crowd funding. Art Tunnel Smithfield had a mission to introduce art into the public realm and through its indigenous planting, create a bio-diverse wildlife habitat and micro-park. Its *Art Tunnel* functioned as an outdoor gallery for emerging artists, its *Art Platform* for community-created artworks. It aimed to activate both civic and placemaking reflection and action and widen the sphere of urban actors and notions of creativity. Research in Dublin with Art Tunnel Smithfield ran September 2013 to February 2014, and then went on to spend time at a second garden, Mary's Abbey [n.5] and other Dublin initiatives, ending in March 2015. In total, Dublin research time comprised of seven extended duration visits, before and after Art Tunnel Smithfield's closure, the number of days in Dublin totalling close to three months.

To place Art Tunnel Smithfield in its wider Dublin context, it was operating in a city with an identified 'vacant land issue'. Three hundred vacant land sites had been identified by DCC of an estimated, and thought underestimated, sixty-three hectares (Kearns and Ruimy, 2014, p.66). In the past twenty years, and in marked effect since Ireland's economic crash, public space in Dublin has been privatised, financialised and regulated with independent cultural production commodified, 'Dublin's "great enclosure"' (Bresnihan and Byrne, 2014, pp.1–4). This has seen a push-back in the form of urban commoning of public, in particular vacant, spaces, re-appropriating such spaces to meet collective needs and in the process, redefining forms of ownership, social, cultural and material (re)production and governance (ibid.). A 'praxis of the commons' emerges from a dissatisfaction with the Dublin urban realm's proscribed cultural production and symbolic economy within it (ibid., p.5). The praxis is a relational social body (ibid., p.12) where the spaces are collectively 'owned' by those that participate in them (ibid., p.9) and financed collectively through crowdfunding and/or non-monetary exchange and circulation, a subversive act 'when we consider how expensive, individualised and disconnected much of the social and cultural activity in the city has become' (ibid., pp.9–10). This emerging Dublin new urbanism breaks the 'liveable-city glass ceiling' of a 'bigotry of low expectation' (ibid., p.48) where Dubliners do not believe the city can become a desirable place for people to live and is attempting to address the contemporary Dublin urban difficulty of a successful, liveable inner-city by rendering its response on a cultural and social reimagining (ibid., p.15). These spaces operate independent of, and often do not seek, formal institutions and are led by communities of 'urban pioneers' (ibid., p.98) – read, urban co-creators [3.6.2] – that 'operate beyond the auspices of public and private management and which have grassroots, DIY ethos'. The urban commoning praxis has created a place in Dublin for informal crticial spatial practice (Rendell, 2006), integrating people, space and place, materials and knowledges-in-exchange (Bresnihan and Byrne, 2014, p.11) which have produced new ways of 'working, playing, and deciding together, and the production of shared knowledges and resources' (ibid., p7). Open participation in the urban commoning process has been cited as a motivating factor to join and continue in the process in contrast to the boundaried participation of enclosed spaces that alienated people from commodified and normative Dublin culture.

5.2 The art practice and process of Art Tunnel Smithfield

This section details the art practice and process at Art Tunnel Smithfield, positioning it as an example of social horticulture and the informal aesthetic, and considers the issue of 'beautification' in such practices. It continues with a consideration of issues of participation, place attachment and place loss and closes with a consideration of thresholds and boundary crossing.

5.2.1 Social horticulture, the informal aesthetic and beautification

The social practice art at Art Tunnel Smithfield functioned primarily through the act of gardening, a *social horticulture* (Anderson and Babcock, 2014, p.141) analogous to a living mural and a site where people could be creative without needing to be 'an artist' as such (Chonody, 2014, p.4). Whilst it was recognised and purposed in practice that Art Tunnel Smithfield 'is not [going] to win a design prize, it's really about the more they get involved, the better' (Art Tunnel Smithfield artist) – the garden functioned as a living mural of social cohesion. The art process at Art Tunnel Smithfield signposted the informal aesthetic concept that underpins social practice placemaking practice. This operated alongside the material aesthetic of a deliberate 'gift and scavenger culture' (Art Tunnel Smithfield artist) of wooden pallets and repurposed containers and materials for raised and bed gardening and artworks. The informal aesthetic came to be articulated as '"Art Tunnel-esque", [it] cannot be part of something planned, it has to just land somewhere' (Art Tunnel Smithfield artist). However, it also drew comments that it looked 'a bit tatty' and like a 'closed shop' (Smithfield community member). Artists here commented that they found the process 'chaotic', 'disorganised' and 'piecemeal' (Art Tunnel Smithfield artists) but they came to realise that it could not have happened otherwise, the chaos being integral in iteratively getting to know the place.

Material art outputs manifested through the design aesthetic of the artists and the confines of the materials used; the aesthetic sensibilities of those in the Smithfield community; and the greening of a derelict space against a grey stone built environment. The garden in Art Tunnel Smithfield was described as a 'colourful palate, in an otherwise grey and dismal setting' (Art Tunnel Smithfield artist) (Figure 5.1); 'they [the public] just wanted a nice place to look at and not get depressed' (Art Tunnel Smithfield artist) and 'it became a very beautiful place to be, a real haven in the middle of the city' (Art Tunnel Smithfield artist). It was equally recognised that the garden had brought a visual vibrancy to the wider area in a hard-landscaped site – 'The colour breaks up a grey corridor of the city' (Smithfield community member); this improved the perception of the area to those travelling through Smithfield on the Luas tram in particular and also, 'people like to have something to look at when they walk past' (Art Tunnel Smithfield community member). At Mary's Abbey, a second community garden space led by neighbourhood community members two blocks along from Art Tunnel Smithfield and opened soon after the Art Tunnel Smithfield closure, 'some said that they didn't want it to be like the Art Tunnel', that they 'they want something nicer' and 'want it to be pretty, they think that structures made out of pallets aren't pretty' and that this prospect 'some found upsetting' (Art Tunnel Smithfield participant). As also observed at The Drawing Shed [n.6], it was perceived by the case-study artists that the practice integrity and intent of the informal aesthetic was missed by many outside of the social practice arts or social practice placemaking field and there was an observed friction between its informality, perceived as unprofessional and lacking rigour, and a friction with the formal funding sector and political administration of conventional arts practice and aesthetics. Several artists stated that professionalism had to be evidenced to DCC for it to be able to trust the artists; artists worked towards this relation out of their own fear that a perceived lack of professionalism would hinder their funding potential, and lead in the wider spatio-cultural milieu to a standardisation of art in vacant sites through the repeated funding of a small select group. Moreover, there was a reflection that the community art practice aspect of Art Tunnel Smithfield suffered for the perception of its informal aesthetic: where the lead artist had anticipated that informality would make the development of the space easier, 'it turns out the opposite is true' (Art Tunnel Smithfield artist) and it felt formality was needed for people to commit to the project, as participant or financial or political supporter.

Material outputs and the rematerialisation of the art object are what differentiates social practice placemaking from social practice arts and this material output may have the outcome of 'improving' the visual sense of place of an area. This outcome does not denigrate the integrity of the social practice arts intent in social practice placemaking though (Bishop, 2006 [a]). In sum, 'Don't underestimate the visual, the beauty' (Art Tunnel Smithfield artist). The garden acted as a re-materialised art object of an ideological (Baker, 2006, p.147) and processual 'critical materialism' (Whybrow, 2011, p.28) to question the function of the urban realm (Lossau, 2006, p.47), moving beyond the 'diverting decoration' (Kwon, 2004, p.65) criticism levelled at such activity, subverting a beautification for critical means. As seen with mural art, such a beautification of an area is a visual and symbolic marker of a community's valuing of art and itself (Chonody, 2014, pp.30–1),

Figure 5.1 TL, CW, BL, Art Tunnel Smithfield, view of *Art Tunnel* and *Art Platform*, Dublin, 2013, Benburb Street graffiti, Dublin, 2015, Smithfield, Dublin, 2014.

a process that acts as an exercise in place identification (Kester, 2011, p.204) and group self-efficacy (Chonody, 2014, p.39). The place of the social practice placemaking activity was thus implicated into being the art object, seen here too as a key oppositional defining factor vis-à-vis community art. As one artist stated – and also at The Drawing Shed – their work was community art, by virtue of its functional (Nicholson, 1996, p.116) placing, but blended with social practice arts, and this was not perceived as conceptually problematic; what was problematic was the perceived tendency of community art to value product over process. For participants, social practice placemaking was 'art, but not as you know it' (Dublin arts sector member) that was material, outward looking, intellectual, political and supported community capacity building.

5.2.2 Issues of and around participation

Artist participation across the projects tended towards the collaborative, either explicit collaboration between two or more artists, or – and this was facilitated by the informal aesthetic space and artists being in close proximity to each other – involved in a dialogue of exploration and feedback on each other's practices and processes and giving hands-on help to each other in the devising and delivery of artworks. Two of the installation commissions at Art Tunnel Smithfield comprised an artist/architect team, creating site-specific installations designed to draw attention to the expanse of wall in the site, and by implication, similar end of terrace sites in Dublin. *Weave* (2013) was a two-storey rope canopy, placed at the entrance to the site; *Loom Seat* (2013) was created using surplus rope and placed at the foot of the canopy and in a gardened section of the *Art Platform*. The joint commission saw the artist and architect work with community volunteers in the build, a process that involved skill swapping, with the artistic and architectural practices being informed by the other; the team found their respective creative practices extended in the process of making, interaction and collaboration, forming an interdisciplinary practice. The architect reflected that their practice was challenged by the fluidity of the artist's practice and the social aspect of work; the artist found working with the architect facilitated a new perspective on and aspect to their art process and object:

> …in that they had that lovely eye of being able to see the possibility of an artistic installations. But then I really enjoyed them making things that were practical as well, so people would be able to use it as a piece of furniture as well. (Art Tunnel Smithfield artist)

Whereas in the case study contexts of London and Indianapolis, the lead artists took on a lead role at times and in contexts appropriate to the development of the project, in Dublin, the lead artist did not want to be seen in any overt leader role: but as the founder of Art Tunnel Smithfield, it was observed that they were looked to for inspiration and guidance, which dispirited some when it was perceived to be lacking. The reluctance of the Art Tunnel Smithfield lead artist to be placed in artist-as-leader position affected the culture of participation in the space – those from the public that were involved were either 'volunteers' or 'participants', not 'collaborators' per se, as people were 'invite[d] to come up with ideas' (Art Tunnel Smithfield artist), the operative word here being 'invited'.

The abstracted participation of the audience gaze emerged throughout the research project as a site of study in its own right as a deliberate act of non-participation. This was acute at Art Tunnel Smithfield with its footfall not necessarily translating into participation: 'They'll walk past, but they won't come in … like when people sit in cars and look at the sea' (Art Tunnel Smithfield artist). For many people in or passing through Smithfield, it was enough that Art Tunnel Smithfield was 'just there' and beautifying the area. An 'Art Tunneler'[1] undertook some informal research on the Luas tram, asking its passengers what they thought of the space and the common response was that they 'liked the colour and that it was there' (Art Tunnel Smithfield participant paraphrasing). Others would stop to ask questions when walking past – primarily along the lines of what the place was and what people were doing in it – which engendered a feeling of connection to the space even though they had not entered it:

> They may not want to get involved, but they feel involved once they know about it. 'It's all good, but we're not doing anything'. Which is fine too, they're all quite happy that something is being done with the space … (Art Tunnel Smithfield artist)

One passer-by commented that they 'stumbled across this place the other day on my way somewhere else' and that 'it's like finding a Narnia-like garden in the middle of the city' (Dublin resident) but were adamant they would not go into the space; another that 'I really love having it there, and look in from the sidewalk at least once a week to see what's new' (Dublin resident). An old man walking past Art Tunnel Smithfield shouted through the fence, 'What you doing in my garden?' and went on to say that he had lived around the corner to the site since 1931 and that 'I think it's beautiful…great for the area, great that it's here' (Smithfield resident). The Art Tunnel Smithfield artists recognised the space's abstracted and visual aesthetic value and some made artworks accordingly:

> The Luas quite regularly stops here so we have a readymade audience from that … it's basically kinda a bit of street performance text in a way … you know, people can see it as they're walking by or on the Luas, so it gets to be seen by a huge section of the, cross section kind of, the population. (Art Tunnel Smithfield artist)

As evidenced in the above quote, between the audience gaze and crossing the threshold to any degree of participation, lies a liminal space of issues of inclusion and exclusion and barriers to participation – communicative, material, conceptual and psychological. At Art Tunnel Smithfield, it was recognised that it was a challenge to find community groups to work with – to firstly find the groups and to then secondly convert a passive interest into an active participation. In Dublin it was remarked that '[local neighbourhood] news does not get out, people don't know about it' (Dublin arts sector member) and that 'You never hear what happens to other places, I guess they don't get the interest outside of the local people, I guess they don't get the same kind of push' (Art Tunnel Smithfield artist). Here, projects like DCC's *Beta Projects* [n.8], an open sourced call to artists and non-artists for pilot urban design interventions ideas, a proportion given seed piloting funding before possible roll out city-wide, was viewed as a necessary communicative tool: "cos you just need somebody [to] start bridging the gap between the conversations' (Dublin arts sector member) across the city.

With Art Tunnel Smithfield, for some it was enough to be connected virtually to the project: 'I am a member on Facebook, but have never entered as a key holder' (Dublin community member). This was underscored by a perception that the land was 'closed' – 'Dublin's "great enclosure"' (Bresnihan and Byrne, 2014, pp.1–4) – gated off and with limited open hours for a limited community group. Many thought Art Tunnel Smithfield was a private concern: 'We went to the garden the other weekend, walking back from town…thought it was grand… didn't think that we were allowed in before that … really glad we did' (Dublin community member). Smithfield itself was an area also that was not perceived of as a cultural destination – though that was seen as changing in the course of the research project. The regeneration of Smithfield some years previous had been on the basis of it functioning as a space for large open air events. It lacked seating and was designed as a 'large holding space' (DCC staff member) and was initially programmed with open air sports and music events and markets that proved unworkable for the capacity and resources of the space. This then leads discussion on to the changing perceptions of space in Dublin observed at the time of research, based in part of galvanised sense of place attachment. The following sections will focus first on place attachment, and attachment as evidenced in the occurrence of place loss, and then informed by this, return in more detail to the crossing of perceptual boundaries to space and place as well as the lived experience of the city.

5.2.3 *Place attachment and place loss*

In Dublin project participation was motived by a desire to feel a local connection to place and its people. Neighbours were observed at Art Tunnel Smithfield and at Mary's Abbey in active communication as a means of community building, and this was the underlying purpose behind the project. Whilst protagonists expressed a surprise that their preconceptions for a 'rough area' – based variously on a maligned reputation of Smithfield in wider Dublin and from previous direct experience of antisocial behaviour for some on Smithfield streets – in the garden space itself, they largely did not face any antisocial behaviour from local residents or passers-by. Firstly, this was understood by protagonists as based on a latent, non-active affective bond with Art Tunnel Smithfield and with the area, residents being pleased to see the space being activated and cared for and beautified, as above. Stories were told to the researcher during social times in the space by Art Tunnel Smithfield protagonists and in conversation with Smithfield residents of the 'former Smithfield' where it was felt the 'community pulled together' in times of adversity (predominately, poverty and intra-city cultural, political and religious attrition) and that Art Tunnel Smithfield and the cultural renaissance of Smithfield that it was integral to, was part of a 'new Smithfield' (Smithfield resident) that had a similar and revived clear and strong place-identity and community bonding.

Secondly, to join the social practice placemaking project micropublic (Amin, 2008) – here, the volunteer community at Art Tunnel Smithfield – was an act of elective belonging, of choosing to be part of a place as an individual that was part of a group, and choosing to be active in it, performing an identity:

> It's a conscious decision to put yourself there, to volunteer, to meet people interested in the same thing, so it's a kind of not just a garden in a scrap of Dublin, it's actually, the types of people that want to come and to put their energy into it, they're the types of people, there's always a sense of coming together. (Art Tunnel Smithfield participant)

Art Tunnel Smithfield was viewed as something that the Smithfield residents had a collective ownership of, whether they were active in the project or not, and it overcame, with the drawing together of the project micropublic, previously absent or unequal levels of familiarity and neighbourliness: there was a period of 'getting to know us' (Mary's Abbey participant) when forming the committee and starting the project. Moreover, for some, this was the first time they had been engaged in a local project and from it came their first communication with their neighbours – 'One of the local volunteers was shouted at by one of the locals, and he said back, "Thank you for speaking to me for the first time in thirty years"' (Mary's Abbey participant) – which was to them, evidence of a previously absent but now activated sense of community. The Mary's Abbey project also surfaced some long-standing community frictions – such as that of the feeding of pigeons by one community member on the street that was quickly forbidden in the garden space – but the proximity of people and a familiarity with site and an emerging sense of neighbourliness was called upon to mobilise and maintain the project. Whilst an aspect of this was seen as potentially divisive, it was also seen as a necessary part of the process and essential in starting a wider conversation on the future of the area:

> Is this creating the biggest fight ever in your community when it should be doing the opposite? It became pretty apparent that they were having their fights anyway, but this was giving them a platform, which I think is a good thing, a physical platform. So I think it's good 'cos it's getting things out. It's really about other things as well. (Mary's Abbey participant)

However, the arts programming at Art Tunnel Smithfield did not engender a place attachment for all: Art Tunnel Smithfield was criticised by one local community member that it had fallen short of its perceived promise at the time of its closure: 'The first ideas, they seemed grander than what is there now, more high-end installations' (Art Tunnel Smithfield community member) but was still largely appreciated by those interviewed for being part of a Smithfield-area 're-curation' (DCC interviewee). Whilst the palisade fencing around the Art Tunnel Smithfield space for some acted to isolate it and effectively gate it off from their participation or concern – as seen with gated residential communities (Seamon, 2014, p.15) and evidence of Dublin's 'great enclosure' – it was not impenetrable and a symbolic attachment to place was evidenced through this barrier.

Part of the process observed was the closure of projects – the closure of a short-term project with The Drawing Shed and significantly a long-term one in Dublin with Art Tunnel Smithfield, place loss here the closure of the whole initiative and its site (Figure 5.2), experienced at a profound emotional level by the protagonists: 'If you looked at it now you wouldn't really know what it was …. When you have been here that long, then this is mine, I live here. This has been our home' (Art Tunnel Smithfield participant). Protagonists spoke of an emotional discombobulation when taking growing plants out of the ground, 'to suddenly be pulling plants out of the earth seems a bit wrong, against everything that we had set up the Art Tunnel for' (Art Tunnel Smithfield participant). This was antithetical to their place identity as linked to the aims and ethos of the project, and there was a psychogeographical distress at passing the site subsequently with many avoiding the area:

Figure 5.2 L, Image of Art Tunnel Smithfield, Dublin, after closure, March 2015; R, An impromptu planning meeting between volunteers at Mary's Abbey garden, Dublin, September 2014.

I try not to [walk past it]. I now usually go up through Smithfield instead, so I go out of my way to not walk past it, which is terrible. But I got so emotionally involved, 'cos it was the first kind of project like that, from the ground up, that I had got involved in, so I just, I found the whole thing upsetting … (Art Tunnel Smithfield participant)

It was commented on that some plants and trees were 'tenacious' (Art Tunnel Smithfield participant) to still be growing on the site in the immediate months after closure, an analogy to the project, and before site-stripping it was hoped by some that this would enable a similar cultural use of the site at a later date or at another Smithfield site – which was proven accurate in the transference of people and plants to the later Mary's Abbey project (Figure 5.2). When Smithfield residents said 'Thank you, and do it again' (Smithfield community member) to the Art Tunnel Smithfield team, this was communicated from a sense of disbelief that what had become a fixture of the Smithfield place identity was going. For other protagonists it enlivened in them place attachment in Smithfield that engendered a deeper active relationship to and working in the place – one participant went on to petition DCC on local concerns, became involved in local food and culture programmes and orientated their social life to Smithfield.

With Art Tunnel Smithfield then, one sees Proshanky et al.'s (in Gieseking and Mangold, 2014, p.77) assertion that place attachment is based on what a place is thought to meant to be like – and that when this changes, discombobulation occurs, effecting a change in the relation to place and place identity. When an established place attachment relation is perceived as under threat, then collective action can occur (Mihaylov and Perkins, 2014, p.71) – but in this example,

Figure 5.3 Images of Art Tunnel Smithfield, Dublin, fence signage, July 2013.

collective action was not to protest closure but to create new opportunities and mobilise in a wider civic arena of Smithfield area politic, working in some way towards mitigating feelings of grief and upset (Twigger-Ross et al., 2003, in van Hoven and Douma, 2012, p.67).

5.2.4 Space/place thresholds and boundary crossing

Threshold crossing or dissolving was seen to be rooted too in an attitude to land use – what could or could not be understood as permitted in a given place. This was particularly noted in Dublin where a change in the agency of the protagonists in social practice placemaking had a mutually constitutive relation to a change in attitude to land use and ownership.

As a first step to changing attitudes to land use, the art practice and process at Art Tunnel Smithfield was, in being an outdoor gallery, a deliberate intention on the part of the lead artist to give space for emerging artists and community members to show work on an equal basis, and commissioned artists saw this outdoor gallery setting as an attraction, 'a space that wasn't a gallery or a sanitized space' (Art Tunnel Smithfield artist). The setting also made demands on the artists and the contextual form of the work: 'an outdoor exhibition is still quite traditional, so the way you have to approach this [siting at Art Tunnel Smithfield] and the consequence of that, of the work being created and its actual presence, you have to consider' (Art Tunnel Smithfield artist). One Art Tunnel Smithfield artist saw the formal gallery setting as a 'permitted space' which was 'to the detriment of art in general' as 'art is meant to kinda push itself and challenge itself if it's going to move or change or evolve or keep momentum' (Art Tunnel Smithfield artist) and commented that the Art Tunnel Smithfield model of working acted to push those artists involved over boundaries to their habituated practice. It was also recognised though that the use of the conventional term 'gallery' for some helped them make sense of the place and acted as a draw to the place:

> Over and above, it was the ART [*sic*] tunnel, it has art installations and exhibitions and events, to bring different people to it, because each artist has a different draw. It's a bit like a gallery, the space gets recognized because of the artists there. It probably gathered a lot more momentum because of that and it had a much wider catchment than a local community garden network. (Art Tunnel Smithfield artist)

An issue of 'How can you make an enclosed space more open?' (Art Tunnel Smithfield artist) was prevalent across all three case studies. At Art Tunnel Smithfield, the palisade fencing that demarcated Art Tunnel Smithfield was observed as a barrier to participation; whilst some passers-by did speak through the fence, for many it was viewed as impenetrable, even with 'welcome' and explanatory signs (Figure 5.3).

Some mentioned the aesthetic of the signs as off-putting; for others the lock on the gate and the perceived effort involved in becoming a key holder – the act of having to ask for this permission – hindered them from crossing the threshold of the site and of further participation. This was a lesson learned that transferred to Mary's Abbey where if someone came into the garden a few times, 'then it's rational to give them the access code' (Mary's Abbey artist).

All three case studies, whilst delivering in the hyperlocal location, also acted to galvanise actors across a wider locale and work with, strengthen or activate a sense of a city terroir (Zukin, 2010, p.xi). This was evidenced in Dublin with the crossing of the culturally entrenched north-south city divide marked by the River Liffey/*An Life* ('It's worse than the Berlin Wall...' [Dublin resident]) of people coming to Art Tunnel Smithfield as a destination experience to participate in and in Art Tunnel Smithfield being part of the renaissance of the Smithfield area that was perceptually foreshortening people's perception of this as far from the city centre, encouraging greater footfall. The psychological barrier to participation in public realm arts in Dublin was noted unanimously by interviewees as being rooted in a cultural attitude to land, its access, ownership and subsequent degree of publicness. It was stated that Ireland has a disproportionate respect for private property, based on its history of colonialism and its legacy of land and ownership and its literal and legal fight for land and that the Georgian enclosure (Bresnihan and Byrne, 2014) of public spaces was redolent in the city's psyche:

> There's an Irish obsession around land, down to historical events and land being taken away from people. I'm making massive generalisations, but people do talk about it. There is this really unhealthy relationship with property here. (Dublin architecture sector member)

Parks and green spaces were often gated during day and night hours, a public and private policy rooted in fear that 'If you open things up then people will destroy them, local community and the council and the parks, you see these parks closed, you just watch it crumble, it's just that "We'll lock the gates, we'll keep people out"' (Dublin artist). An architect noted that Dublin was built on a Protestant public space model, lacking the open plazas for social meeting and protest manifest in Catholic-designed cities: in Irish urban culture it is 'The bars [that] act as other city's plazas or piazzas, where people can speak what they want' (Dublin arts sector member) and Dublin's public land use has been influenced by this cultural factor. Landlords and land owners were seen to be reluctant to renting or opening up space for use, for fear of a creeping claim to security of tenure on the part of the tenants. One artist commented that Dublin 'has a problem with responsibility of land, not the duties of it' (Dublin artist), meaning that landlords have a concern that tenants would achieve an emotional ownership of, and claim to, land so landlords would rather let the land lie vacant. Land was also known to stay within families for generations, so rarely changed hands. Lastly, it was remarked – and evident when walking Dublin – that there is no shared space policy for streets in Dublin with cars having the priority and pavements and pedestrian areas not being user-friendly.

Together, these factors were seen to result in the Irish having a reverence of land, a large volume of un- or under-used land, and an underdeveloped street protest or pedestrianised culture:

> That's an historical thing for Ireland. There was a time when we didn't own this country and it's gone to the other extreme, where nobody but the landlord owns that land. People are very protective of their own piece of land, to the neglect of public space. (Dublin artist)

A sense of ownership, closed and fixed, of land was felt from DCC Parks and Housing departments by architects and artists in Dublin, the departments viewed as considering themselves as incumbent land owners that were fearful of a change in land use:

> You'd ask the Housing Department, 'Well who owns this land?', and their answer was 'Well, we do'. Well, the real answer is 'Well you don't, you don't own it, you the Housing Department are just looking after it'. But that's how they think, they think that they own the land and the Parks think that they own the land. (Dublin architect)

Any change in land use was also known to be a slow process, working against the capacity of grassroots and some professional schemes. The Planning Department was also not thought to be universally entering into conversation outside of its professional sector and at the level of arts funders, many commented that bodies such as Arts Council Ireland/*An Chomhairle Ealaíon* (ACI) [n.8] 'failed to display cohesion, to maximise opportunity for benefit' (Dublin artist) or lobby to include artists in planning and urban design debates. The case of Granby Park [n.9], a temporary park set up by arts collaborative Upstart [n.10] on a vacant site in the north inner city, a project designed to bridge a gap in time and space before a social housing development on the land, was for one artist was emblematic of the lack of momentum and commitment to realising long-term change in DCC:

> I can see it being a stalemate for some time. It was a public-private park with no private any more. Housing, that's not going to happen. For the residents, I can't help feeling that it's great that we came for a month and transformed something into this lovely space for the community, and then you know, what's changed? Sure, we inspired others, but that site, what's happened to the residents of Dominic Street? They're no further on. But they had a lovely month. That's quite sad. It raised more awareness, but like everything else, if there's not momentum, then things don't get done here. (Dublin artist)

Learning from the park context in London with The Drawing Shed can also be applied across to the outdoor nature of Art Tunnel Smithfield. It was commented that in Ireland the leisure focus is 'ad hoc at best, and all in pubs or at home, not as street life' (Dublin artist) and that 'We use our streets here for traffic, not for play' (DCC staff member), so the public realm in this area was not conceived as being people-centred or open access. The 'vacant land issue' in Dublin and the re-appropriation of these spaces was illustrative of a latent creative and civic agency, moving the city from being 'stuck in pubs' to 'opening up ideas to people, that they can do stuff in public space, gets it in people's mind that it can be a park' (Dublin artist). The following section will turn to 'the vacant land issue' and issues arising of active citizenship and temporary land use and the 'Dublin new urbanism' (Kearns and Ruimy, 2014, p.48).

5.3 'The vacant land issue', active citizenship and generative planning

As stated in 1.5.1, at the time of research Dublin was experiencing a high and visible volume of areas of vacant land in the central city area. As was observed during the research itself, this was a conversation and concern for artists and communities with various initiatives and projects along these lines forming in response to the issue. This process was seen to generate an active citizenship – the focus of the first part of this section – and a change in attitudes to temporary land use, which was informing a Dublin new urbanism – the focus of the second part of this section.

5.3.1 Active citizenship

Social practice placemaking was seen to have a positive and generative effect on the levels of interest those participating had in their locale as both a physical and political entity and through which they enacted an active citizenship and citizen agency. Participation in projects occasioned a grassroots political awareness in those involved, whether artist or non-artist, provoked by the performative nature of the art practice – as one participant termed this, it made them 'an active citizen' (Art Tunnel Smithfield participant). This was a process of subtle to overt political conscientisation, fostered from an enhanced social capital manifested in the individual through participation and a feeling of connection to the group, the wider community and place: activity in place brought about conversations about place, that created in many a curiosity about place and from this, deeper conversations about the issues in place that were motivating the project at hand; for some protagonists, there was a consequent increase in levels of political awareness and activity in the wider city context.

In Dublin, participation in Art Tunnel Smithfield precipitated an active citizenship process for a number of its protagonists and it was part of a cultural shift in the city with regards to its attitude to and policies concerning vacant land. All of those interviewed from the Dublin arts and political sectors commented that the city's enclosed urban form heightened respect for private property and led to a tacit acceptance of a land ownership and enclosure status quo. Thus, whilst there was a voiced anger about the paucity of the Dublin urban environment and its vacant land, there was little demonstration: Ireland was described by one of the arts sector members as a 'culture that complains a lot, but does little about it'

(Dublin arts sector member), and moving out of a culture of complaining, into an active one, was noted as motivation to join the Mary's Abbey activity for one of its protagonists. The research period however was placed in a time of change – with the arts as a driver for questioning the function of land, its moral ownership as common land, or certainly for common use, and the potential of the citizenry in this as custodians: 'Why not give platform to local initiatives? Give ownership and the community polices itself, it will happen' (Dublin artist). The number of projects operating at the grassroots in vacant spaces in Dublin was a literal and metaphorical representation of a political statement about land rights that was resulting in a change of attitude amongst its constituents:

> You've got to keep on saying, to all these community groups that we work with, 'No, Dublin City Council don't own that land, people own that land'. And you're [DCC] just looking after it for them. It eventually seeps in … 'cos its true. (Dublin artist)

Leadership change though was located at grassroots level, artists looked to as leader in a new, active citizenship. With awareness of projects came an awareness of possibility in the urban realm, projects leading by example. Artists used the vacant space activation as an opportunity to ' … reclaim public space in an artful way, without permits' (Art Tunnel Smithfield artist) and to challenge public space and events rules and regulations head on, or to avoid them altogether: 'There's less stopping you than you think. Everywhere has red tape. Get on with it' (Smithfield arts sector member). There was a growing sentiment-into-action that the 'Smithfield programme needs to be taken out of DCC hands' (Smithfield arts sector member), showing a desire for the arts programming in the area to be of and from the area, the Smithfield citizens, through groups such as Art Tunnel Smithfield indicative of a change in the Dublin arts ecology, proactively locating power in the local:

> [Art Tunnel Smithfield is] a very positive place … Ireland has a great culture for complaining about things, that people should be doing a better job, but it's wonderful to actually see people being very proactive and channelling, instated of just complaining about things, to channel their energy into something very positive for the community, and long may it continue. (Dublin artist)

It was stated that Art Tunnel Smithfield achieved an 'open[ing] up of ideas to people, that they can do stuff in public space' (Dublin artist), this activity, by situating and activating a citizenship of place, 'remind[ing] us how important public space is, the last vestige of our democracy, where we all gather regardless of race, gender, socio-economic status, wherever do we gather' (Dublin artist). Thus, vacant land in Dublin had a civic agency of the countersite (Holsten, 1998, p.54, in Watson, 2006, p.170) activated through a relational arts practice occurring in liminal sites. The grassroots was seen as an agent in urban and awareness of and participation in grassroots projects such as Art Tunnel Smithfield was an informal route into the arts and a creative culture, which was itself creating more participative design models. Dubliners were 'becoming less fearless' (Dublin artist), demanding more of the function of places, asking for a different art from that of monumental public art or community-centred art, and an active role of sense of place creation:

> It's not about dropping in Public Art. It's not about building a community centre and saying 'Hey, look, this is where you can be creative'. Let's just build an open environment in which people can create their own sense of space. If people want a space to function a certain way, they will make that happen, that's a creative process. (Dublin artist)

This shift in agency, in the grassroots-led placemaking of Dublin, had a resultant conscientisation of a civic mentality manifested in various ways. Art Tunnel Smithfield itself was viewed as an activation of a counter, liminal site on the Luas corridor that would have been designed against such use, 'The Luas … they created it as a corridor, not a place for pedestrians, they don't want anyone there, but [Art Tunnel Smithfield's] efforts are the opposite of course' (DCC member). Art Tunnel Smithfield found too that those with a community of interest in the space as an arts or gardening destination would also travel, and this simple act of travel was a psychogeographic boundary changer; it was a received – but very generalised – cultural notion that Dublin south- and north-siders are culturally and socio-politically different from each other and do not mix, yet Art Tunnel Smithfield drew participants from both sides of the River Liffey. For one Art Tunneler, their civic participation took the form of joining more community groups in Smithfield, getting involved later as a committee member in some, and becoming more aware of the position of Smithfield in the wider Dublin politic. Participation and joining the Art Tunnel Smithfield collective place identity also led one to complain about Art Tunnel Smithfield land owner Bargaintown's illegal signage in the area, getting others to do the same, which resulted in the signs being removed, and 'act of small revenge' (Art Tunnel Smithfield participant) for the closure of the garden, but one that 'was still the right thing to do, [as] what they were doing was illegal after all and an eyesore' (Art Tunnel Smithfield participant).

5.3.2 Temporary land use and 'Dublin new urbanism'

The culture of temporary land use (Lehmann, 2009, p.32) in Dublin was seen to be contributing to urban life through the generation and encouragement of new urban activities, a bespoke 'Dublin new urbanism'. Art Tunnel Smithfield was an 'area catalyst task' (Kearns and Ruimy, 2010, p.206), a simple and low cost intervention outside of municipal planning and the professional design sector, with disproportionate individual, community and place gains including increased community confidence and control. This section addresses this first from the perspective of the bottom up – that of the grassroots – and then from the top down – that of the city administration – and then draws both together in the observed practice of Dublin's new urbanism.

5.3.2.1 *From the bottom up*

Art Tunnel Smithfield brought a daytime lease of life to adjacent Benburb Street and directly inspired further creative commissions in the area, such as the photographic installation of Smithfield people, *Complexions: An Exhibition of Character* (The Complex with Jarlath Rice), with funding secured by a DCC officer who spent time at Art Tunnel Smithfield and wanted to add more to the street. Such animation was a challenge Kearns and Ruimy (2010, p.93) saw as not being met all over Dublin and its protagonists saw themselves as urban pioneers (ibid.). Art Tunnel Smithfield was not alone in activating vacant land and Dublin-wide projects were coalescing under collectivised project banners such as Connect the Dots [n.11], a networking and vacant land project activating initiative, and What if Dublin? [n.12], a public-engagement installation running throughout the St Patrick's Day festival in 2015, that used visual imagery to propose questioning of what the city should and could be.

The relation of artists to DCC though was complex and a topic of much debate between artists. Throughout conversation in Dublin with artists and creative practitioners around the relation to the administration, a paternal and intergenerational analogy was often invoked, of a paternalism in describing how they experienced the efforts of macro institutions at placemaking and working with the grassroots. Dublin artists interviewed spoke of DCC as coming 'from a different generation, different background, different conditioning' (DCC artist) and that:

> Sometimes I feel we need to go to family counselling, 'What he meant was this' and 'What he meant was that', and 'You both are trying to achieve this', and 'These are your methods', and 'This is how to do it' ... you know, a mediator in-between. (Dublin artist)

A breakdown in or misunderstanding of language was even part of this intergenerational miscommunication, 'They [DCC] basically don't speak the street language and the street doesn't speak the DCC language' (DCC artist):

> It's like a parent from a different generation that doesn't know how to talk to the kid, and the kid gets really angry and rebellious 'cos 'Why am I so misunderstood'. But in actual fact, both parties are speaking about the same thing. (DCC artist)

DCC spoke paternalistically of wanting to 'to protect volunteers from themselves' (DCC member), which however well-meaning in the face of its weighty public realm regulations and exposure to legal action, was also used as a moralising and financialised device against the grassroots, DCC wanting in effect a return on its investment of support:

> We [DCC] will do this if you want it and you're willing to put yourself behind it, because we're investing in you, there's a cost per person you know, and if you want this to be successful then you need to put something in too. (DCC member)

One artist stated Ireland's policy to be one of a parent to 'help people to help themselves' but also that this had to follow on with allowing 'people to develop their own opportunity' (Dublin artist). The paternalism of DCC was thus viewed as being outmoded, and a participative not collaborative relationship, the grassroots being a 'youth voice of new citizenship' (Dublin artist).

Dublin artists articulated a strong sense of disquiet and lack of efficacy with regards to DCC. Not only was there 'no point in contacting DCC, they're not interested, they're not going to fund it' (Dublin artist) as DCC's 'Default answer is "No", not "How can we make this happen"' (Dublin artist) but also 'DCC couldn't care. The less they can do the better as [they're] less exposed to risk. Therefore nothing gets done, if it is done, its done very safely, [with] no creativity' (Dublin artist). If projects DCC was involved in were successful this was due 'to nothing that DCC has done' (Dublin artist). Some saw themselves in an predicament with regards planning and vacant land – with the increased use of vacant land it was seen that DCC could then claim that 'the vacant land issue' is diminishing, and therefore become inert in this issue. The wider public's hopes had been dashed from over-consultation and dialogue with DCC was seen as circuitous: 'The community, they've had their hopes kicked up. People best not get their hopes up, because it's still Dublin City Council, you know' (Dublin artist). One saw DCC's interest in grassroots projects as that to be seen to be fulfilling quotas, but as such projects weren't revenue projects they were not the sincere concern of the council; this artist saw the twofold situation of there being 'no green quota to be filled' and 'the economic group in the council is still the most powerful group' resulting in creative urban design being comparatively devalued and thus rendered powerless: 'It's not the visionaries or the designers, it's the people who get the money in that have the most power' (Dublin artist). When designers were brought into the picture, it was a 'tokenistic, superficial design engagement' (Dublin arts sector member). Bids for the purchase of vacant buildings were also seen to 'favour those that are wealthy' (Dublin arts sector member), a property portfolio required to enter negotiations – 'What arts or urban farming or community set up is going to have that?' (Dublin arts sector member) – or the National Asset Management Agency/*níomhaireacht Náisiúnta um Bhainistíocht Sócmhainní Eilliú* (NAMA) [n.13] having a stranglehold on property use and ownership: 'When NAMA is involved, they take priority over everything, and they have no remit of any social agenda and are crude caretakers of places. They let them rot' (Dublin arts sector member). Nothing short of a 'radical change in the city council and of the ideology of the country' (Dublin artist) would change this situation:

> None of this vacant land stuff is going to have any kind of impact until there is a change in the economic ideology so that it spreads and starts to understand the social value and cultural value. And that's to do with leadership that's to do with how the entire city council operates. (Dublin artist)

The planners and planning departments were also viewed as working against the interests of an appropriation of vacant land in Dublin. 'It's going to planning. It's gone to the other side, the dark side ... [laughs]' (Dublin artist). Whilst DCC had a perception that it had an open door to suggestions and advice giving – 'If you need some guidance, we are here' (DCC member) – artists commented that 'Why bother, they're not interested anyway' (Dublin artist) and that DCC's offer of support needed to be more explicit. DCC was no longer seen as a necessary partner in projects. In this 'new form of citizenship, the council is not even on the radar of people to work with' (Dublin artist). With the number of temporary or short duration projects in Dublin at the time of research and this being the widespread model of urban interventions in the city, artists questioned whether the ephemeral nature of this, coupled with an assumed short attention for social practice arts of the public and the cultural desire for the new, worked contra to a need they felt for a longitudinal residency in place. The attitude of the macro was seen in Dublin and Indianapolis as being to utilise what it saw as the 'entrepreneurial spirit of the artist' to bring arts into the city and its commercial and vacant spaces. This creative placemaking process was seen as a positive in Indianapolis, 'representative of growth and change attitude in Indy' (Big Car artist), but as a negative in Dublin. Those artists interviewed in Dublin saw this as DCC deploying the arts in precarious community liaison schemes, a 'we need this to happen' desperation to bridge a gap between DCC and the community and overcome mistrust of the council. In such cases, the arts were deployed in place or of to allay community (over) consultation. In this respect, public participation in such projects was viewed as a form of active citizenship by DCC, but one that was not perceived to have a commensurate experiential level of power. Social practice placemaking projects pushed through by DCC were seen too as a way of shifting the responsibility of placemaking out from the council.

The research shows an ambivalent positioning of the micro vis-à-vis the macro though. One the one hand, there was a clear – and often cynical and angry – opposition to the macro. Planning, Housing, Parks and Arts departments at DCC were singled out for scorn by many and in general terms DCC was perceived as being disengaged with any grassroots cultural activity and there was a high level of mistrust between the public and DCC due to a perceived lack of transparency of the body politic.

5.3.2.2 From the top down

In Dublin, social practice placemaking was informing DCC of a new way of working, a first step being open to 'learning from other cities ... there are some [in DCC] that have been trying very hard to prove this' (Dublin arts sector member); the groundswell of social practice placemaking activity evidenced in Dublin was forming a cultural shift in attitude to land amongst the art and cultural sectors, based on a co-produced active citizenship ethos. A shift in a paternalistic disposition of some council members and departments was evidenced by the macro openly seeking out the ideas from the micro, a move towards agonistic generative planning (Mouffe, 2005; Munthe-Kaas, 2015, p.30) and Lefebvrian (1984) spaces for representation, such as DCC's *Beta Projects* initiative. For DCC this was 'a good way of getting ideas in from the public, and having voiced activity around what we're doing' (DCC member). Thus, in *Beta Projects*, DCC was displaying a level of permeability in collaborative planning that itself had a mediating agency, as seen as required above, in bridging the gap between the macro and micro. Successes in the rezoning of land were noted as having an effect on planners and public alike in changing perceptions on how land can be used.

Dublin faced a similar issue to that found in London and Indianapolis, 'That's the debate for Dublin today: how do you go about urban regeneration when the forces against you are so strong' (DCC member); this was something that it and those from the grassroots realised required a collaborative and consistent approach, 'then its impact is going to be much greater' (Smithfield community member). It emerged that there was a need for planners and policy makers, as a professional class, to enter into dialogue and work collaboratively with the grassroots to grasp a developmental relationship model (Benington, 1996, p.162) of placemaking to help all realise their placemaking policy goals (Rekte, 2011). The political agency in Dublin was seen in some significant respect as located not within the council, but at the grassroots, 'Regeneration starts by creative people' (Dublin council member), and with projects such as Art Tunnel Smithfield as 'To change cache of an area, people have to manifest it' (DCC member) and lessons learnt from the failure of Smithfield regeneration, including that a solely top-down approach will fail. Some Dublin architects in interviews stated that they had been surprised with what they had been able to achieve in terms of the design and function of land with DCC; others stated that they saw DCC learning from cultural programming mistakes, especially ones in the Smithfield area around large-scale public events. In interviews, DCC respondents also recognised a measure of their own limitations in engagement in the city's vacant land: that it cannot be top-down, that 'it is time for a new debate' (DCC interviewee) and that Art Tunnel Smithfield and others had been instrumental in both galvanising an interest in vacant land in the public of Dublin and also in fostering a space and time for this debate. The Government of Ireland/*Rialtas na hÉireann* Construction 2020 (May 2014) report recognised the need for public engagement in the discussion around vacant land in the city and DCC interviewees recognised that 'regulation inhibits innovation'. DCC is part of the EU-funded urban sustainability and resilience network, TURAS [n.14], and created *Beta Projects*, a scheme that actively canvasses for artist and resident suggestions for the alternative use of vacant land, and pilots these where possible. Interviewees saw this as an example of DCC treating the creative community as valued 'cultural translators' (artist interviewee) to bridge the gap between the public and the administration, but, essentially, coming from the level of the street to begin with.

Exemplification and learning was seen across Art Tunnel Smithfield to Mary's Abbey too. The closure of Art Tunnel Smithfield was positioned as a necessary signal to other landlords of vacant sites in Dublin that such projects could be successful for a mid- to long-term duration without a threat to land rights. As one DCC member commented on the Art Tunnel Smithfield closure and the granting of the Mary's Abbey site in receivership from NAMA, 'To have the precedent of it, as, if we can apply it here, we can apply it later on [elsewhere]' (DCC member). Once at Mary's Abbey, the DCC liaison also took the approach of 'Lets explain why we might do that, rather than not do something from fear or do things the same way again' (DCC member), a lesson learnt in community relations from Art Tunnel Smithfield amongst other projects. Planners also saw the hurdle of the 'miles and miles of red tape' (DCC planner) needed to sanction projects and proactively advised projects to help them get around it; Art Tunnel Smithfield was advised by DCC planners to have a key code entrance system, not an open access one, to avoid hefty public liability insurance for example and Granby Park to be open for under six weeks as this kept it within the parameters of more flexible planning regulations as temporary and commercially exempt. Furthermore, planners saw the success of such projects lying in not controlling them, 'The question is really, what do you need to control, ignore or help with?' (DCC planner). In Smithfield too there was a perceived 'edginess' that should 'be allowed to happen' (DCC planner), with community-initiated programming in direct response to, and created for, the area, such as local history and spoken word and music festivals, favoured over macro cultural programming of the DCC.

5.3.2.3 The new urbanism

It was identified – said with some trepidation by artists however – that the micro and macro were moving in the same direction of travel, albeit with different starting points:

> This may be unorthodox, but I feel the will is definitely there, the interest is definitely there. I think when it comes to, you know, wishing Dublin well and wanting things to happen, I think everybody is on the same page. (Art Tunnel Smithfield artist)

It was evident too that the perceived paternalism was to some degree being overturned by the existence of the grassroots projects – the macro was seen to be learning from the micro in the setting of the case studies. In Dublin for example, DCC arts-based placemaking initiatives such as *Beta Projects* acted as a conduit for sustained dialogue, 'they're very much necessary 'cos you just need somebody [to] start bridging the gap between the conversations' (Dublin art sector member). The urgency felt in Dublin to address 'the vacant land issue' was also appreciated as too big for DCC to attempt alone, and that change had to be led by different factions and in the face of pragmatic parameters, 'There's no time anymore to say you can't do things' (Dublin artist). The role of artists here was to exemplify (McCormack, 2013, p.12) and to employ their skills as networkers to bring a larger cohort of people together in this endeavour that could be achieved by the council and its siloed, and resented by many, departments. In regards emplaced arts, 'It's about vacant land and collaborative practice, about collaborative practice around vacant land as the issue' (Dublin arts sector member). One of the roles of an artist witnessed was that of a cultural translator for city administrations, the artist not used as an externally facing commissioned mouthpiece for the city, but as a 'listening trumpet' (Dublin artist) from which the city authorities could learn. The re-activation of vacant land was seen by some as signalling just this cultural shift that was required in the mind-set of DCC. The considerable success of Granby Park also revealed a public appetite for community areas. Projects such as Art Tunnel Smithfield introduced a critical dialect between cultural policy and that of urban design and planning. It was a concern of many that success in Dublin was still being seen in terms of the property market 'to get us back on track, that's a very Western point of view, and does that work?' (Dublin arts sector member). Smithfield regeneration was spoken of as problematic as it was of 'All the big developers, and nothing behind them. A classic capitalist development, that one, so monumental' (DCC member).

Art Tunnel Smithfield was asked to leave its site at the point of two years residence. No reason was given for this by the landowner, furniture discount chain Bargaintown, but many commented that it was a two-fold reasoning: firstly, of fear of the affection felt for Art Tunnel Smithfield, and thus a fear of tenure; and secondly, to realise the value of the land to a higher rental or development value (whilst the plot was too narrow for building on, it was surmised that it could be used a parking or storage space for an adjacent mooted housing development, and thus rented at a higher value – and this is in fact what happened soon after Art Tunnel Smithfield closure). Whilst some protagonists and Smithfield community members looked to the lead artist to protest this decision, stating that 'they', here meaning the landlord and DCC, 'didn't realise the value of what was there' (Art Tunnel Smithfield participant) it was decided to close Art Tunnel Smithfield without remark. This was done to secure the longer-term and wider future for such use of sites in the future – as an act of its own exemplification (McCormack, 2013, p.120). Furthermore, whilst the closure of Art Tunnel Smithfield was seen by some as sending a positive message to landlords and DCC that vacant space use would not threaten long-term tenure, by others this was viewed as a failure in compromise, it feeding a commercial gentrification that would result in grassroots arts being priced out of an area and supporting the political status quo of land ownership in the city.

Many artists said of DCC that 'it still won't immediately say yes [to an idea], but it'll now try to find a way to say yes, to work within regulations to make something happen' (Dublin artist). Planners were known to 'turn a blind eye' to some instances of land use – the advertising board at Art Tunnel Smithfield, and a source of its income, being one such instance. Such 'organised disobedience' (DCC planner) as one planner termed it, could be tolerated for the longer-term gain, particularly so on Dublin's northside, as opposed to its Georgian tourist destination of the southside. Within DCC, a 'coalition of the willing' (DCC member) was formed of those across departments with an interest in vacant land and a desire to work differently in facing its issues. Whilst some in and out of DCC viewed this as a powerless, and therefore redundant group, as a first inter-departmental gathering, it was a sign of changing mind-set. The shifting ecology precipitated by the grassroots in 'the vacant land issue' was also bringing about a change in some of the roles that DCC was taking on. With DCC's Vacant Spaces Scheme [n.15], the council was acting as an intermediary leaseholder 'to give the landlord comfort that the tenant could be got out, to monitor the tenant, to make sure they were paying all the bills all that kind of stuff' (DCC member). This was said in relation to NAMA, and whilst this scheme 'never really got off the ground' (DCC member) its model was adapted in future projects such as Mary's Abbey for example. Here DCC underwrote the initial risks and responsibilities at the start of the project but these soon moved to the community group. DCC was there to offer 'the land owner comfort that they can get it back'; DCC was 'just in the sandwich this time', 'growing the community into a position where they can take it over and run it, and contain it' and the responsibilities in short time 'being rested on the group now' (DCC member). This was also seen as precipitating a cultural shift in the community: where the community first looked to DCC to steer the way, in short time too, it felt empowered to take on the contractual and licensing agreements. Thus, it was seen that a binary and an oppositional stance was being deconstructed, due in large part to the active citizenship and DCC showing its learning curve in the public realm. DCC had also been consulting the city on 'the vacant land issue', attempting to bring artists into the city conversation through a series of open meetings. It started the Vacant Spaces Scheme to help populate vacant shop spaces with creative enterprises on a short-term basis and began to licence its own spaces and act as intermediary, as in the case of Mary's Abbey garden, between the arts and community sector and NAMA on land in its jurisdiction; at the time of research it was mooting a Vacant Land Levy to disincentivise the hoarding of un-used land; and was linked at a European level via membership of urban resilience and sustainability initiatives URBACT [n.16], USEAct [n.17] and TURAS.

In Dublin, one can see how the grassroots momentum in re-appropriating vacant land was leading to a cultural and policy shift in DCC towards generative planning – though this process by no means 'complete' in the course of the research time, but observed certainly with a stated intention and aspiration of creating in Dublin the 'new political paradigm based on communal values' and a new relational model of citizen-city administration based on a positive place attachment that Benington (1996, p.152) talks of. The interiority of place identity – physical, social and autobiographic insideness (Dixon and Durrheim, 2000, p.32) – and attachment to Dublin was seen as a generative process in itself (Friedmann, 2010, pp.154–6; McClay and McAllister, 2014, p.8) that through in many instances, creative, social practice placemaking encounters, precipitated a process of assemblage that operated in the Lefebvrian 'generative relation' of affective spacetimes (McCormack, 2013, p.2; Munn, 1996, in Low, 2014, p.20). This in turn was forming an agonistic and exemplifying planning relation (McCormack, 2013, p.12; Tabb, 2012, p.202) from the grassroots towards the city administration as the 'user-generated urbanism' or 'collaborative city-making' (Marker, in Kuskins, 2013) of generative planning. This was a way to also manage the expectations the public was perceived to have of DCC, ' … of what the council can and can't do … some of the people would expect us to be there all the time, well, we can do things to start things moving' (DCC member). Dublin's new urbanism was seen to have the intention to illustrate to the city populace how it could be reimagined and refunctioned, so whilst a project such as Art Tunnel Smithfield would leave a memory of it being 'a really successful pop-up park' it would also, 'shift awareness of people to understand how this [vacant land] all can be reimagined in the city' (Dublin arts sector member). This active citizenship was experienced as informal in that it did not necessarily, at the level of the project delivery, have an interaction with formal political institutions or processes. Rather it operated outside or to the side of these, and was experienced as an organic, iterative process of conscientisation. Therefore, as a social practice placemaking approach demanded collaboration and co-production from all in Dublin, oppositions were being re-evaluated and created anew in practice.

The re-appropriation of vacant space was seen as a challenge to the 'huge tectonic process that grates and rubs' (Art Tunnel Smithfield participant) and which was in its process and outcome, redefining land and people, and surfacing a conversation around not just vacant space but the larger issue of land in the city. This undermined vacant spaces' 'discursively weak' (Lilliendahl Larsen, 2014, p.319) affective potential through its arts-based activation, and by networking in Smithfield, Art Tunnel Smithfield positioned itself in the local physical and socioeconomic spatial systems (Foo et al., 2014, p.175) to place the local community in the local urban morphology (Sennett, 2007). The growth of arts and local-led cultural programming and infrastructure in Smithfield, of which Art Tunnel Smithfield was a part – together with principally artist's hub Block T [n.18], the Generator Hostel [n.19], Lighthouse Cinema [n.20] and the café Third Space [n.21] – was emblematic and symptomatic of a Dublin-wide axis turn of art in the public realm. As Kearns and Ruimy (2014, p.48) state, the Dublin new urbanism altered perspective is attempting to redress the contemporary Dublin urban difficulty of a successful, liveable inner-city by rendering its response on a cultural and social reimagining (ibid., p.15), and that this – as seen in the research – is being led by communities of 'urban pioneers' (ibid., p.98), a group that would be composed of urban co-creatives [3.6.2]. From the example of Art Tunnel Smithfield as social practice placemaking and its agency within and without its urban creatives, participation in the project can be seen as firstly reflexive and from this, transformative, as based on empowerment to be able to decision-make and coalition build, separate of external organisations (Bishop, 2006 [a]; Kester, 2004).

This chapter has presented social practice placemaking as conducive to collective efficacy of sense of belonging in place, place operating as a unifying secure base in the urban locale, especially in times of change (Anderson, 1986; Castells, 1983; Lynch, 1981; Manzo, 2005; Unwin, 1921, in Nicholson, 1996, p.114). With this unifying function, the data challenges the notion that place attachment operates at a local level; whilst intra-city communication was in some instances problematic, social practice placemaking activity was seen to foster a sense of a wider citizenship and place bondedness (Scannell and Gifford, 2010) in the local and city-wide context. Place attachment was seen to motivate people to spend active time in place through processes of neighbourliness (Mihaylov and Perkins, 2014, pp.68–9) and community-minded place protective (Carrus et al., 2014, p.156) and material improvement (Mihaylov and Perkins, 2014, p.61) behaviours; thus, place attachment has social cohesion and capital aspects. The mechanism to promote this in the research project was social practice placemaking and its art processes. These processes were seen to have outcomes of individual and community level conscientisation, manifested by increased social awareness, expression (Chonody, 2014, p.2; Cozolino, 2006, p.147; Hou, 2010; Kelly, 1984; Legge, 2012, p.5), efficacy (Chonody, 2014, p.39) and the dissolving of subject positions that maintained a cultural horizontality (Carmona et al., 2008, p.14; Chandler, 2014, p.42; Jackson, 2011, p.52; Petrescu, 2006; Sherlock, 1998, p.219). Social practice placemaking then creates a shared, collectively performed urban public realm, the project spaces of which are opened up for public debate and controversy (Munthe-Kaas, 2015, p.17) where the countersite (Holsten, 1998, p.54, in Watson, 2006, p.170) agency of space allows its multiple uses. Social practice placemaking therefore can be seen as part of a grassroots social criticism (Miles, 1997, p.188) emanating from a dissatisfaction with the political structures and material form of the urban built environment; when enacted, this forms a 'parasitic takeover' (Klanten and Hübner, 2010, p.103) of an active citizenry taking the built environment function into their own hands. Art in this process is of agonistic (Mouffe, 2005) dissensus surfacing where conflict is viewed as an explorative, and positive, process. When city administrations take note of the social practice placemaking message and act to inform decision-making, social practice placemaking has a 'trickle-up' (Burnham, 2010, p.139; Silberberg, 2013, p.10) agency. Social practice placemaking from this perspective has a critical spatial practice function vis-à-vis the neoliberal condition, countering its instrumentalisation of creativity by exemplification (McCormack, 2013, p.12) and parasitic (Serres, 1980/2007) role. However, the degree to which this is manifest varies and projects have no or little control over whether their narrative is co-opted into that of the neoliberal, nor the resources to aggressively counteract this. The transformative potential of social practice placemaking in this regard is also limited as activity in the public urban realm does not automatically equate to efficacy with regards to democratic power; the re-appropriation of space does not grant automatically right to the city or its control.

5.4 Note

1 Vernacular demonym for anyone involved in Art Tunnel Smithfield.

5.5 Bibliography

Amin, A. (2008) 'Collective culture and urban public life' in *City* 12, 1.
Anderson, K. A. and Babcock, J. R. (2014) 'Horticultural therapy: the art of growth' in Chonody, J. M. (ed.) *Community art: creative approaches to practice*. Champaign, IL: Common Ground.
Anderson, S. (1986) 'Studies towards an ecological model of urban environment' in Anderson, S. (ed.) *On streets*. Cambridge, MA: MIT Press.
Baker, M. (2006) 'Afterword' in Warwick, R. (ed.) *Arcade: artists and placemaking*. London: Black Dog Publishing.

Benington, J. (1996) 'New paradigms and practices for local government capacity building within civil society' in Kraener, S. and Roberts, J. (eds.) *The politics of attachment: towards a secure society*. London: Free Association Books.
Bishop, C. (2006) [a] 'Introduction – viewers as participants' in Bishop, C. (ed.) *Participation*. London: Whitechapel Gallery and The MIT Press.
Bishop, C. (2006) [b] 'The social turn: collaboration and its discontents' in *Artforum*, 178–83 [Online]. Available at: www.gc.cuny.edu/CUNY_GC/media/CUNY-Graduate-Center/PDF/Art%20History/Claire%20Bishop/Social-Turn.pdf (Accessed: 28th November 2015).
Bresnihan, P. and Byrne, M. (2014) 'Escape into the city: everyday practices of commoning and the production of urban space in Dublin' in *Antipode* 47, 1: 36–54.
Burnham, S. (2010) 'Scenes and sounds: the call and response of street art and the city' in *City* 14, 1–2.
Carmona, M., de Magalhães, C., and Hammond, L. (2008) *Public space: the management dimension*. London: Routledge.
Carrus, G., Scopelliti, M., Fornara, F., Bonnes, M. and Bonaiuto, M. (2014) 'Place attachment, community identification, and pro-environmental engagement' in Manzo, L. C. and Devine-Wright, P. (eds.) *Place attachment: advances in theory, methods and applications*. Abingdon: Routledge.
Castells, M. (1983) *The city and the grassroots: a cross-cultural theory of urban movements*. London: Edward Arnold.
Chandler, D. (2014) 'Democracy unbound? Non-linear politics and the politization of everyday life' in *European Journal of Social Theory* 2014, 17, 1: 42–59.
Chonody, J. M. (2014) 'Approaches to evaluation: how to measure change when utilizing creative approaches' in Chonody, J. M. (ed.) *Community art: creative approaches to practice*. Champaign, IL: Common Ground.
Cozolino, L. (2006) *The neuroscience of human relationships: attachment and the developing social brain*. New York: W. W. Norton and Company.
de Certeau, M. (1984) *The practice of everyday life*. Berkeley: University of California Press.
Dixon, J. and Durrheim, K. (2000) 'Displacing place-identity: a discursive approach to locating self and other' in *British Journal of Social Phycology* 39, 27–44.
Foo, K., Martin, D., Wool, C. and Polsky, C. (2014) 'Reprint of "The production of urban vacant land: Relational placemaking in Boston, MA neighborhoods"' in *Cities* 40, 175–82.
Friedmann, J. (2010) 'Place and place-making in cities: a global perspective' in *Planning Theory and Practice* 11, 2: 149–65.
Gieseking, J. and Mangold, W. (eds.) (2014) *The people, place, and space reader*. New York: Routledge.
Hou, J. (2010) *Insurgent public space: guerrilla urbanism and the remaking of contemporary cities*. London: Routledge.
Jackson, S. (2011) *Social works: performing art, supporting publics*. Abingdon: Routledge.
Kearns, P. and Ruimy, M. (2010) *Redrawing Dublin*. Kinsale: Gandon Editions Kinsale.
Kearns, P. and Ruimy, M. (2014) *Beyond Pebbledash and the puzzle of Dublin*. Kinsale: Gandon Editions Kinsale.
Kelly, O. (1984) *Community, art and the state*. [n.p.]: Comedia.
Kester, G. H. (2004) *Conversation pieces: community and communication in modern art*. Berkeley: University of California Press.
Kester, G. H. (2011) *The one and the many: contemporary collaborative art in a global context*. Durham: Duke University Press.
Klanten, R. and Hübner, M. (2010) *Urban interventions: personal projects in public spaces*. Berlin: Gestalten.
Kuskins, J. (2013) 'Love or hate it, user-generated urbanism may be the future of cities', 23rd September 2013, *Urbanism*. Available at: http://gizmodo.com/love-it-or-hate-it-user-generated-urbanism-may-be-the-1344794381 (Accessed: 4th December 2013).

Kwon, M. (2004) *One place after another: site-specific art and located identity*. Cambridge, MA: The MIT Press.
Lefebvre, H. (1984) *The production of space*. Translated, Nicholson-Smith, D. Malden, MA: Blackwell Publishing.
Legge, K. (2012) *Doing it differently*. Sydney: Place Partners.
Lehmann, S. (2009) *Back to the city: strategies for informal intervention*. Ostfildern: Hatje Cantz.
Lilliendahl Larsen, J. (2014) 'Lefebvrean vagueness: going beyond diversion in the production of new spaces' in Stanek, Ł., Schmid, C. and Moravánszky, Á. (eds.) *Urban revolution now: Henri Lefebvre in social research and architecture*. Farnham: Ashgate Publishing Ltd.
Lossau, J. (2006) 'The gatekeeper – urban landscape and public art' in Warwick, R. (ed.) *Arcade: artists and placemaking*. London: Black Dog Publishing.
Low, S. (2014) 'Spatializing culture: an engaged anthropological approach to space and place (2014)' in Gieseking, J. and Mangold, W. (eds.) *The people, place, and space reader*. New York: Routledge.
Lynch, K. (1981) *A theory of good city form*. Cambridge, MA: MIT Press.
Manzo, L. (2005) 'For better or worse: exploring the multiple dimensions of place meaning' in *Journal of Environmental Psychology* 25, 1, 67–86.
Massey, D. (2005) *For space*. London: Sage Publications.
McClay, W. M. and McAllister, T. V. (2014) 'Preface' in McClay, W. M. and McAllister, T. V. (eds.) *Why place matters: geography, identity, and civic life in modern America*. New York: New Atlantis Books.
McCormack, D. P. (2013) *Refrains for moving bodies: experience and experiment in affective spaces*. Durham: Duke University Press.
Mihaylov, N. and Perkins, D. D. (2014) 'community place attachment and its role in social capital development' in Manzo, L. C. and Devine-Wright, P. (eds.) *Place attachment: advances in theory, methods and applications*. Abingdon: Routledge.
Miles, M. (1997) *Art, space and the city: public art and urban futures*. London: Routledge.
Mouffe, C. (2005) *The return of the political*. London: Verso.
Munthe-Kaas, P. (2015) 'Agonism and co-design of urban spaces' in *Urban Research and Practice* 8, 2: 218–37.
Nicholson, G. (1996) 'Place and local identity' in Kraener, S. and Roberts, J. (eds.) *The politics of place attachment: towards a secure society*. London: Free Association Press.
Petrescu, D. (2006) 'Working with uncertainty towards a real public space' in *If you can't find it, give us a ring: public works*. [n.p.]: Article Press/ixia.
Rekte, S. (2011) 'New urbanism and pocket neighborhoods: developing stronger communities', 18th August 2011. *Global Site Plans* [Online]. Available at: http://globalsiteplans.com/environmental-design/new-urbanism-and-pocket-neighborhoods-developing-stronger-communities/ (Accessed: 4th December 2013).
Rendell, J. (2006) *Art and architecture: a place between*. London: I. B. Tauris.
Scannell, L. and Gifford, R. (2010) 'Comparing the theories of interpersonal and place attachment' in Manzo, L. C. and Devine-Wright, P. (eds.) *Place attachment: advances in theory, methods and applications*. Abingdon, Oxon: Routledge.
Seamon, D. (2014) 'Place attachment and phenomenology: the synergistic dynamism of place' in Manzo, L. C. and Devine-Wright, P. (eds.) *Place attachment: advances in theory, methods and applications*. Abingdon: Routledge.
Sennett, R. (2007) 'The open city' in Burdett, R. and Sudjic, D. (eds.) *The endless city*. London: Phaidon.
Serres, M. (1980/2007) *The parasite*. Minneapolis: University of Minnesota Press.
Sherlock, M. (1998) 'Postscript – no loitering: art as social practice' in Harper, G. (ed.) *Interventions and provocations: conversations on art, culture and resistance*. Albany: State University of New York Press.

Silberberg, S. (2013) Places in the making: How placemaking builds places and communities. *MIT Department of Urban Studies and Planning* [Online]. Available at: http://dusp.mit.edu/cdd/project/placemaking (Accessed: 13th August 2015).

Tabb, W. (2012) 'Beyond "cities for people, not for profit": cities for people and people for systemic change' in *City* 16, 1–2.

van Hoven, B. and Douma, L. (2012) '"We make ourselves at home wherever we are" – older people's placemaking in Newton Hall' in *European Spatial Research and Policy* 19, 1: 65–79.

Watson, S. (2006) *City publics: the (dis)enchantments of urban encounters*. Abingdon: Routledge.

Whybrow, N. (2011) *Art and the city*. London: I. B. Tauris.

Zukin, S. (2010) *Naked city: the death and life of authentic urban place*. Oxford: Oxford University Press.

5.6 List of URLs

1. Dublin City Council/*Comhairle Cathrach Bhaile Átha Cliath*: www.dublincity.ie
2. Art Tunnel Smithfield: http://arttunnelsmithfield.com/
3. Fieldwork and Strategies: www.fieldworkandstrategies.com/
4. Luas: www.luas.ie
5. Mary's Abbey: www.facebook.com/MarysAbbeyGarden/?fref=ts
6. The Drawing Shed: www.thedrawingshed.org/
7. *Beta Projects*: https://dubcitybeta.wordpress.com
8. Arts Council Ireland/*An Chomhairle Ealaíon*: www.artscouncil.ie/
9. Granby Park: www.granbypark.com/
10. Upstart: http://upstart.ie/
11. Connect the Dots: www.connectthedots.ie/
12. What if Dublin?: https://twitter.com/what_if_dublin?ref_src=twsrc%5Egoogle%7Ctwcamp%5Eserp%7Ctwgr%5Eauthor
13. National Asset Management Agency/*níomhaireacht Náisiúnta um Bhainistíocht Sócmhainní Eilliú*: www.nama.ie
14. TURAS: www.turas-cities.org/
15. Vacant Spaces Scheme: www.dublincity.ie/main-menu-services-recreation-culture-arts-office/city-council-matches-landlords-empty-buildings-and
16. URBACT: http://urbact.eu/
17. USEAct: http://urbact.eu/useact
18. Block T: www.blockt.ie/
19. Generator Hostel: http://generatorhostels.com/en/destinations/dublin/
20. Lighthouse Cinema: www.lighthousecinema.ie/
21. Third Space: www.thirdspace.ie/

6 Big Car, Indianapolis

> '...makes me want to paint the whole city. I love it, I love doing it too...' (Big Car [n.1] *participant*)

The Big Car case study is of key significance to issues of the artist as leader and the role of social practice placemaking in regeneration. The work of Big Car, such as with its *SPARK: Monument Circle* [n.2] project, a month-long largely creative placemaking-programmed residency on the central downtown Indianapolis Monument Circle, are all examples of social practice placemaking: all act as the metaphorical breathing space to give air to place issues and work to dissolve issues of boundaried space and delineated space functions. This chapter will first consider the role of the artist as leader, extending thinking on this position from previous chapters, and then move into aspects of collaborative practice. It will then consider social practice placemaking: first, with regards to social cohesion; second, with a discussion of place identity; and third, as arts-led regeneration.

6.1 Introducing Big Car

Big Car is located in Indianapolis, Indiana, USA. It was formed in 2004 by Jim Walker, Shauta March and a collaborative of artists, working in and from the Fountain Square area of the city. It is a 501c3 non-profit arts organisation and collective of artists – staff and freelance – that uses creative placemaking and socially engaged art with communities and in collaboration with other artists from across the USA and internationally. At the time of research, Big Car operated along three lines: cultural programming and social practice projects; citywide arts-led placemaking; and the Garfield Park Creative Community development project [6.6.1]. Research with Big Car took place throughout September–October 2014 and April 2015.

6.2 The artist as leader and the Staff Artist

> 'He's like the Jeff Koons of social practice art. Though he'd hate to be called that'. (*Indianapolis arts sector member*)

With Big Car, its founder Jim Walker was resoundingly looked to as a leader and Indianapolis-wide as a figurehead for place-based arts. Like Koons [n.3], Walker brought a team of artists together in the 'fabrication' of the art work, but moreover, their artistic practice was encouraged through the collaborative leadership style that Walker employed. Walker was seen as someone that had a skill at 'curating a group of individuals who each had slightly different skills and strengths', bringing them together in a cohesive whole and matching their skills to community need, 'seeing a diverse group of artists that were doing different things in the city different ways, and that's part of Jim's brilliance, is that he has the ability to identify that and he also has the ability to identify challenges in the community' (Big Car Board member). It was at the intersect of the artist-as-leader and of collaboration that Big Car was realised and it was important with Big Car that all in the organisation had 'the ability to have your own idea and involve people in it and let what they suggest be in the mix and being willing to trade off on skills and what they want' (Big Car artist). Walker's leadership style was one where he delegated – 'the delegating in our group is the number one job' – according to peoples strengths to excel – 'you have to delegate to people's skill sets and things will happen, versus saying why can't you just do that, they aren't going to if they can't' (Big Car artist).

Overarching this was a constant 'trying to figure out how to give more guidance to those that need it, to bring out their capabilities, give them more specific role' (Big Car artist). It was this duty of care across the team that was seen as the success in the navigation of Walker's role changing from that of 'Big Car artist' to 'Big Car director' with the growth of the organisation and its increased layers of administration and management in its ten years: whilst Walker may be able to speak of himself as an initiator, this was done with modesty, 'there's no ego to Jim' (Big Car board member). With Big Car maturing to its tenth year at the time of research, its journey to an increased level of administrative professionalism was a matter of conversation. This was viewed as maintaining Big Car's ability to keep arts at its core, with artists employed as 'Staff Artist' on fulltime and project terms. In playing to the individual strengths of the team, the Big Car staff body was empowered to be active in the group, bartering and trading relative skills on projects intra-group, proactively asking for or offering help, as well to suggest their own ideas within or for projects. This was seen as an essential trait of a Big Car Staff Artist, 'the ability to have your own idea and involve people in it and let what they suggest be in the mix and being willing to trade off on skills and what they want' (Big Car artist) – and this will be discussed more in the following section. Thus, with Big Car, we see relative expertism [2.5.1] in action in the artists cohort – the creation of a horizontal creating field with each artist valued and empowered to enact their skillset whilst simultaneously engaging in knowledge exchange with others.

6.3 The arts practice and process of Big Car

The arts practice and process of Big Car involved a number of similar interventions as seen across the Dublin and London case studies – the making of objects, the

Big Car, Indianapolis 155

dialogic of storytelling, social horticulture for example, and specific practices and processes will be mentioned throughout this chapter – and this following section extends from this anticipated knowledge to explore the modes of the organisation's approach to its placemakings, explored through the collaborative dynamic, access to the arts and social justice concerns, and the disposition and role of the artist.

6.3.1 Collaborative participation and the temporary aesthetic

This 'mix' and 'trading' amongst Big Car Staff Artists was indicative of the ethos of the intra- and inter-group collaborative dynamic, observed as a natural ebb and flow between artists and non-artists: 'people galvanise around an idea, and it happens, everyone gets on board' (Big Car artist). The Staff Artists saw this as an opportunity for them to engage in a mutual beneficial working relationship with Big Car and their colleagues, 'some of it's for me, some of it's for the organisation' (Big Car artist). Big Car referred to itself as 'a collaborative for city engagement' (Big Car artist) and this was embedded through its projects, its staffing and its role in the Indianapolis arts ecology. As with Art Tunnel Smithfield [n.4] and The Drawing Shed [n.5], Big Car worked through an informal aesthetic, cited as the 'temporary as an aesthetic' (Big Car artist) and the social practice art of Big Car was a reflexive questioning for the organisation. It kept an open and flowing questioning dialogue on whether if its activity was overtly named 'art' would this stop people from being involved? 'Does it matter if the public don't see what we do as art? Would calling it art put them off? Make them feel they can't participate, criticise it, if they know its "art"?' (Big Car artist). It was observed by Big Car artists that the challenge for many came from notions of creativity, the public member not perceiving of themselves as artistic; consequently the joining of a community of artists or an arts activity was thought intimidating – the conceptual and psychological barrier to first step arts participation seen in the other case studies too. Put simply, art was seen as an 'othered' activity to be participated in only by trained artists, not an open, fluid and social activity as posited with social practice arts and placemaking. It was equally observed that this can be countered by creating a common experiential platform, 'I've got pretty good at encouraging people and shooting down their excuses of why they can't, 'cos, I've learned to understand more that a lot of people have anxiety when it comes to creativity' (Big Car artist). Thus, as with The Drawing Shed [4.5], we see the same gradual progression in an individual from a non- or abstracted participation and audience gaze to a mediated, engaged or co-produced participation. Transformation of self was also seen in the individual: where participants had acted as co-creators, the affect this had ostensibly altered self-perception, if only for the moment, such as one painting day volunteer with Big Car articulating, with some emphasis, that 'I am an artist today' (Big Car participant).

Involvement in Big Car's social practice placemaking process was seen to have a reflexive-to-transformative capacity, through the exemplification process (McCormack, 2013, p.12) that showed both what was within and without the spacetime conditions (McCormack, 2013, pp.12–3; Munn, 1996, in Low, 2014, p.20).

Big Car artists saw that participation in just one activity could have a transformative effect, 'You may only do one thing but that could open your mind to whatever you're into, you may only have to do it once and it could transform who you are' (Big Car artist). Non-participation was also an act of participation through its individual choice activation, 'Engagement is what it's about, someone saying yes as well as someone saying no' (Big Car artist).

6.3.2 Access to arts and social justice

Big Car's working out of the formal gallery setting, as with all case studies, was a deliberate tactic to generate access to the arts to a wider constituency and to foster a culture of travelling to arts projects as a regular activity, over and above the destination travel of a visit to a museum for example. This strategy was seen as one of social justice and redressing cultural–economic power imbalances, 'You have to be from a certain background to go through those doors and have these certain privileges, and we want to go out to communities where there isn't that sense of privilege', to bring arts to places where there was no cultural offer and through an emplaced social practice arts, 'have these conversations based around freedom, security, whatever, with people that don't go inside an art gallery', consciously thinking outside of 'white box terms' (Big Car artist) with regards to art-making and participation. The performative 'human interaction system' (Big Car artist) that was the social practice placemaking model at Big Car featured a variety of relational objects, including the mural site. One artist commented of a Big Car Chicagoan mural that the mural object was rendered less significant to that of the relationality of the activity of functional community formation and dialogue precipitated by the action of mural-making, this comment moving mural art from community arts to social practice arts in its performative intent and decreased focus on the mural as product.

6.3.3 The artist disposition and role

This was further related to the belief in the disposition of the social practice placemaking artist. During *Art in Odd Place Indianapolis* (AiOPIndy) [n.6], it was noted that some artists in the programme did not participate for the duration of the event, and this was felt contra to the meta ethos of Art in Odd Places (AiOP) [n.7] and to social practice placemaking and social practice arts in more general terms. Walker referenced Fletcher (2014) [n.8] as an 'ignitor' and someone that was 'turning it around' and as an artist that was 'talking about others, not themselves', 'reflecting the world and they become invisible' (Big Car artist): it was understood that the social practice arts and social practice placemaking artist has a particular character:

> Ultimately what it comes down to is that artists that are interested in others are different to artists that are interested in themselves, and the subject of their art is themselves and that's a different type of artist. (AiOP artist)

This artist was thus acting in the mode of the New Situationists (Doherty, 2004), the 'radically related '(Gablik, 1992, pp.2–6) artist whose role is to invite the audience into the artwork (Kwon, 2004, p.66) by creating a countersite (Holsten, 1998, p.54, in Watson, 2006, p.170) where the site of the artwork acts as a counter to dominant narratives and is purposed for relational and dialogic creative interrogation (Kwon, 2004, p.66). With a questioning of the impact of splash interventions in particular, the artist's role was recognised as largely limited – or specialist – in creating the opportunity for art experience, 'I just have to know that's my work to open those doors, I can't have the wherewithal to stay and create community, that's not my work, my work is to facilitate opening a door in places and minds' (Big Car artist). Artists here also commented – as seen similarly in Dublin and London with comments on perceived knowledge deficit of social practice arts and placemaking and its rigorous practice disabused as informal – that there is a perception that social practice arts artists lack a level or scope of professional knowledge and etiquette, but drawing on Whitehead's (2006) *What Do Artists Know* as one example:

> When you're in that conversation, you have to stand up for yourself and be smart enough to understand others aspects, the politics, the engineering, the funding. … People expect you to be a flaky arts thing. (Big Car artist)

The wide horizon of knowledges held in the social practice placemaking artist across many different professional sectors however was seen as a benefit of working outside of the gallery system and in the public realm: these artists were seen to be able to think strategically and tactically and make connections and understand multi-step processes (Whitehead, 2006), seen as an integral skill for the social practice placemaking artist: 'those people are the most valuable to an organisation like ours, they're able to see what is right in front of them and also able to make connections down the line' (Big Car artist), whereas 'studio artists aren't prepared for that kind of stuff, its just white noise to those other artists' (Big Car artist). The data shows the informal aesthetic to be contextually performative (Whybrow, 2011, p.35; Yoon, 2009), a temporary and improvised method of space production (Lynch, 1981, p.21; Tonkiss, 2013, p.108) as a site of both social aesthetic, embodied (Rendell, 2006, p.52; Lippard, 1997, p.267) and material response, as integral to differentiating social practice arts from social practice placemaking. It is not routine or accidental per se (Tonkiss, 2013, p.8) but haptic, rigorous and intentional.

The following sections now turn to consider the social practice placemaking of Big Car in respect to social cohesion, place identity, and arts-led regeneration.

6.4 Social practice placemaking and social cohesion

> *'One word that sums up what we're talking about is "transformation". It's creativity. Working with people, some sort of transformation takes place, transform through your sensibility and outlook' (Big Car artist)*

The work that Big Car undertook was about 'building the community that's on your doorstep, and building the arts community, creating a groundswell of an arts scene here' (Big Car artist). Mural painting (see Figure 6.2, discussed later in the chapter) was an activity deployed by Big Car with a symbolic value and relational role in group identity formation and place-bondedness, the activity designed as a group creative exercise, the mural reflecting aspects of its locale, its painting occurring in-situ and its installation public and participative. This accords with Chonody's (2014, p.32) assertion that participatory murals are a form of public art that create a sense of place attachment and identity to further intra-community connections through its participative function. Whilst Big Car 'never intended to do so many murals' (Big Car artist) and was aware of this practice's participative leverage rather than its co-produced agency, the practice was continued as it had a close 'mission fit' with Big Car's concern for community and social interaction. This is where this practice's, as community arts rather than social practice arts, challenge and sometime-friction lay: 'it's not just about creating [art] pieces, it's about what you are creating as far as a community grows ... it's not the goal, it's all about engaging people because of the art' (Big Car board member).

Big Car artists were explicit that the arts were used as a 'medium between lots of different things to create connections' (Big Car artist) and an evident neighbourhood and functional community (Nicholson, 1996, p.116) gathered around the projects, forming in the instant, a micropublic of resident and non-resident protagonists. It emerged that a key motivator to join in social practice placemaking projects was to activate social cohesion, the 'sense of community' and of bringing members of the functional neighbourhood perceived as desperate, together, that was mentioned by respondents across all case studies. In Indianapolis, there was a sense that 'The whole of the modern world is to atomise. But when we have street parties they go every well. People want to talk to their neighbours, even if it's just to gossip' (Big Car artist), and that through social practice placemaking projects, a 'tipping point' had been reached and that consequently 'there's neighborhood communities now, I feel like part of a community' (Big Car artist). The feeling of care to a place also affected a feeling of community self-care in some:

> My hope is that others would be inspired to take all such similar spaces for similar use. It is a real delight to me as a neighbour. It's nice to have someone care. I feel like I am cared for as well as the space. (Big Car participant)

This was then begetting of social capital, with a sense of a place-attached duty of care to neighbour and to place, 'It is a real delight to me as a neighbor. It's nice to have someone care. I feel like I am cared for as well as the space' (Big Car participant), but with a concern too about such activity diminishing, 'but it's still really important to keep it going otherwise it just goes into the distant past and what's that going to be in the future?' (Big Car participant) – here, the aesthetic dislocation of the social practice placemaking experience is seen to have an effect on the longer-term thinking on this interviewee. To join the social practice placemaking project micropublic (Amin, 2008) was an act of elective belonging, of choosing to be part of a place as an individual that was part of a group, and choosing to be active in it, performing an identity; the following section turns to place identity.

A challenge to galvanising or fostering social cohesion was, as one Big Car artist commented, that city-wide communication channels and attitudes were problematic in Indianapolis: 'Lots have no idea what is going on in Indy [*sic*], people stay in [their] township area, go to [the] local mall, to work, and that's it, no as yet way of mass publicizing an event across the whole of the city' (Big Car artist). Big Car experienced language barriers at the *Galería Magnifica* storefront gallery and workshop project at Superior Market (Figure 6.1) on the Far East Side: 'ultimately it would be great that it becomes a place where people go to more regularly, with programming, but [Spanish-speaking language] communication is difficult out there' (Big Car participant). These communicative barriers were one of the factors that stymied the longer-term strategic plans for the project as the initial 'pushing of that conversation' (Big Car artist) towards a cultural, heath-based programme in the store could not start. But moreover, this was from not knowing cultural gatekeepers in the area and thus not being able to communicate through them: 'in this neighborhood [Garfield Park], it's really easy as we know all the communicators, you can bring them in, get them connected with others, there's a place to start, but out there [Far East Side], we don't know anybody' (Big Car artist).

With Big Car, its social practice arts approach begat social cohesion and social capital, acting counter Durkheim's (1893) *anomie* and an adverse urban experience, the place attached behaviour that of the self-reliant, autonomous and mutually trusting individuals (Bowlby, 1979/2010, p.137) with developed peer-to-peer proximal relations (Ainsworth, 1991, p.38; Gosling, 1996, p.149; Pahl, 1996, p.98; Marris, 1991, p.70). These relations were based on both bonding and bridging ties

Figure 6.1 Big Car (2014), *Galería Magnifica* [gallery and workshop], Superior Market, North High School Road and West 38th Street, Indianapolis, 2014.

(Mihaylov and Perkins, 2014, pp.69–70); they were galvanised around common issues and participated in communally (Ainsworth, 1991, p.43; Carrus et al., 2014, p.155; Friedmann, 2010, p.155) – begetting yet further social capital and cohesion and place attachment. The art in this process had the function of being a creative act of social capital (Nowak, 2007), individuals in this fused into a group identity – a 'group-in-fusion' (Lilliendahl Larsen, 2014, p.326) – as a creative co-produced process (Lehmann, 2009, p.18) wherefrom people can become civically engaged in a wider operational context. The following section turns further to the issue of place identity in this art practice and process.

6.5 Social practice placemaking and place identity

The turn above here to include notions of the self in place attachment brings the text more explicitly to place identity and this section will focus on the case of Indianapolis as an example of a manifestation of and shift in place identity. Big Car arts projects fostered an approach that demanded the travel through and between protagonists and audiences non-residential neighbourhoods: this was seen as an 'out of the ordinary' city exploration but was evidence of an increasing sense of symbolic place attachment and elective place-identity across the whole city that worked contra to the myriad neighbourhood psychogeographic and cultural boundaries.

6.5.1 Travel and the car

The car and the act of driving was seen as integral to the territorialised experience of Indianapolis, both supporting it, by enforcing a sense of distance and afar destination, and undermining it, by being the means to foreshorten distance. In the latter regard, it was seen that the space of the car and the roadscapes of the Interstate became habituated places that produced a taken-for-granted topographical comprehension of cities; Big Car was pushing to dissolve these familiar territorial boundaries by opening up the city through an arts-based cartography. Driving came with its own set of social symbolism however and it should not be assumed that all Americans have a car, with those reliant on public transport effectively disempowered from fluid intra-neighbourhood travel. This was keenly understood by Big Car, and was the impetus behind its negotiation for bus rapid transit to the Garfield Park area.

6.5.2 Territory crossing and the city terroir

Indianapolis has a great many named geographic neighbourhoods which solidifies a sense of demarcation between these areas at the intra-city level, neighbourhoods which may not be more than a five-block geographical area in some cases, and hinders intra-city interchange:

> [The neighbourhood] has a name and that's important for the identity of the place. But, just a hunch, I wonder to what extent the naming of the

neighborhoods and the return to the name has a negative effect on this cross-city interaction. It's like you're in your own little city where you live. (Indianapolis arts sector member)

These neighbourhoods however have 'no sense of place in this area, no gathering place, no hub, no centre, no main drag' (Big Car artist), Big Car thus encountering a further intra-neighbourhood geographic conceptual barrier to arts participation. The city was also recognised to have a spatial stratification of populations of wealth and class-based identities, these demographics being arts attenders: those others, not. Thus, the conceptual understanding of the city held by its residents precluded some from participation. It was often commented that 'areas of the city don't talk to each other' and that 'people won't move out of their geographical area, they won't travel to be involved in the cultural life of the city' (Indianapolis arts sector member). Whilst those in the arts sector, as workers or attendees, were noted to travel to cultural events across the city, for many, unless there was a food retail destination motivating inter-neighbourhood car travel, the car itself:

makes it very neighborhood based, makes you reluctant to go do things … to see other parts of town and see how things are connected and see what's so close by in these places, get off the main street corridors of the city. (Indianapolis arts sector member)

Whilst Big Car worked singularly at specific locations on projects, and whilst these projects may have been differently funded, it employed a cohesive city-wide approach to its work, working against perceptually delineated neighbourhoods to foster a sense of a city terroir (Zukin, 2010, p.xi) that intersected these. It used the 'structural motivators' (Big Car artist) of ancillary retail destination stores for example, where it rented adjacent vacant storefronts, to foster a desire to travel intra-city. With endemic preference to drive the city on the Interstates, which residents did from neighbourhood to neighbourhood traveling over the city roads, swathes of the city were hidden from view from the car, bisected at street level too by the Interstate material infrastructure (Figure 6.2) – with the Interstate in this image having divided a mixed ethnographic neighbourhood of residences and shops, where it was perceived intra-ethic interactions were reduced or not evident at all since, and here being made a more walkable space and joined from one aspect to the other through mural painting. Whilst neighbourhoods were small and commonly within walking and cycling distance, it was felt that the use of the Interstate and its particular psychogeographical affects resulted in a similar social and cultural territorialisation, with news of events not travelling intra-neighbourhood. Arts-destination intra-city travel was observed as positively impacting people's perception of the topography of the city as one that was easily navigable between neighbourhoods – the same foreshortening of distance between home area and destination as seen in Dublin – to join the city in a more easily traversed whole.

Figure 6.2 L, Big Car (2014), *Indy Do Day* [n.9] mural painting [event], Indianapolis, 2014. R, Example of Interstate infrastructure bisecting pedestrian route that linked two retail parades, South Meridian, Indianapolis (2014), image taken during Big Car (2014) *Indy Do Day* mural painting.

This in turn was observed as creating a micropublic and microecology of a networked social practice arts/social practice placemaking in the city. In Big Car group settings there was much talk seemingly of difference, people attributing in a joking and light-hearted fashion stereotypical characteristics to others from where they were from outside of the state and the US; this potentially divisive talk in fact acted as a positive means to articulate an Indianapolis place identity and inclusion by way of comparison, people reflecting themselves in the 'otherness' of people around them and explaining themselves in this process, as well as the conversation acting as a means of intragroup understanding and bonding. Big Car artists articulated and exhibited a sense of pride in being an Indianapolis artist and defined themselves in that way, both within the city ecology and external to it, as observed at the Open Engagement 2015 [n.10] conference where Big Car artist delegates self-identified as Indianapolis artists to others; and whilst Big Car had a national mural project on its roster and a demand to work out of the city, it made a conscious decision during the time of research to stay working in Indianapolis: 'our real impact is here, and there's certainly still work to be done here' (Big Car artist). The manifestation of place identity in the individual and in the group was also seen to manifest in a notion of the authentic place (Seltzer, in Fletcher and Seltzer, in Finkelpearl, 2013, p.162; Zukin, 2010, pp.6–17, 2013, p.220). For one artist, this time was a conscientisation of both their artist and 'Hoosier'[1] identities fused as one, rather than disparate, their self-actualised artist identity integral to their felt collective identity of the city of Indianapolis and the wider Midwest:

> If you've got [artistic] ambitions to do stuff, and I think where Indy is now, I'm really excited about it from where its been and personally, to have just a

little part of that, and with Big Car, to have maybe even a little bit bigger part of that. Its really exiting ... I've lived here twenty-three years and I've only just now said that I'm a Hoosier. I guess it seems like home, to have known people this long, to have that longevity. (Big Car artist)

Territory crossing was a major theme and observation with Big Car. In its storefront location, *Showroom* and the first *Listen Hear* [n.11] space next to it at Lafayette Place (Figure 6.3), the materiality of the infrastructure was a barrier to participation – such sites are driving destinations without a pedestrian walk-up. The *Galería Magnifica* neighbourhood was not linked by a regular public transit service to the rest of the city so, and as with Lafayette Place, footfall was dependent on those that were prepared to drive and/or that were dedicated and regular arts attendees. This had an impact on the arts programming: 'It's hard to do the art out there, there's not that kind of pole to draw people in, no one is going out there for any reason so it's a lot harder to programme that space' (Big Car artist). However, there was a 'sense in Indy that if you want to do something, then just do it, the city becomes a project space' (Indianapolis artist) and Big Car worked to counter a psychogeographic territoriality, evidence again here of a reconfiguration of the city terroir.

Thus, whilst a change in the city terroir was evolving, at the time of research, no greater claims could be stated of this. Changes were noted, but were nascent and as stated above [5.2.4], this was far from a linear or smooth process and deep cultural, social and demographic forces were active in maintaining a status quo of boundaried city lived experience. Rather than people identifying with a fixed functional community or neighbourhood identity (Nicholson, 1996, p.116), place

Figure 6.3 TR, CW, BR, Big Car (2014) *Showroom* (centre, left, TR) [storefront gallery, events and workshop space] and *Listen Hear* (centre, right, TR) [sound art production, exhibition, installation and performance space], Lafayette Place, Indianapolis, 2014.

identity was, through participative activities, beginning to be expressed rather as fluid and of a wider city-geographic, symptomatic of a grounded existential feeling across the place of the city: 'I'm definitely a west-sider, or I was until I came out here' (Big Car artist on involvement in its east-side *Galería Magnífica* project). This is contra to place identity thinking that views the localised base as its operational site (McClay, 2014, p.8). A sense of elective belonging and of a shared group level symbolic meaning attribution to place (Scannell and Gifford, 2010) in forming a cohesive place identity was keenly observed in Indianapolis, where the individual's identity as an artist and/or resident was articulated as connected to a sense of place identification with the city. Whilst attachment to the hyperlocal site or the neighbourhood was still prevalent and articulated as the first site of a respondent's place identity, a symbolic and orientated identification (Norberg-Schulz, 1971, in Sime, 1986, p.50) with the city was also noted. Big Car's role and social practice placemaking activity in the overall sense of Indianapolis' cultural renaissance was seen as a key component in greater levels of interest in the city, 'people want to move to Indy now' (Big Car artist), whereas previously Indianapolis was nicknamed 'Indy-no-place' (Big Car artist). This 'creative agency' was emphatically 'a good thing for Indy' (Big Car artist) with the arts both creating a new scene in the city and this becoming woven into the place identity of Indianapolis as a creative city – both in a general sense and in the Floridian (2011) one. Thus, even though projects were not experienced continuously, on a day to day basis, or in some instances at all, rather, experienced through word-of-mouth reportage for example, place identity was functioning symbolically (Scannell and Gifford, 2010, p.26) and electively through this activity.

6.6 Social practice placemaking as arts-led regeneration

> *'Big Car invests in these places, with money and activity, but also with emotion'.*
> *(Big Car Board member)*

Big Car's model of artist participation had a scope of reach outside of the arts sector and of the immediate project setting, and this had a necessary co-produced ethos and manifestation. A positive place attachment and active citizenship will promote place protective activities (Carrus et al., 2014, p.157; Saegert, 2014, p.400) and Big Car's instances of *arts-led regeneration* can be seen as examples of this, their motivation largely to stave off a commercial gentrification, in favour of a development of place that was arts and artist-led.

6.6.1 Arts-led regeneration and the material and cultural

Social practice arts was utilised this way with Big Car to take protagonists 'just that little step outside of your normal comfort zone' to affect a placemaking change, 'using the dialogue of art and social practice as a way for us to say we can improve our communities this way' (Big Car artist). Big Car's first home was in the Fountain Square neighbourhood of Indianapolis, and was synonymous with people from all sectors in Indianapolis with the regeneration and gentrification of that neighbourhood. Other creative industries aggregated around this and as a result, Fountain Square is now much improved: it is linked to downtown and the rest of the city by the Indianapolis Cultural Trail cycle route [n.12] and a regular bus service; has improved roads and sidewalks and a wide variety of stores and entertainment; and is home to the city's film and music industries. Big Car – and others – used the vacant spaces in and of storefronts of Fountain Square to 'fill the material and cultural gap' (Big Car artist), but with the inevitable rise in prices for space, Big Car moved to other sites in the city, with the dual motivation of taking its practice to areas in need. One arts sector professional commented:

> I mean this affectionately, but I think of Big Car as when you're going on a road trip and you don't have to be discerning about what you pack, you just put it all in the car ... which is great 'cos it means they can do a project over on this side of town, they do a project in collaboration with this organisation, and they can do a project somewhere else and it all kind of intersects. (Indianapolis arts sector member)

At the time of research, Big Car was undertaking its *Garfield Park Creative Community* regeneration scheme, centred around the re-use of two vacant former factories (a 12,000 sq. ft. and 30,000 sq. ft.), one a former tyre factory (hence the working title of *The Tube*) which it was renovating into meeting and office workspaces, exhibition and performance space, and substantial maker and fabrication workshops. These were all for Big Car and community use, sitting alongside a third commercial property acquisition of a former white-goods store to which it relocated the *Listen Hear* sound art project space and gallery and that it is using as a base for its forthcoming radio station. Big Car also entered into a mixed-ownership purchasing of circa ten houses on the surrounding Cruft and Nelson Streets (Figure 6.4) to turn into permanent and rotation artist residencies; and it was in talks with the city administration about funding community co-produced traffic calming measures along the arterial route from downtown, Shelby Street.

Regeneration was recognised by Big Car as needed in the city to address the 'ring that goes all around the city that is this commercial wasteland, tonnes of foreclosures' (Big Car artist). Integral to this approach was Big Car's embedded knowledge of the area, 'where the placemaking part comes in [to *The Tube*], is that we understand the connection to the broader strategy of this neighbourhood, where a social practice artist that doesn't care about that kind of stuff would just stick their thing there and it would be about that thing' (Big Car artist). 'The city' was recognised as wanting the Garfield Park regeneration to succeed 'to be a catalyst for actual change that will lead to economic development, a better neighbourhood ... help attract talent to live here, boost the tax base' (Big Car artist). It was also stated that 'the city' realised the value of Big Car's arts-led regeneration model to achieve this. Big Car's motivation was anchored in the neighbourhood, and it was working systemically to 'keep the artists in' and where they have security of affordable

Figure 6.4 TL, Exterior aspect of one of *The Tube* warehouses (pre-renovation) from Cruft Street, Indianapolis, 2014. TR, Interior aspect of one of *The Tube* warehouses (pre-renovation) from Cruft Street, Indianapolis, 2014. BR, Example house for renovation around *The Tube* site (pre-renovation), Indianapolis, 2014. BL, Site of Garfield Park *Listen Hear* sound art gallery (pre-renovation), Indianapolis, 2014.

housing would 'stay places for artists to be leaders in the community' (Big Car artist). The Garfield Park project was an example of place attachment precipitating a sense of place stewardship, in these beginning stages, of the artists to the area and its community, forming a communal place identity (Brown and Perkins, 1992, p.284, in Mihaylov and Perkins, 2014, p.62; Giddens, 1992; Nowak, 2007, p.98; Wilmott, 1987). It was proven in this that place-attached individuals were more likely to enact place-protective, neighbourly and civic behaviour (Carrus et al., 2014, p.157; van Hoven and Douma, 2012, p.74) and were becoming resilient to change by encountering that change on their own – communal – terms. The social practice of Big Car operated along secure attachment and micropublic behavioural lines – of sensitivity, responsiveness and mutual understanding and negotiation (Cozolino, 2006, p.14; Marris, 1991, in Holmes, 1993, p.205). Again, this was seen to be counter to Durkheim's (1893) *anomie* and an adverse urban experience. The group act of this arts-led regeneration was to act more in favour of the place, enacting both bonding and bridging ties once more towards a deeper exploration of citizenship (Holmes, 1993, p.208) and cognitive, affective and behavioural aspects of place attachment (Manzo and Perkins, 2006, in Mihaylov and Perkins, 2014, p.63). The six facets of place attachment – interaction, identity, release, realisation, creation and intensification – were also seen in this project (Seamon, 2014, pp.16–9).

The Garfield Park scheme, and the Big Car organisational model of operating in vacant spaces, was working to release the assets of the built environment of the neighbourhood, matched with a community desire for sympathetic and locally based regeneration and a need for affordable housing and arts opportunities for artists. As with The Drawing Shed's commissioning of artists from housing estates to work on its housing estate projects, the Big Car team was largely resident in the Garfield Park and neighbouring Fountain Square neighbourhood and this local knowledge and place bondedness was galvanised with the skills of the artist to model its arts-led regeneration scheme:

> It's like Sim City, you can see everything building up around here, you can feel the connection, and that's not being driven by investors and for profits, who can drive it and who can initiate stuff like that, we can, that's what we do. (Big Car artist)

The Tube development was primarily funded by Big Car's own monies, a Community Development Trust [n.13] grant and investment from the Riley Area Development Corporation [n.14], a community development corporation (CDC). The project formed a land trust 'for the artists working hard with neighbours to improve the area' (Big Car artist). Its aim was, through an arts-led regeneration, to improve the quality of life for the residents, with fewer vacant sites and an improved pedestrian and transit infrastructure, fostering the area as a destination for Indianapolis by creating a cultural driver for economic development that helps existing businesses thrive as well as attracting new concerns to the area. The goal too was to improve people's lived perceptions of place so that 'residents feel proud of their place and consider art and creativity to be integral to its culture' (Big Car artist). Again, we see here the art practice and process of the regeneration as productive of a group-in-fusion *communitas* through the creative act of social capital and action (Saegert, 2014, in Gieseking and Mangold, 2014, p.397; Carrus et al., 2014, p.156), *The Tube* a literal 'workshop of civic engagement' functioning as a source of place identity, stability and growth (Nowak, 2007). Turn here can be made to the creation of 'authentic places' as a social product of placemaking (Zukin, 2013, p.17). The arts-led regeneration was asserting the moral right of the authentic place to be able to remain *in place* and connect the hyperlocal of Garfield Park with the wider structural forces of Indianapolis.

The placing of these spaces as adjacent in Garfield Park was central to Big Car's social art practice and the facilitation of the agency of arts in space:

> a connection between more artist, contemporary highbrow stuff and trad [*sic*] stuff that does bring a community with it. I don't care so much about the art on the walls as the social impact of having these shows. (Big Car artist)

Big Car's arts-led regeneration was illustrative of the social practice placemaking approach to link placemaking and social practice arts in a community turn (Hou and Rios, 2003) of placemaking, differentiated from creative placemaking in its social intentionality and non-culturised (Zukin, 2010, p.3) use of the arts in placemaking to heterogeneous spatial ends (Deutsche 1991, in Miles, 1997, p.90; Kwon, 2004, p.65):

> Placemaking is this other thing that organisations do. And then there's arts organisations, and people don't connect them, the job of landscape architects and planners, different to what a studio artist would do but conversely placemaking is absolutely a social practice and we're doing both. (Big Car artist)

The arts-led regeneration of Garfield Park though was also enacting change on the area, however well-minded it was in comparison to a feared corporate development-led gentrification. Mindful of the thinking that any change in place, no matter how ostensibly positive, can affect feelings of place loss – of grief, loneliness and diminished well-being – it remains to be seen in time how Garfield Park responds to its re-creation, however community-driven it may be.

6.6.2 The role of the artist and arts organisation

The subject of arts-led regeneration brought into the frame the question of the role of the artist once more. Big Car self-defined itself as an organisation that had created systems centred on human interaction, the artist having to be aware of their role in relation to others – relative expertism once more, as this artist's position respected that of the community as individuals and group as equal stakeholders and sites of knowledge. One artist used the metaphor of the artist role being that of a sports referee, being away from the limelight of the activity at hand, which comes from a place of confidence in their practice:

> You should not see the referee or remember them at the end of the game. And it's the same with this kind of art, who did it, its fine for you to know that, but if it's a public project its good for that person to be invisible, that's about confidence, I'm confident enough in myself not to have to have my stamp all over it. (Big Car artist)

Others accorded with this position of the artist: 'they're invisible but they're making everything work, they set the clockworks in motion, they wind it up and then it happens, you're setting everything in motion but you have to have the right gears' (Big Car artist). As place-based artists too, their role was to both work with and across areas of expert knowledge, echoing Whitehead's (2006) *What Do Artists Know* and relative expertism. From this, Big Car was in a position to lead in an arts-led regeneration. Big Car's embedding artists in an arts-led regeneration was working to prove to the resident community and to the city administration that artists are an employable and vital part of the city culture, offering the artist's perspective on cultural, economic, political and planning matters (Whitehead, 2006) – itself working towards a sustainable model of social practice placemaking for the city as a community-driven placemaking that was revaluating and repositioning planning and design and the role of citizens in the making of the public realm

(Hou and Rios, 2003, p.19). Big Car's model of arts-led regeneration rejected consultation in favour of co-production as a model for planning and urban design based on 'consensus building, conflict resolution, and organizational participation' (Hou and Rios, 2003, p.20). This process is that of generative and developmental planning (McCormack, 2013, p.2; Till, 2014, p.150), 'To be able to offer, have input on a city wide level, an artist's viewpoint as part of the conversation' (Big Car artist) – that embeds art practice as a source of learning, devising and delivery of urban design, moving beyond a creative placemaking approach that utilises the arts as an object of beautification or commercial prosperity.

The Tube space and neighbourhood approach was a continuum – as quoted below and as seen with working through and between community art and social practice – of practice for Big Car, 'It's not starting over, it's a continuation, an answer to not abandoning the area' (Big Car artist). By building the 'community on the doorstep, that's also building the arts community', by taking a systemic approach to 'keep the artist in' (Big Car artist) in community development. The knowledge of the social practice arts artist *as a placemaker* – as with Whitehead's (2006) *What Do Artists Know* again – was viewed as specialist and expert. This attitude is making change at the city level with, for example, the programming of *SPARK: Monument Circle*, a project that used temporary cultural experiences to inform the function and planning of the central downtown Indianapolis space, a commission granted to Big Car by the city authorities. Cumulatively, Big Car's social practice placemaking activity has 'turned [artists'] practice around … they see themselves as in parallel practice to placemaking, or placemaking see them as parallel back' (Big Car artist). At an organisational level, the pushing of proscribed and normative boundaries of art in Indianapolis by Big Car was a challenging of what an arts non-profit should be. Bringing *Art in Odd Places* to Indianapolis was a deliberate move on Big Car's part to educate its city administration and business community on what art can be in the public realm:

> They're entrenched here in what they think is art. Like, this is what I wanted [motioning to The Duty Free Ranger in performance on Monument Circle (Figure 6.5)] but some said 'Let's paint down by the skate park', like, they don't get it, that this is art too, that art is not just decoration. It's getting there though. (Big Car artist)

AiOPIndy (Figure 6.5) was presented by Big Car with other cultural partners, one being the Indianapolis Museum of Art (IMA) [n.15], its (then) Curator of Audience Experiences and Performance, Scott Stulen, also acting as a curator of the *AiOPIndy* programme. This was seen as 'an all-pronged attack on the Indy arts scene' (Indianapolis arts sector member) but with Stulen's role to make the IMA permeable to local artists and also being part of the *AiOPIndy* programme, together this also 'incorporated the bottom up to the top down of the big art institution, working into being a lot of engagement activities to the museum setting … makes people question what the role of the museum in Indy is' (Big Car artist).

The impact and degree of transformation and reflection engendered by this work and the blanket acceptance of the notion that 'people will reap the benefits of it regardless, whether they directly engage or not' (Indianapolis artist) has to be

Figure 6.5 Big Car and Ed Woodham (2014), *Art in Odd Places Indianapolis* [event], Indianapolis, 2014.

questioned however. In Indianapolis, it was felt that at times the city administration was happy to let Big Car do the job of regeneration activity where the city was falling short – and such criticism was levelled at the city authorities in Dublin by a number of research protagonists and the London boroughs of Walthamstow and Wandsworth by The Drawing Shed artists. All the case study organisations questioned their role in gentrification, Big Car posing the difficult question to itself of Fountain Square, 'Did we make it too good that we drove all the artists out? The artists did all the work, and now they're not here' (Big Car artist). This ambivalence was being heeded at Garfield Park, 'At The Tube, we're aware of that, and we're ready to head it off, making sure that we know what's going to happen and we're ready for it' (Big Car artist) and strategies and tactics of mitigation were seen across all case studies. The following chapter acts as a critical synthesis of this and other issues raised in the data and seen across the case studies.

6.7 Note

1 'Hoosier' is the vernacular demonym for a resident of the state of Indiana.

6.8 Bibliography

Ainsworth, M. D. (1991) 'Attachments and other affectional bonds across the lifecycle' in Parkes, C. M., Stevenson-Hinde, J. and Marris, P. (eds.) *Attachment across the life cycle*. London: Routledge.

Amin, A. (2008) 'Collective culture and urban public life' in *City* 12, 1.

Bowlby, J. (1979/2010) *The making and breaking of affectional bonds*. Abingdon: Routledge.

Carrus, G., Scopelliti, M., Fornara, F., Bonnes, M. and Bonaiuto, M. (2014) 'Place attachment, community identification, and pro-environmental engagement' in Manzo, L. C. and Devine-Wright, P. (eds.) *Place attachment: advances in theory, methods and applications*. Abingdon: Routledge.

Chonody, J. M. (2014) 'Approaches to evaluation: how to measure change when utilizing creative approaches' in Chonody, J. M. (ed.) *Community art: creative approaches to practice*. Champaign, IL: Common Ground.
Cozolino, L. (2006) *The neuroscience of human relationships: attachment and the developing social brain*. New York: W. W. Norton and Company.
Doherty, C. (2004) *Contemporary art: from studio to situation*. London: Black Dog.
Durkheim, E. (1893) *The division of labour in society reprint*. Basingstoke: Palgrave Macmillan, 2013.
Fletcher, H. and July, M. 'Hello' in Fletcher, H. and July, M. (eds.) *Learning to love you more*. Munich: Prestel.
Friedmann, J. (2010) 'Place and place-making in cities: a global perspective' in *Planning Theory and Practice* 11, 2: 149–65.
Finkelpearl, T. (2013) *What we made: conversations on art and social cooperation*. Durham: Duke University Press.
Florida, R. (2011) *The rise of the creative class revisited*. New York: Basic Books.
Gablik, S. (1992) 'Connective aesthetics' in *American Art* 6, 2: 2–7.
Giddens, A. (1992) *The transformation of intimacy*. Cambridge: Polity Press.
Gieseking, J. and Mangold, W. (eds.) (2014) *The people, place, and space reader*. New York: Routledge.
Gosling, J. (1996) 'The business of "community"' in Kraener, S. and Roberts, J. (eds.) *The politics of attachment: towards a secure society*. London: Free Association Press.
Holmes, J. (1993) *John Bowlby and attachment theory*. London: Routledge.
Hou, J. and Rios, M. (2003) 'Community-driven place making: the social practice of participatory design in the making of Union Point Park' in *Journal of Architectural Education* 57, 1, 19–27.
Kwon, M. (2004) *One place after another: site-specific art and located identity*. Cambridge, MA: The MIT Press.
Lehmann, S. (2009) *Back to the city: strategies for informal intervention*. Ostfildern: Hatje Cantz.
Lilliendahl Larsen, J. (2014) 'Lefebvrean vagueness: going beyond diversion in the production of new spaces' in Stanek, Ł., Schmid, C. and Moravánszky, Á. (eds.) *Urban revolution now: Henri Lefebvre in social research and architecture*. Farnham: Ashgate Publishing Ltd.
Lippard, L. (1997) *The lure of the local: senses of place in a multi-centered society*. New York: New Press.
Low, S. (2014) 'Spatializing culture: an engaged anthropological approach to space and place (2014)' in Gieseking, J. and Mangold, W. (eds.) *The people, place, and space reader*. New York: Routledge.
Lynch, K. (1981) *A theory of good city form*. Cambridge, MA: MIT Press.
Marris, P. (1991) 'The social construction of uncertainty' in Parkes, C. M., Stevenson-Hinde, J. and Marris, P. (eds.) *Attachment across the life cycle*. London: Routledge.
McClay, W. M. (2014) 'Introduction: why place matters' in McClay, W. M. and McAllister, T. V. (eds.) *Why place matters: geography, identity, and civic life in modern America*. New York: New Atlantis Books.
McCormack, D. P. (2013) *Refrains for moving bodies: experience and experiment in affective spaces*. Durham: Duke University Press.
Mihaylov, N. and Perkins, D. D. (2014) 'Community place attachment and its role in social capital development' in Manzo, L. C. and Devine-Wright, P. (eds.) *Place attachment: advances in theory, methods and applications*. Abingdon: Routledge.
Miles, M. (1997) *Art, space and the city: public art and urban futures*. London: Routledge.
Nicholson, G. (1996) 'Place and local identity' in Kraener, S. and Roberts, J. (eds.) *The politics of place attachment: towards a secure society*. London: Free Association Press.
Nowak, J. (2007) *A summary of creativity and neighbourhood development: strategies for community investment TRFund* [Online]. Available at: www.sp2.upenn.edu/siap/docs/

cultural_and_community_revitalization/creativity_and_neighborhood_development. pdf (Accessed: 4th December 2013).

Pahl, A. (1996) 'Friendly society' in Kraener, S. and Roberts, J. (eds.) *The politics of attachment: towards a secure society*. London: Free Association Books.

Rendell, J. (2006) *Art and architecture: a place between*. London: I. B. Tauris.

Saegert, S. (2014) 'Restoring meaningful subjects and "democratic hope" to psychology (2014)' in Gieseking, J. and Mangold, W. (eds.) *The people, place, and space reader*. New York: Routledge.

Scannell, L. and Gifford, R. (2010) 'Comparing the theories of interpersonal and place attachment' in Manzo, L. C. and Devine-Wright, P. (eds.) *Place attachment: advances in theory, methods and applications*. Abingdon, Oxon: Routledge.

Seamon, D. (2014) 'Place attachment and phenomenology: the synergistic dynamism of place' in Manzo, L. C. and Devine-Wright, P. (eds.) *Place attachment: advances in theory, methods and applications*. Abingdon: Routledge.

Sime, J. D. (1986) 'Creating places or designing spaces?' in *Journal of Environmental Psychology* 6, 49–63.

Till, K. E. (2014) '"Art, memory, and the city" in Bogotá: Mapa Teatro's artistic encounters with inhabited places' in Sen, A. and Silverman, L. (eds.) *Making place: space and embodiment in the city*. Bloomington: Indiana University Press.

Tonkiss, F. (2013) *Cities by design: the social life of urban form*. Cambridge: Polity.

van Hoven, B. and Douma, L. (2012) '"We make ourselves at home wherever we are" – older people's placemaking in Newton Hall' in *European Spatial Research and Policy* 19, 1: 65–79.

Watson, S. (2006) *City publics: the (dis)enchantments of urban encounters*. Abingdon, Oxon: Routledge.

Whitehead, F. (2006) *What do artists know* [Online]. Available at: http://embeddedartist project.com/whatdoartistsknow.html (Accessed: 16th June 2015).

Whybrow, N. (2011) *Art and the city*. London: I. B. Tauris.

Wilmott, P. (1987) *Friendship networks and social support*. London: Policy Studies Institute.

Yoon, M. J. (2009) 'Projects at play: public works' in Lehmann, S. (ed.) *Back to the city: strategies for informal urban interventions*. Ostfildern: Hatje Cantz.

Zukin, S. (2010) *Naked city: the death and life of authentic urban place*. Oxford: Oxford University Press.

Zukin, S. (2013) 'Whose culture? Whose city?' in Lin, J. and Mele, C. (eds.) *The urban sociology reader* (3rd ed.). Abingdon: Routledge.

6.9 List of URLs

1 Big Car: www.bigcar.org/
2 *SPARK: Monument Circle*: http://circlespark.org/
3 Jeff Koons: www.jeffkoons.com/
4 Art Tunnel Smithfield: http://arttunnelsmithfield.com/
5 The Drawing Shed: www.thedrawingshed.org/
6 *Art in Odd Places Indianapolis*: www.artinoddplaces.org/indianapolis/
7 Art in Odd Places: http://recall.artinoddplaces.org/
8 Harrell Fletcher: www.harrellfletcher.com/
9 Indy Do Day: http://indydoday.org/
10 Open Engagement: http://openengagement.info/
11 *Listen Hear*: www.bigcar.org/project/listenhear/
12 Indianapolis Cultural Trail: http://indyculturaltrail.org/
13 The Community Development Trust: www.cdt.biz/
14 Riley Area Development Corporation: http://rileyarea.org/
15 Indianapolis Museum of Art: www.imamuseum.org/

7 Conclusion
Towards a deeper understanding of the arts in placemaking

This chapter offers a conclusion to the research project by way of synthesising the case study data according to the three research aims into three thematics. To recapitulate, the aims were: firstly, to examine the practice and process of performative arts–informed placemaking and its effect on the emplaced arts experience; secondly, to investigate what existing social practice arts and place thinking can contribute to performative arts–informed placemaking and this artform as a means of reinterpreting the urban public realm; and thirdly, to explore the role of emplaced performative arts practice in shaping social cohesion, arts and civic participation and citizenship. The data thus far has been presented by case study, but here, by way of conclusion, these are now extrapolated as findings into three thematics: firstly, the practice and process of social practice placemaking and its effect on the emplaced arts experience; secondly, reinterpreting the urban public realm through the arts in (social practice) placemaking; and thirdly, social practice placemaking and social cohesion and active citizenship. Here data from each of the case studies is presented concurrently for the first time by way of critical comparison and new data is introduced to build narrative and theory, as the projects deal with common issues from their nuanced cultural context and comparisons.

7.1 Syntheses of findings

The thematic presentation of findings as conclusions begins with attention to the arts practice and process found and its effect on the emplaced art experience of social practice placemaking. It develops this narrative in the second thematic of how the social practice placemaking arts experience is an agent in the reinterpretation of the urban realm. In the third thematic, this section closes with attention on the outcome of this experiential reinterpretation as social cohesion and active citizenship.

7.1.1 Thematic one: the practice and process of social practice placemaking and its effect on the emplaced arts experience

Through the case studies, different drivers of social practice placemaking work were observed; different methods of knowledge production through practice; and different layers of tactical and strategic urban interventions as situated

174 Conclusion

arts practice. This section first presents where the case studies saw themselves in the social practice arts heritage of practice and then focuses on the core aspects of practice observed in the field, that of the arts heritage, performativity and duration, and the role of the artist as leader.

7.1.1.1 Social practice arts heritage

> *'We're going to use the Kaprow thing...' (The Drawing Shed* [n.1] *artist)*

Artistic influences were explicitly talked about, with practice anchored in the post-Modernist, sometime avant-garde, relational and social arts lineages from both the UK and USA. Direct reference was made to Kaprow [n.2] in particular, in Big Car's [n.3] ten year anniversary exhibition, *No Brakes* (2014), and in The Drawing Shed's reference to the Happenings in devising *IdeasFromElse[W]here* [n.4], this also based on artist Jim Haynes' (1984) arts lab model of artistic enquiry. With Big Car, a number of artists and art movements were spoken about regularly, including (in alphabetical order): the Beat Generation; Breton; Bukowski; Dada; Deschamps; 'hippies' and 'all that stuff of "let's just do something"' (Big Car artist); Max Ernst; Surrealism; and William Carlos Williams (poet). Contemporaneously, The Drawing Shed used the texts of Richard Sennett, in particular, *Together: The Rituals, Pleasures and Politics of Cooperation* (2012) and Big Car referenced its USA peers: curator and writer Hans Ulrich Obrist (2014) and the artist Harrell Fletcher [n.5], the latter as an 'ignitor' and someone that was 'turning it around', 'talking about others, not themselves' (Big Car artist), and Rick Lowe (Project Row Houses [n.6]) and Theaster Gates (Rebuild Foundation [n.7]) in respect to the motivation to materially, socially and economically change place through an arts-based placemaking.

The terms 'performative' and 'social practice' were also part of the daily vocabulary of The Drawing Shed and the Big Car artists and managerial staff, the latter describing its own practice as 'maximalism', as opposed to minimalism. Art Tunnel Smithfield [n.8] used the 'community art' term, using the word 'community' in its naming of sites within the space, the *Community Platform* and its art content created solely by 'any member of the community' (Art Tunnel Smithfield artist), which was sited physically opposed to the *Art Platform* which was described as a curated and lead artist–commissioned space. Such an artform link was problematic in London with The Drawing Shed and Big Car. In London, The Drawing Shed artists viewed their work as to 'push that envelope a bit more...they're also practicing artists, they want the work to operate as an art piece, not just a piece of community engagement' (The Drawing Shed artist). For one artist here, community art was regarded as of a lower status to that of social practice arts as it was not thought to challenge those from the community it purports to work with, 'worthy, not high calibre enough, people can handle this kind of work [social practice], [they] do not have to be coaxed into it' (The Drawing Shed artist). In Indianapolis, whilst mural activity was seen as located in the community art praxis, this methodology was refuted, 'the term "community art" conjures up something very different, it makes it sound a little down-homely studio somewhere, painting roses on a fruit plate' (Big Car Board member). Social practice arts and community arts were not necessarily viewed as in aggressive conflict or as a binary, rather part of a 'blended' (Big Car artist) continuum whereby any difference was determined by the intention and any perceived friction was a noted conversation topic between protagonists and was both worked through in conversation and practice, or was left to one side and the self-determined social practice arts activity continued.

Across the case studies, whilst the neighbourhood and the functional community (Nicholson, 1996, p.116) was implicitly understood and talked of, the core of the concern manifest was of a *relationship*, over that of *community*, this articulation reflective of an understanding of the fluidity and malleability of the notion of community. The case studies shared an ideological drive that obliquely and directly informed their practice of 'collective planning' (Art Tunnel Smithfield, Big Car and The Drawing Shed artists). It was also observed, and a stated aim especially of The Drawing Shed, that whilst there was a simultaneous adoption of a social practice arts practice and process, equally, these methods were interrogated in the live body of work, as noted too by Rounthwaite (2011, p.92).

These references locate these organisations firmly in the social practice arts/ socially engaged arts praxis and discussion emerged from this self-awareness across the case studies to their commonly externally perceived co-location with community art: 'Does community art create different work or just have different values?' (Big Car artist). In the setting of the neighbourhood and functional community (Nicholson, 1996, p.116) spheres, which variously intersected within and across projects, it was 'hard to keep the balance there between the art stuff and the community stuff, there you can't do the art stuff' (The Drawing Shed artist). This tension was not necessarily resolved, sometimes viewed as a provocation to work, at other times as a 'blended continuum' (Big Car artist) of practice where the differential was in critical intent (Bishop, 2006 [b]; Bishop, in Jackson, 2011, p.48; Finkelpearl, 2013, p.49; Froggett et al., 2011, p.9; Kester, 2004, pp.90–101; Schneekloth and Shibley, 2000, p.138).

It should also be noted here that the artists interviewed and observed in this research project showed a high degree of reflexivity and ethical awareness of their position as artists. Whilst in the development of the typology and the research as focused on a certain type of art practice, that of social practice, the artists inhabited a contradictory and complex position of self and practice. The artist's profiles incorporated various traits from across art practices and acted upon these to a greater or lesser extent, context contingent. Thus, the artist practice was one of a fluid hybridity, though anchored in a social practice: this challenges the theoretical mapping of artist practice given here by way of a positioning of the research project.

7.1.1.2 Performativity

The performative practice of the case studies operated to, as Bishop (2006 [a], p.10, 2012, p.11) suggests, 'bring art closer to everyday life': it also worked to 'co-mingle' (Critical Art Ensemble, 1998, p.73) art and everyday life by being situated in and operating from the spatial context of the project at hand. In these examples, the art activity of the case studies can be seen as an activation (Froggett, 2011, in Doherty et al., 2015, pp.15–6) of the audience away from spectacle to social exchange (Debord, in Bishop, 2006 [a], p.12; Deutsche, 1992, p.164; Minton, 2009; Mitchell, 2003, p.130; Zukin, 1995, p.259). The art intervention and relational object acts as exemplifier (McCormack, 2013, p.12) of social change (Till, 2014, p.165; Stern, in Lowe and Stern, in Finkelpearl, 2013, p.146) through the aesthetic dislocation (Kester, 2004, p.84) [2.4.1] of making the everyday strange (Froggett et al., 2011, p.97) and in the creation of 'socially cooperative experiences as art' (Finkelpearl, 2013, p.27). In the co-mingling (Critical Art Ensemble, 1998, p.73.) people became inter-related through 'empathetic identification' (Kester, 2004, pp.77–8) with difference in a process of distanciation from the everyday, activated by performativity, bonded through seeing both the similarities and differences levelled of their and others lifeworlds.

7.1.1.3 Duration

Case studies were engaged in projects of varying durations dependent on myriad contextual factors that are outside the academic capacity of this research project, such as competing workloads of the practitioners, of their third party stakeholders and competing logistical demands of these and other projects.

However, an aspect of project duration of relevance is that of the relative impact of shorter or longer duration, impacting as it was seen to on the degree of 'cohabitational time' (Latour, in O'Neill, 2014, p.198) in the project locale and situation. With splash projects it was accepted that the interactional experience of the art was transitory and of passing-by: the sight of the artwork may have been in the peripheral vision, intermittent or fleeting (if it occurred at all), and comments made more likely to be asides than conversational. It was also seen though that splash interventions were a playful opportunity for passers-by to go off-script from their usual day-to-day, to be hands-on with art materials and to be in conversation with art and artists.

Duration in space was also deployed operationally and creatively and worked with to the best of its parameters. In one respect, a longer duration was clearly seen to facilitate longer-term programming, with a slower, naturally unfolding pace of activity, resulting in deep relations between the artist and the residents and place and the unfolding of new ways of producing space (Froggett et al., 2011, p.95; Jackson, 2011, pp.68–9; Kester, 2004, p.171; O'Neill, 2014, pp.198–9; Stern, in Lowe and Stern, in Finkelpearl, 2013, p.145). On the other, limited and short duration was seen to force an intensity of focus and splash impact of activity. Just as with splash art interventions, work here acted to open up space within the dialogic aesthetic for reflection of the everyday habitat and its critique (Hamblen, 2014, p.83; Kester, 2004, p.83; Kwon, 2004). It is important to note here though, that the shorter term activity was still founded on a high frequency of visits and their long duration, factors that off-the-shelf processes such *Lighter, Quicker, Cheaper* (PPS [b]) urbanist interventions can learn from – there are 'no short cuts' (Cleveland, 2001, p.21) to such arts activity, no matter the duration. Whilst this position does not give an final answer to the question of the necessity, or not, for the social practice arts/social practice placemaking artist to have to be a resident of the place in which they are working (a question itself that is perhaps diversionary to and undermining of the social practice arts practice based as it is on absolute binaries), it does give nuance to an understanding of how such artists can operate in place as either resident or non-resident and across different types of project, using the durational and spatial parameters of any given situation to the best advantage of their work and the community in question.

7.1.1.4 The informal aesthetic

The perspective of the informal aesthetic, as a mode of practice observed in the field, gives a new lens through which to view placemaking as a whole and social practice placemaking in particular, in addition to and alongside relational and dialogical aesthetics. The informal aesthetic practice is both experimental and rigorous and active in place as spacetime refrain (McCormack, 2013, p.2) and countersited agency. As an embodied critical spatial practice, social practice placemaking is contingent on the effectuation of the conjoining of art and place through process and is a manifestation of a co-produced 'thinking through making' approach to urban issues which does not have proscribed entry or exit points for protagonists, nor outcomes or outputs. Parallel to Dewey's (1958) 'art as experience' and 'doing is knowing', The Drawing Shed worked with an ethos of 'knowing through making'. The 'seemingly worthless "non-objects"' (The Drawing Shed artist) of the go-karts made at Nine Elms during *Some[w]Here Research* [n.9] were re-valued and re-purposed in their making, 'grappling with the powerful relationship between imagination, survivability and resistance' (The Drawing Shed, 2015). In social practice placemaking, the artist's relative expertism is as creative thinker, disruptor and/or negotiator, operating place and local knowledge as the artist-in-place and situated knowledge of artist-as-expert (Schneekloth and Shibley, 2000; Whitehead, 2006). The artist employs, through a perpetual learning loop, an *a posteriori* knowledge, that knowledge formed from direct experience in the field (Bonjour, 2011; Landesman, 1997; Sosa, 2011) to inform future practice.

7.1.1.5 Artist as leader

A sometime friction was observed in the artists when positioned as leader, stemming from this emulation being contra to a collaborative, co-produced social practice placemaking practice and the disposition of the social practice artist as being that of a collaborator; however, in some instances, consensual and mutual leadership styles were employed to dissolve this uncomfortability (Froggett et al., 2011, p.103). This was enacted to the best success of the projects when coupled with the special character disposition of the social practice placemaking artist, their social role (Hope, 2010, p.69; Lossau, 2006, p.47) equally important as lead artist with artist and non-artists alike, contributing further to the production of a new discursive framework of aesthetic knowledge of social practice placemaking, the artist as producer of situations, the art work as the ongoing project, the viewer as the co-producer (Lowe, in Lowe and Stern, in Finkelpearl, 2013, p.147; Sennett, 2012, p.53). Artists were aware of contradictions or conflict in their position in social practice placemaking practice and process with regards to proscribed models of participation prevalent in the arts and urban planning sectors, and acted outside or beyond these, and often to subvert them. The position of the social practice placemaking artist as embedded or not had context-specific relative merits: what was agreed as imperative though was a disposition in the artist to the relational and performative social practice placemaking practice. Spatial and temporal relations are explored through the social practice placemaking arts practice as a spacetime (Munn, 1996, in Low, 2014, p.20): the performative practice both acted as a mirror to the hyperlocal site and its issues, as well as a provoker of 'reflection into action' (Kester, 2004, pp.90–101). As a material art practice, social practice placemaking then is a 'user-generated urbanism' or 'collaborative city-making' (Marker, in Kuskins, 2013) and of a unitary urbanism (Debord, 1958, p.95). As such, placemaking theory, to date, of the two umbrella terms of placemaking and creative placemaking, is inadequate alone to describe and advocate for the variety of practices, which employ a variety of arts and civic mechanisms.

7.1.1.6 Participation

Participation was a site of inherent, and unresolved, conflict for the artists across the case studies. Artists were aware of the ontological assumptions of participation and their role with or against this, acutely mindful of their position in the spectrum of thinking and power differentials of participation in the neoliberal dilemma: whether art is purposed to question the hegemonic system or ameliorate social and political issues (Beech, 2010, p.15–7; Grodach, 2010; Kwon, 2004; Rounthwaite, 2011, p.92) and the assumption that participation is emancipatory (Bishop, 2006 [a], p.10; Colombo et al., 2001, p.462). Rancière's (in Jackson, 2011, p.52) 'rupture' levelling process of social and aesthetic hierarchies was seen in the informal aesthetic of collaboration and the deeper levels of participation to co-production and sole authorship, and in the regard held of participants as experts in their own lives (Chonody, 2014, p.2). When co-produced, social practice placemaking located agency in city inhabitants and the art process (Till, 2014, p.168; Tonkiss, 2013, p.10); it gave literal and metaphorical space for people to discuss issues of identity and marginality for example (Bosch and Theis, 2012, pp.69–70). The process transformed the aesthetic experience into one of an embodied meaning, by placing the participant into the art practice and process. The motivation for this for the artist was to aid an empowering, and possibly active, political conscientisation in the participant. The artist relinquished authorial control (Bishop, 2006 [a], p.12; Brown, 2012, p.1; Grodach, 2010, p.476; Kravagna, 2012, p.254) in the informal aesthetic co-produced process, concerned more as they were with this transformational experience and the restoration of social bonds (Bishop [a], 2006, p.12).

It was clear that the case study groups were close-knit, from the self-presenting core group of volunteers at Art Tunnel Smithfield to the cohort of practitioners with The Drawing Shed that ran through the project portfolio during the time of research (and have continued since) and the 'family' (Big Car artist) that Big Car formed. One artist, Ed Woodham, actively and deliberately linked both London and Indianapolis in practice by being part of organisation's projects – working as an artist with The Drawing Shed, and bringing Art in Odd Places [n.10] to Indianapolis with Big Car (*Art In Odd Places Indianapolis* [n.11]). This gave a sense of an international global practice community, with the artist cross-referencing projects in both places to each other. New practice across and beyond these two organisations was also formed from the researcher's role of *Thinker in Residence* – the researcher being a critical friend of the projects through the form of blog writing – this starting with The Drawing Shed [n.12], continuing with Big Car [n.13] and then being transferred with other writers with AiOP [n.14], through the link of the common denominator artist. Big Car had an extended family too of arts organisations in Indianapolis that it had a close relationship with.

It was evident that the motivation amongst the urban co-creators to join and stay in projects was to form close and psychologically positive community connections, even if these were temporary or liminal: the bonding process in some cases, such as the mural painting, was quickened due to the time limit, and in some cases, as with the long-term nature of The Drawing Shed's residency on the Attlee Estate (*LiveElse[W]here*, [n.15]), becoming layered and nuanced over the longer duration. Thus, the artist cohort and the organisational banner under which they formed became a secure unit – akin to the family in Bowlby's (1979/2010) thinking, and also talked of explicitly as a family and of paternalist relations, not always positive, just as in family life. Projects formed a cohort of positively socialised individuals (Giddens, 1992; Pahl, 1996, p.96; Wilmott, 1987), with positive attachments to place through the forming of a secure base of micropublic (Amin, 2008), social practice placemaking activity acting to confirm each other's existence as an extension or substitution (depending on the individual) of familial bonds, countering any felt *anomie* (Durkheim, 1893).

The barriers to participation observed – communicative, material, conceptual and psychogeographic – as with degrees of participation, were both distinct and interlinked in a moment for a person and project, and also fluid. As one artist observed, this transformation process was about 'permission' (The Drawing Shed artist): the contraction of the *a priori* social practice placemaking event was dependent on viewer to participant to protagonist role, the permission granted in the moment through intra-group cohesion and reciprocity (O'Neill, 2014, p.201). Across non-artist participation can be seen incremental stages of participation towards the non-artist participant becoming the producer of artworks in their own right (though they may not refer to themselves as an 'artist' still). Non-participation can be viewed as part of the participation process through its presence as an oppositional force – and this will be discussed in more detail below. The nature of the activity for artists predisposed projects to a collaborative and generative process that was mutually supporting and of knowledge exchange

180 Conclusion

and capacity building. Artists though problematised their position as artist, in reference to participative models of the arts sector and claims of collusion with neoliberal gentrification process: the artists were self-aware of this difficult, contradictory position and acted either outside of these structures or to a degree within them, aiming to subvert them from within. It was recognised though that this was an ongoing conversation with self and others, and the community the artist was active within, and a conversation that was had on a repeated basis with practice behaviours adapted accordingly.

7.1.1.7 Transformative outcomes

The performativity of social practice placemaking is located in the space and place of the everyday, its practice and process activation for social exchange and transformation in the first instance. Where artists observed transformation in the participants, and whilst a change in some regards was seen in the course of the research – the elements of play and happy-making, the sense of connection to the group and to the place – there was no remit in this research of quantifying or qualifying this with the participant: it could not be known if the art experience had a long-term and/or profound effect. What was observed was a sense of increased individual agency in the moment which was explained as a personal conscientisation, an awareness or an increase in a group level identity formation, and, for some, an ongoing (in the term of the research project at least) engagement in the urban form and its cultural and political setting. A transformation was seen though in participants of what was thus thought to be art and for the artists, of what was thought to be community. In this latter regard, a transformation of practice was observed in the artists through the nature of the collaboratory, reframing and a reorientation of their practice to a co-produced one or one with reconsideration of community agency and authorship.

Art in social practice placemaking is a tool and a function to produce a built environment critique sustaining democracy in the urban realm by antagonising it (Bishop, 2012, p.264), the social practice placemaking artist working as a facilitator aiming to empower and build a platform with people to give voice to people in the decision-making process. The independent spaces in focus here revert abandoned spaces and subvert disinvested commons as a productive urban space and process (Tonkiss, 2013, pp.21–2) and this has significant sites of learning for planning. Social practice placemaking, as an informal design process, accords with progressive politics as it does not fit squarely into formal models of land and property ownership (Roy, 2005, p.148). Informality is instead a 'differentiated process embodying varying degrees of power and exclusion' (ibid.). In the socially and professionally horizontal (Sherlock, 1998, p.219) relative expertism spectrum of expert positions, the social practice placemaking artist acts as a change agent. In this, as an ideologically driven practice (with commensurate material outcomes nonetheless), the artist, whether resident in the hyperlocal site or not, also has a parasitic role, as that change agent. This was to, as repeatedly articulated across the case studies, 'hold things open', to critique and agitate through exemplification (McCormack, 2013, p.12), through aesthetic dislocation, activating a politically subversive potential – and signalling social practice placemaking's intentional and material differentiation from participatory arts, and from other forms of placemaking. The informality of social practice placemaking is a challenge to a regulated production and consumption of space (Franck and Stevens, 2007, p.272; Minton, 2009; Sennett, 1970; Tonkiss, 2013) and, through the informal aesthetic, creates new, alternative urban forms (Lynch, 1981, p.21; Roy, 2005, p.148; Tonkiss, 2013, p.108).

Both short and long project duration were used to the benefit of the social practice placemaking practice; duration of any length was viewed as having the impact as aesthetic dislocation, though longer duration was seen as premising of deeper practice and greater social engagement. With regards the experience of art that social practice placemaking precipitated, this was of the relational micropublic (Amin, 2008) and micro-community (Kester, 2011, p.29) that, through performative exemplification negotiated difference through relational and dialogical interaction as a social aesthetic that gave voice to individuals and communities and was agentive of an understanding and representation of urban lifeworlds. Thus, whilst social practice placemaking is informed by social practice arts, the practitioners across the case studies overtly placing it in this arts lineage and directly and obliquely using its processual mechanisms, social practice placemaking has specific material and place-led concerns, outcomes and outputs that differentiates it from social practice arts as a placemaking practice. The following section will continue the exploration of social practice placemaking vis-à-vis social practice arts by focusing on issues of participation found in the data.

7.1.2 Thematic two: reinterpreting the urban public realm through the arts in (social practice) placemaking

This thematic refers to the second aim, to investigate what existing space and place thinking can contribute to performative arts-informed placemaking, and this artform as a means of reinterpreting the urban public realm.

7.1.2.1 Aesthetic dislocation

Social practice placemaking works with the involved aesthetic dislocation effect (Kaji-O'Grady, 2009, p.108; Kester, 2004, p.84), where through performative agency, everyday habituation of place is made strange (Froggett et al., 2011, p.95) and disrupted. This precipitates a reflection on that place and exemplification of its potential, different, lived experience, a cognitive and behavioural transformative process of 'engagement through alienation' (Klanten and Hübner, 2010, p.3; Rendell, 2006). As seen in the data, there is a limit to this dislocation: a familiarisation with these interventions may take place and they consequently join the habituated lived experience of the urban dweller and become normalised. Memories are formed in the moment and reflection on the social practice placemaking intervention, which goes on to conjoin with the everyday habituation, which also dilutes the special dispensation of the artist [4.3.1]. Habituation with social practice placemaking interventions is still not though a negation of the transformative agency of social practice placemaking, and its longitudinal impact on individuals, the community and the wider political context in which they are sited. This transformational agency is founded on the social practice placemaking interventions relational capacity as the parasitic (Serres, 1980/2007) aesthetic third operating in a spacetime countersite (Holsten, 1998, p.54, in Watson, 2006, p.170). Here, Subject/Object positions are subverted in the dialogic aesthetic and subsequent critical thinking, the 'problem-posing pedagogy' (Freire, in Finkelpearl, 2013, p.30) of a dominant, hegemonic space production ideology critique.

7.1.2.2 Spatial boundaries

Across the case studies, a sensing of boundaries and relative boundary positons was crucial to the social practice placemaking process: artists were working with the cognitive aspects of place, using processes of social practice placemaking to subvert this. A social practice placemaking experience of the cityscape and neighbourhood-scape gave rise to alternative, and embodied, mappings of the space and place that made past histories and the contemporaneous everyday forces of urbanism visible and malleable through the arts-based animations of liminal spaces. Boundary-crossing of space was seen in the hyperlocal context between artists and non-artists in a complex mutual relationship that recognised both the agency of the artist and non-artist alike, and of participants and non-participants.

7.1.2.3 Performed place attachment and exemplification

A positive place attachment and active citizenship will promote place protective activities (Carrus et al., 2014, p.157; Saegert, 2014, p.400) and instances of arts-led regeneration can be seen as examples of this, their motivation largely to stave off a commercial gentrification, in favour of a development of place that was arts-led and place context-responsive and responsible. The social practice placemaking of The Drawing Shed in Nine Elms was articulated as ideologically motivated to 'throw up fault lines' (The Drawing Shed artist), using a social practice arts questioning process to unlock issues. *Some[w]Here Research* was active in creating a space for residents critical discourse, 'There's a critical edge. Even in passing people are talking about their life here...[we're] creating gaps in which this can pass through, creating space for it' (The Drawing Shed artist).

Across the case studies, social practice placemaking activity was seen to support the theory that group activity in place was a process that created an emotional affective bonding to that place (Lewicka, 2014, p.351), the social practice arts practice and process acting as a polylogic organising framework of person–process–place (Scannell and Gifford, 2010) that functioned to create a sense of place-belonging and place identity through an actualisation of self-identity (Hernández et al., 2010, p.281, in Mihaylov and Perkins, 2014, p.66; Proshansky et al., 1958, in Gieseking and Mangold, 2014, p.77). The social practice placemaking group experience fostered dynamic and discursive interpersonal relations that nurtured reflexivity and self-actualisation, the protagonists getting to know themselves though others, to paraphrase Cozolino (2006, p.342), generating a self-to-group feedback loop that functioned to create a sense of place-bondedness and belonging. Evident in the data was place attachment as a holistic, dialectic and generative phenomenological experience (Seamon, 2014, p.12) and physical, social and autobiographic insideness (Dixon and Durrheim, 2000, p.32). As Lewicka (2014, p.351) states, this organising framework and its affective outcomes creates or bolsters existing social cohesion and social capital, which subsequent sections will detail. All case studies wilfully presented a topophillic (Tuan, 1974, p.4) relation to place – the projects they were involved in were all working to positive goals and came from a positive motivation to improve place and social interrelations. Debate and disagreement was seen as a positive part of this process too, as a part of an explorative arts-based approach to questioning, devising, problem-solving and solution-finding. These processes were viewed as a behavioural enactment of interest and care which promoted positive affectual place bonds (Carrus et al., 2014, p.154). Non-participation and ambivalence were accepted and seen as an organic state of relation to project and place. To link place attachment to positive urban theory: in the creation of the micropublic (Amin, 2008) in the social practice placemaking practice/process, with the creation of a secure base in the moment of the project activity, discussion and encounters of difference were given safe platform and worked through in open dialogue (Marris, 1991, in Holmes, 1993, p.205; Cozolino, 2006, p.14), nurturing a space for, and increase in, both emotional and interpersonal expression (Cozolino, ibid., p.147), the foundation of the social practice placemaking polylogic organising framework (Scannell and Gifford, 2010).

The performed space of performed identities is the relational capacity of the aesthetic third [2.4.2] and this is the role of art in the place attachment process. The processes employed by the case studies were multiple – anything from clearing a river bank (Big Car) to conversing over a communal lunch (The Drawing Shed) to building a multi-storey sculpture (Art Tunnel Smithfield) – but all involved a relational and dialogic concern of 'practice as process', i.e. that the performative concerns of the practice informed and were mirrored in process. Located in the social aesthetic, social practice placemaking-formed micro-community (Kester, 2011, p.29) membership precipitates a 'group-in-fusion' (Lilliendahl Larsen, 2014, p.326) communitas (Turner, in Tuan, 2014, p.106) of a larger sense of other and a wider sense of place through intra- and inter-group contact with a cooperative action function for problem solving (Dewey, 1958; Freire, in Finkelpearl, 2013, p.30). The 'co-mingling' (Critical Art Ensemble, 1998, p.73) of the micropublic (Amin, 2008) and art practice and process of social practice placemaking has a conscientisation function of a collective and political efficacy engendering affective place-protective (Till, 2014, p.167) activities, a performed commitment to place, where place is responded to and in turn, (re)shaped. As an informal aesthetic practice, social practice placemaking is based on an experimental pedagogy – as above – that both organises and transforms space (McCormack, 2013, p.29) and is contextually performative, of gestured, temporary or improvised space productions. This urbanisation is a fluid synergy of both adverse and positive urban experiences as motivating factors, that works with both mixophobic (Bauman, 2003, in Watson, 2006, p.168) and mixophillic and meta-psychological dynamic and adaptive positions (Rapaport and Gill, in Bowlby, 1979/2010) to foster 'meaningful social worlds' (Fischer, 1975; Key, in Krupat, 1985, pp.133–4) through participative relational performances (Anderson and Nielson, 2009, pp.305–6).

The role of social practice placemaking, as an artform in place of the informal aesthetic was as haptic, navigational and somatic exemplification (McCormack, 2013, p.12) of urban form change, an embodied site of experience and tactical response process and of a co-created process where agency is located in the urban dweller (Till, 2014, p.168; Tonkiss, 2013, p.10). In the frame of relative expertism, artists also navigated across areas of expert knowledge in their own practice and that of other urban protagonists. The outcomes of social practice placemaking, at individual and community level conscientisation and as part of the place attachment process are of increased social awareness, expressive conscientisation and efficacy and the dissolving (but not flattening) of subject positions. The site of the social practice placemaking project setting evidenced the potential of participation to co-production. Projects aimed to break down such barriers to participation and of arts as the 'other' by interfering with and reforming the cultural and spatial geography of their operative settings – the agency of the countersite (Holsten, 1998, p.54, in Watson, 2006, p.170) in effect. The countersite has a parasitic role to break the signal-noise (Serres, 1980/2007) of hegemonic urbanism, enacted as a socialised (Carmona et al., 2008, p.14) and mutually generative embodied (Lefebvre, 1984, p.170) placemaking (Sen and Silverman, 2014, p.4) by a conscientised micropublic (Amin, 2008) or micro-community (Kester, 2011, p.29) that were reproducing self-territorialised (Friedman, 2010, p.154; Sime, 1986, p.60) space. Transformation of what was considered art, and therefore funded and programmed as art, also affected at the macro level of the city administrations the city's relation to artists, which changed in many aspects through the social arts in place activity.

The embedding of arts into place and its affective place attachment outcomes were seen to encourage a sustainable place-based ethics of care (Till, 2014, p.151) enacted from enhanced socially cohesive and high-capital communities. In Dublin in particular it was seen that the countersite (Holsten, 1998, p.54, in Watson, 2006, p.170) of its vacant land acted as an informal critical spatial practice: it is a diverted space of conceptual indeterminacy that signifies new spatial codes (Lilliendahl Larsen, 2014, p.336) bespoke to their hyperlocal and city context (Foo et al., 2014, p.175). Other placemakings can learn from this in the development of stakeholder relations for the immediate and long-term care of place, utilising the skills of place professionals that are grounded in professionalised knowledge as well as the community in question and creative thinking (Roberts, 2009, p.446). For that community in question, the 'un-working' (Mozes, 2011, p.11; Whybrow, 2011, p.19) of the city is done and communicated through the social aesthetic structure of the urban event space (Hannah, 2009, p.117) that creates multiple performative assemblages of alternative urban forms. It is this people-centred co-produced intentionality that differentiates social practice placemaking from other placemakings, which does not hold that place is acted upon to be 'made' or complete, but that it is already in existence and to be worked with processually.

7.1.2.4 Dissensus and agonistic agency

The case studies in generalist terms expressed a topophillic (Tuan, 1974, p.4) relation to place and placemaking outcomes and outputs. But integral aspects of the social practice placemaking process actively engage with negative and ambiguous feelings towards place as integral to its experimental and pluralist practice and process. It regards failure as a process and not an ends and is a means of creating a space for intersectional spatial discussion and activity that dissolve – in a positive regard – shared cultural assumptions (Finkelpearl, 2013, p.21). Social practice placemaking has an agonistic agency where the use of arts is functioned for dissensus surfacing and emplaced in countersites (Holsten, 1998, p.54, in Watson, 2006, p.170) as political space to side of hegemonic space and political discourse. In social practice placemaking, communities are organised to take advantage of political opportunities and use the arts for pervasive change, including, as seen with Big Car, social practice placemaking artists with the community gaining control of real estate community assets and setting their own regeneration agendas.

7.1.2.5 Duration

Varying durations, as refrains (Guattari, 2006), and their ritualistic 'invention of tradition' (Sennett, 2012, p.88) within a project, were appreciated for their relative 'aesthetic palpability' (Jackson, 2011, p.69) and operational and agentive merits. Value was evidently felt in splash interventions as aesthetic dislocation (Kester, 2004, p.84), as above, casing rupture in the 'habit-memory' (Connerton, 1989, in Sen and Silverman, 2014, p.4) of daily routine and differentiating that place anew in time and geography, and in the moment of delivery, all arts interventions acted with this agency. The virtue of a short duration was maximised to both stimulate interest in passers-by through to protagonists and utilise the rupture to create a 'breathing space' countersite (Holsten, 1998, p.54, in Watson, 2006, p.170) to reflect back to people the potential or current changes of a place (Lippard, 1997, p.287).

However, case studies leaned to regarding a longer duration in place as increasing the depth of social engagement: whilst short duration was part of the project-by-project funding landscape for all, this was not seen as a strategy for pervasive social or material change, thus the tactical aspect of such interventions was thought to have limited exemplifying (McCormack, 2013, p.12) agency. Rather, an increased timeframe was thought to precipitate an increased depth of social engagement and change (Froggett et al., 2011, p.95; Jackson, 2011, pp.68–9; Kester, 2004, p.171; Stern, in Lowe and Stern, in Finkelpearl, 2013, p.145).

186 *Conclusion*

The stopping of a project was seen as having the 'mediating change function' (Proshansky et al., 1958, in Geiseking and Mangold, 2014, p77) that caused upset to the place attachment process. Such issues were seen with Big Car. On leaving its *Service Center* [n.16] venue after a five-year residence, the group was given two months' notice and had to decide in that short timeframe if it was to continue working in the Lafayette area of the city. It appreciated the durational and organisational advantages that a fixed site gave the organisation over that of its pop-up model seen across the city and the longer-term model was being actively pursued to build on this impact pop-up model in the arts-led regeneration in Garfield Park [6.6.1] [n.17]. Similarly, the presence of The Drawing Shed on the Atlee Estate over a number of years was aiming to become part of the culture of the estate and have a legacy that resonated accordingly.

7.1.2.6 *Place efficacy and change*

The transformative potential of social practice placemaking can be placed on a spectrum of Freire's 'problem-posing pedagogy' (Finkelpearl, 2013, p.30) where the art practice operates in a parasitic (Serres, 1980/2007) spacetime (McCormack, 2013, pp.12–3; Munn, 1996, in Low, 2014, p.20) of dominant space production ideology critique that makes the process of change socially horizontal (Sherlock, 1998, p.219) – and note should be made here too of the role of relative expertism [2.5.1] in this as productive of the same horizontal process. Thus, reflexive and transformative potential was spoken of with a notion of caution, modesty and some tempering. Many spoke of projects 'planting a seed of an idea', 'that they'll continue with other programming, so things will start to happen there' (Big Car artist) – McCormack's (2013, p.12) exemplification again – for the functional and/or neighbourhood community to continue; or of reaching just a few, 'maybe it's enough that a kid in the houses opposite remembers it in ten years' (Art Tunnel Smithfield artist), based on an assertion that an aesthetic dislocation (Kester, 2004, p.84) of/in place will leave an impression on an emotional place memory (Giuliani, 2003; Lewicka, 2014, p.51; Rubinstein and Parmelee, 1992; Scannell and Gifford, 2010); and whilst accepting an impact in general terms, in particular terms, questioning what the longevity of that would be – in cognitive and material form – of splash interventions. Degrees of impact were also mentioned, i.e. did those directly involved as collaborators or protagonists experience more of an impact than those that might be voyeurs to the same, or could this either not be recorded or uniformly predicted? The function of this practice, through a reflexive and transformative lens, is to realise non-artist protagonist's agency as 'authors of their own lives' (Chonody, 2014, p.2), social practice arts diverging from community arts here with the enhanced degree to which such thinking is provoked (Jackson, 2011, p.290), with the added material-action change outcome of social practice placemaking. The performative experience leads to a suspension of the individual's identity (Kester, 2004, p.157) to join the group dynamic via the 'empathetic feedback loop' (ibid., p.77), enhancing peoples efficacy in place, projects allying to social activism and via performativity to aid the community conscientisation process of uncovering their priorities for change (Legge, 2013, p.78).

Place loss can also be positioned as a process of place change – whilst loss is discombobulating and emotionally wrought for participants, it can also lead to new creative actions and interventions and a deepening of place attachment and (more) active civic participation. The experience of place loss across The Drawing Shed and Art Tunnel Smithfield generated some ambivalent feelings, a 'mediating change function' (Proshansky et al., 1958, in Geiseking and Mangold, 2014, p.77) of a change in relation to place and thus place identity. Contra to place loss thinking, and informed by practices of duration of social practice arts, closure as experienced with The Drawing Shed was seen in a positive regard, part of a natural social practice arts process and aided through the process of exemplification (McCormack, 2013, p.12) and symbolic interactionism (Bulmer, 1969, in Milligan, 1998, pp.1–2) to envisage a future, positive, scenario. The art process in the social practice placemaking practice here had a 'mediating change function' (Proshansky et al., 1958, in Geiseking and Mangold, 2014, p.77). With Art Tunnel Smithfield, place loss also generated a positive outcome of deeper community engagement for protagonists, in other gardening or social projects. But place loss at Art Tunnel Smithfield was also seen to support the thinking that it destabilised place attachment (Livingston et al., 2008) and feelings of displacement (Bulmer, 1969, in Milligan, 1998, pp.1–2). There the forced 'emancipation' (McAllister, 2014, p.191) was experienced as psychologically negative (Twigger-Ross et al., 2003, in van Hoven and Douma, 2012, p.67; Weiss, 1973, in Marris, 1991, p.70), the place of Art Tunnel Smithfield having become an attachment figure (Scannell and Gifford, 2010, p.27) and its symbolic function still resonating in popular Smithfield discourse.

7.1.2.7 Located co-production

The awareness of non-participation was also indicative of the artist awareness of who was not participating, either wilfully or from lack of invitation (Beech, 2010, pp.25–6). In this role the artist acted from their position as artist to the ends of catalysing co-production (Gablik, 1992, p.6; McGonagle, 2007; White, 1999, pp.16–7). Whether or not the artist was the instigator of the project – or, in the case of The Drawing Shed in Wandsworth for example, the Council and developer – the artist's integrity created an open-loop 'socially responsive' (Gablik, 1992, p.6) co-produced system. In the case of The Drawing Shed in Wandsworth this was employed to deliberately undermine the intention of the funders for arts to act as a salver to the social issues of gentrification, an act of resistance and of social change (Madyaningrum and Sonn, 2011, p.358). This latter point was further qualified however, seen as the intention of social practice placemaking which may not be possible to realise fully from arts alone, the position of the artist remaining differentiated (Kwon, 2004, pp.139–40; Kravagna, 2012, p.242) from that of participants and the duration of projects not lending itself to deep change, in sum, the artists being 'deployed' in Kravagna's (2012, pp.240–1) 'aesthetically digestible' bites of limited longitudinal significance. The outcome of a broader understanding of community, as a micropublic (Amin, 2008) or micro-community (Kester, 2011, p.29) from social practice placemaking though changed how communities were approached and worked with, and it was hoped, agreeing with Silberberg (2013, p.52) that this would go some way to avoid the box-ticking of community participation in some instances. Social practice placemaking then was purposed to act as the intermedial (Jackson, 2011, p.14) to dissolve boundary and power positions, which in the urban place context, acted as a countersite (Holsten, 1998, p.54, in Watson, 2006, p.170) and spacetime (McCormack, 2013, pp.12–3; Munn, 1996, in Low, 2014, p.20) of creative activity and political resistance (Buser et al., 2013, p.624). Whilst these issues were left unresolved in an absolute sense for the artists – and deliberately so as the issues were seen as fluid and bespoke to context so generalisations were to be avoided – it is still a debate that the placemaking sector needs to be engaged with; here it can learn from the nuanced understanding of participation of social practice placemaking and widen the debate to other forms of placemaking.

7.1.2.8 Material change

Material change was not just to be an expected output in projects that were concerned with social practice placemaking, but also integral to the process, and seen especially in the work of Big Car and Art Tunnel Smithfield. This ranged from the design and installation of street furniture and proposing traffic calming measures (in the case of Big Car in Indianapolis' Far East Side and Garfield Park activity, respectively) to the 'issue of beautification', a common conversation across Big Car and Art Tunnel Smithfield, this relating to a conflation around social practice placemaking to beautify that place and the relative artistic and social merits of this. In this regard, the re-materialisation of the art object in social practice placemaking and its material-change concern was appreciated and seen as integral to this place-based and led intervention.

The relational objects had a currency of political resistance, from the dialogical aesthetic conversations between artists and residents that created a horizontal field and that led to the devising of the objects and on to their invocation of a time past before the reality of gentrification and a security in social housing, to the making of them in the street as a highly visible creative exercise in a countersite (Holsten, 1998, p.54, in Watson, 2006, p.170). Nonetheless, street level participation was experienced as politically counter to the scale of the multi-storey developments of the area's gentrification and the re-materialisation of objects in the social practice placemaking approach engendered a playful and also functional quality to the arts experience. Importantly, place-based participant agency was recognised as something that an arts-based approach could activate *through place*, but also as something that was already *in place*. Big Car described its process as being place-based of 'artist-led work in each location' (Big Car artist) and this accorded with the approach of the other case studies – not of place-*making* per se, but of place-*shaping* or a place-*led* approach to placemaking. Duration again was an issue in this context. The short period of time on the estate was worked with through high frequency of visits and their individual long duration, during which creative acts were repeated to become familiar to viewers, who then variously become protagonists, made secure in the experience by its then familiarity. Thus, whilst the activities were essentially ephemeral in the lifeworld of the estate, artists worked within these remits to become part of the temporary culture of the estate, enabling through their approach a crossing of barriers to participation for some.

The social practice placemaking of the case studies, as relational art and of a dialogic aesthetic, bonded self-to-self, self-to-others and self-to-non-human-agents, both the relational objects of the projects but also, of place itself as a spacetime generative-relation embodied experience of place (Munn, 1996, in Low, 2014, p.20). The interactionist social practice placemaking process was discursive, the individual placed in dialogue with people and place, their place attachment informed through collective social practice placemaking praxis (Dixon and Durrheim, 2000, p.32). The mutual place orientation and identification process (Sime, 1986) was contingent on peer-to-peer proximal relations integral to a social network of shared experience and need (Ainsworth, 1991, p.38; Gosling, 1996, p.149; Marris, 1991, p.70; Pahl, 1996, p.98), where, counter to adverse urban experience thinking, the 'collective self' (Carrus et al., 2014, p.156) was put before the individual self. This operated at cognitive, affective, behavioural levels in the micro-community (Kester, 2011, p.29) and in the case of the USA especially, this was observed as the place-group functioning as an extension of familial ties. Across all case studies, respondents exhibited Dixon and Durrheim's (2000, p.32) 'three senses of "insideness"', an embodied physical insideness of place, a social insideness of community connection, and an autobiographic insideness, an 'idiosyncratic sense of rootedness'. Social practice placemaking acted to counter a rootlessness or adverse urban experience *anomie* (Durkheim, 1893) to place (Sennett, 2012, p.257) and form over duration Tuan's (1974) accretion of sentiment to self, other and place through embodied (Merleau-Ponty, 1962) social practice placemaking. The 'authentic place', as a moral right to the city (Lefebvre, 1984), was both a driver and an outcome of the place attachment process in the case studies and further acted to counter *anomie* (Durkheim, 1893) and connect the hyperlocal to wider cultural forces.

The habituated nature of the projects, of regular and immersed participation, formed a phenomenological habituated experience of place that precipitated meaning making, the 'symbolic interactionism' of Bulmer (1969, in Milligan, 1998, pp.1–2). Social practice placemaking, as informed by social practice arts, was catalysing and galvanising of place attachment (Nowak, 2007) via engendering increased self-esteem and sense of self from place (Twigger-Ross et al., 2003, in van Hoven and Douma 2012, p.66) and transforming an individual identity to that of a collective one that generates new place cognition and attachment (Lilliendahl Larsen, 2014, p.326; Seamon, 2014, p.13). Increased duration in the project space increased proximated habitation, which acted to connect protagonists to the symbolic and functional place of past, present and future and to each other (Seamon, 2014, p.13), a place-bonding experience and process (Altman and Low, 1992; Rowles, 1983; Sixsmith, 1986). However, it was evident that whilst place attachment is attributed to being contingent on a sense of routine that generates a sense of 'rootedness' (Carmona, 2010, p.120, in Sepe, 2013, p.11), the visual and experiential splash of the case study projects, anything from the embeddedness of Art Tunnel Smithfield to the fleeting interventions of Lloyd

Park performances during The Drawing Shed's *IdeasFromElse[W]here*, broke the habituated experience of place (Froggett et al., 2011, p.95) but did not break attached bonds. Rather than a temporary intervention such as artist Mckenzie's (2014) *Fence 2014* upsetting a sense of proximated 'existential insideness' (Relph, 1976, p.55, in Seamon, 2014, p.14), it generated interpersonal exchange amongst those present to it, which in turn generated new memories of place, a constituent factor in positive place attachment behaviour (Giuliani, 2003; Lewicka, 2014, p.51; Rubinstein and Parmelee, 1992; Scannell and Gifford, 2010). Thus, social practice placemaking as a navigational practice and as part of a community turn (Hou and Rios, 2003) in placemaking functioned as a relational tool for complex social networks to traverse through place issues, fostering a group identity and consequently cohesion, a 'group-in-fusion' (Lilliendahl Larsen, 2014, p.326) which had adaptive potential.

Just as Lefebvre (1984) changed the focus of place from physicality to process, social practice placemaking does the same with placemaking, opening up the sector to social, embodied intersectional and fluid meanings of place, forming a new discursive framework of aesthetic knowledge for the placemaking sector that, in the claiming of social practice placemaking discipline, recognises the nuance of arts practices across the sector. This is contingent in the wider reflexive turn in the placemaking sector as evidenced too by critiques of practice, such as that of creative placemaking (Markusen and Gadwa, 2010 [a, b]), the recognition of practice turns such as community-driven placemaking (Hou and Rios, 2003) and the recognition of informal knowledges of place that places the community as leaders and is cognizant of the right to the city (Lefebvre, 1984) discourse. Thus, this text can be placed in a new placemaking approach of a constellated practice as seen in the placemaking typology [3.7.2] that crosses formal and informal urban arts practices with a complex understanding of how both inform the other, opening the sector to a wider social role. This is the 'serious complex work' (Shirky, 2008, p.47) that is free of proscribed and institutionalised formulae of practice and that locates agency in the micropublic (Amin, 2008; Lehmann, 2009, p.31; Silberberg, 2013, p.9; Sorenson, 2009, p.208). This places social practice placemaking in a parasitic role to the placemaking sector as a whole, and the placemaking sector in the same relation to the wider urban design and planning sector.

The case studies were concerned with the social and somatic experience of the urban lived experience and the phenomenological experience of co-production of that lived experience in a further feedback loop between people, place and action. For the artist, the process is one of non-apathetic knowledge production where they allow themselves to be changed by the co-producers of that experience that questions current and informs future practice, the 'reflexive co-mingling' of Critical Art Ensemble (1998, p.73) suited to the condition of the micropublic (Amin, 2008) formation of place identity. The social practice placemaking activity of the case studies was seen then to change the city terroir (Zukin, 2010, p.xi), drawing residents out of their habitual patterns of travel and experience to new places, the social practice placemaking activity being an arts destination like any gallery or museum may be. Social practice placemaking activity increased the cultural offer of an area, elevating it above purely retail functions for example, by employing a place-based approach that added to existing culture (Roberts, 2009; Silberberg, 2013, p.51) and that included community members in its production (Lepofsky and Fraser, 2003, p.132; Lydon and Garcia, 2015). This is the 'user-generated urbanism' or 'collaborative city-making' (Marker, in Kuskins, 2013) aligned to community-driven placemaking (Hou and Rios, 2003) and emplaced arts being functioned as the 'relational glue' (Nowak, 2007) of complex social networks that underpin civic engagement. Furthermore, the re-appropriation of space and its production anew self-defined and personalised that space, which affected the city terroir (Zukin, 2010, p.xi) by re-presenting that space materially and visually (Friedmann, 2010, p.154; Lilliendahl Larsen, 2014, p.336; Sime, 1986, p.60; Zukin, 1995, p.24). Here the arts in social practice placemaking were used towards the creation of a unitary urbanism (Debord, 1958, p.95) and as a site of political, social and aesthetic learning (Franck and Stevens, 2007; Kaprow, 1996/2006; Burnham, 2010).

7.1.3 Thematic three: social practice placemaking and social cohesion and active citizenship

This thinking refers to the third aim, to explore the role of emplaced performative arts practice in shaping social cohesion, arts and civic participation and citizenship. Social practice placemaking was seen to be emblematic and arising from an 'authentic' 'cultural phase' (Sepe, 2013, p.81) of self-determining active citizens.

7.1.3.1 Pluralist place politics

A number of theorists and practitioners believe that there is a positive causal relation between experience of emplaced social practice arts or social practice placemaking (and emplaced arts in more general terms) as an encounter art (Rancière, 2004/2006) and an effectual outcome. This problematises the assertion that social practice placemaking practices in particular produce a new form of 'insurgent' citizenship that activates the participant from passive to active citizen (Crawford, 1999; Franck and Stevens, 2007; Hou, 2010; Larsen, 1999/2006) which leads to a critical path of urban revitalisation and conscious-raising community development that has the power to agitate the macro and dislodges the dominant discourse (Beyes, 2010; Grodach, 2010; Lehmann, 2009, p.17; Miles, 1997; Sherlock, 1998; Suderberg, 2000) and which signifies an altered power relationship and the artist subject as a vehicle of liberation (Finkelpearl, 2013). Social practice placemaking though can counter this position by the political agency of its event space 'disturb[ing] the annulment of politics' (Beyes, 2010, pp.242–3) and maintain its capacity to articulate dissensus and consequently its ideological tools to argue that a creative work is inherently political (Ostwald, 2009, p.97). With participation, there is a need for interrogation of practice to validate its claims, but note should be made too that critiques may also be contradictory. Despite multi-vocal claims, there is an issue with a dialogic aesthetic practice in that it is contingent on all parties sharing a common meaning of discourse (Kester, 2004, p.85) which may have emerged from the subsuming of social relations in relational aesthetics, leading to standardised, homogenising practices (Miles, 2008) and experiences.

The micropublic (Amin, 2008) here may counter this. Rather than being based on an empathetic connected knowing between urban creatives, community is created by the recognition of difference, or the absence of a 'substantive identity' (Nancy, 1986/2006, p.56), individuals are connected to others through an accrual of singularities, an 'ontological sociality', a dialogical encounter that facilitates a partial, accumulative change in subject representation (Kester, 2004, pp.155–7). This is an intersectional process that recognises the social content from which individuals speak and does not seek consensus, redefining 'discursive interaction to empathetic identification' based on a sense of otherness (ibid., pp.113–8). It is an act of 'discursive violence' for any one person to speak for another (ibid., p.130) and there has been an accompanying 'fetishisation of authenticity' (ibid.) in that it is deemed that only artists with a prior intimate knowledge of the community at hand, i.e. that are resident in it, can have the right to work with that community. This is divisive, ignorant of a common consciousness and of empathetic connection and does not allow a position of outsider to be of active reflection to a community (ibid.; Kwon, 2004). However, social practice placemaking may support this fetishisation whereby artists will often be resident in the city or neighbourhood and will co-produce with the community. This 'authentic' position can be left unquestioned, or an 'outsider' artist dismissed immediately on this basis alone, as if a common practice was not qualification enough to work contextually with the community. The community-based catalyst of change artists of Mancilles (1998, p.339), if actually from the community in question though, 'are uniquely positioned to initiate community policy or programming that has far reaching effects'.

7.1.3.2 Active citizenship

> *'People think that places change but often it's your relationship to a place that changes'. (The Drawing Shed artist)*

The quote above illustrates how participation in social practice placemaking, whilst often having a physical output, has an agency that rests on the participant's relation to the place that they are active in. Place here incorporates socio-spatial relations as process and outcome and this following section will focus on active citizenship as an outcome of social practice placemaking and a positive place attachment relationship. Theory schematically views positive place attachment and civic participation in a linear and causal relation (Carrus et al., 2014, p.156) that has social cohesive and social capital outcomes (Nowak, 2007), and where high social capital begets further social capital in a mutually increasing and constitutive process. On the premise of place attachment being of a cognitive community identity, it has attributed behavioural neighbourliness characteristics (Manzo and Perkins, 2006, in Milhaylov and Perkins, 2014, p.63), and when activated through an arts-based process is a literal participation in 'creative action'

(Saegert, 2014, in Gieseking and Mangold, 2014, p.397; Carrus et al., 2014, p.156) of active citizenship (Lydon and Garcia, 2015, p.10; Schneekloth and Shibley, 2000, p.138) that has a place framing (Martin, 2013) activation that materially and culturally changes the urban realm (Bresnihan and Byrne, 2014, p10; Uitermark et al., 2012, p.2549).

Active citizenship sees citizenship as enacted, transducive, reflexive, discursive and collective, a wide definition of citizenship that social practice placemaking has a role in as constitutive of the ludic city (Whybrow, 2011, pp.14–5). It was evident that social practice placemaking participation as arts process and place framing (Martin, 2013) resulted in a sense of unity in an area (Unwin, 1921, in Nicholson, 1996, p.114); a territorial specificity (Anderson, 1986) from gathering as a micropublic in place led to a spatial meaningfulness (Castells, 1983) and a sense of connection (Lynch, 1981) to people and place, and this activated an active, and collective (Sepe, 2013, p.81) citizenship. This active citizenship took the form of a desire to socialise with neighbours and to talk about local issues (Mihaylov and Perkins, 2014, p.61) and in some cases this talk led to affirmative material action to improve place.

But this reading of a place attachment to active citizenship belies the complexity of the conscientisation to action process. Degrees of place attachment will differ from person to person and the ability to act on any degree of place attachment will depend on myriad social, cultural, economic and political concerns to the individual and the group. Seamon's (2014, p.16–9) place interaction through to place intensification is not necessarily a linear or continuous path – people may join or leave at different stages, or not progress past one stage at all; in the case studies, for some, joining a community mural painting project for a day, as in the case of Big Car *Lilly Do Day* (2014) project commission, was enough to state a desire to be doing something to help the city and to feel active in *place realisation* and *creation*; for those at The Drawing Shed's *Live Lunch* (2014) [n.18], conversation was activating as place interaction, identity and release; and in Art Tunnel Smithfield, its 'capital P' politics vis-à-vis vacant land led to individually and as part of a city-wide momentum, activist-located place intensification (ibid.) Thus, active citizenship does not have to be action orientated but can take more subtle and nuanced forms (Ainsworth, 1991, p.43; Carrus et al., 2014, p.155; Friedmann, 2010, p.155) with cognitive aspects of networked community identity, a sense of community enacted as sense of place, and behavioural aspects enacted as neighbourliness (Cattel and Evans, 1999, in van Hoven and Douma, 2012, p.74; Livingston et al., 2008; Manzo and Perkins, 2006, in Mihaylov and Perkins, 2014, p.63; Shumaker and Taylor, 1983). Rather, this active citizenship 'process', if it is a process for all, is closer to Mihaylov and Perkins' (2014, pp.68–9) psycho-behavioural components of collective efficacy as empowerment; a sense of community as social bondedness; giving and receiving neighbouring help; and citizen participation as involvement in volunteer, community or political groups for example.

The political dimension of place attachment lies with an active civic subjectivity that is capable of reflection and transformation through creative activity. Wherein social practice placemaking acts as an 'area catalyst task' (Kearns and Ruimy, 2014, p.48) of neighbourhood strategies (Guattari, 2000), its political dimension lies with an active citizenry that is capable of alternative urbanisms (Carrus, 2014, p.156; Saegert, 2014, in Gieseking and Mangold, 2014, p.397). Site is part of the social aesthetic too, a form to work through urban issues as a place-framing practice (Lehmann, 2009, p.18; Klanten et al., 2012) and the discursive production of place as a basis for local politics (Martin, 2013, pp.91–100) and that subverts the power/weak dialectic of formal/informal space through active citizenship and the material social practice placemaking practice. The operative dimension of bonding and bridging ties (Mihaylov and Perkins, 2014, p.69) navigates a complex socio-spatialecology and a meaning of citizenship (Holmes, 1993, p.208), performed through neighbourly and civic place-protective behaviour (Carrus et al., 2014, p.157–60).

The social practice placemaking arts ecology can inform the progressive politics ecology, understanding as it does, that power and knowledge will be held in different forms and places than previously, and conventionally, thought. There is a perceived and operational binary of top-bottom, planner/non-planner in the planning sector, a binary that compromises and devalues all constituents, and social practice placemaking demands new ways of working. Where notions of creativity in planning are (deliberately) 'fuzzy' (Lilliendahl Larsen, 2014, p.330), the enhanced knowledge and awareness of types of art across placemaking is a means to subvert this, and extend practice. Social practice placemaking processes question who determines and is involved in the creation of the urban form and on what cultural valued benchmarks these are premised on (Froggett et al., 2011, p.91; Sennett, 2012, p.53; Stern, in Lowe and Stern, in Finkelpearl, 2013, p.146; Till, 2014, p.168) and is proactive in the articulation of the benefits of a socially engaged and collaborative urban realm design and planning (Miles, 1997, p.189). Here, social practice arts and social practice placemaking hold citizen tactical power as the response to the strategic power of the government (Lydon and Garcia, 2015, pp.9–10), subverting the idea of a citizenry as occupying weak space (Lilliendahl Larsen, 2014, p.319). As seen in the data, planners will 'go rogue' and force through projects via loopholes or turn a blind eye to interventions. As a tactic, this may have worked in the examples cited, but this is not a sustainable strategy to develop the city, politically or materially.

7.1.3.3 Generative planning

As an embodied experience, the production of space is mutually generative between body and space and is consequently re-producing (Lefebvre, 1984, p.170), a facet of the process that lends itself to a generative, co-produced, planning (McCormack, 2013, p.2) and the production of the authentic place of right to the city (Lefebvre, 1984) and place attachment discourse that, with its reflexive component, would create the ludic city (Whybrow, 2011, pp.14–5). Social practice placemaking widens the social and intellectual concepts of planning, thought by Lefebvre as narrow and trivialised (Boyer, 2014, p.175), and diffuses specialisation. There is a role dissonance (Sennett, 2012, p.203) in this planning milieu of strict professional classes; instead all those in the process, community protagonists included, are urban co-creators acting on a platform of relative expertism (Brown, 2014, p.175; Nicholson, 1996, p.110). For planners, an understanding of place attachment will offer a new lens on community and its values, and how to work with these through urban design, and work through a developmental relationship model for city administrations, where positive relation to place and active citizenship is understood and worked with. This text, and the author (Courage, 2015), calls for planners to engage with place attachment to aid urban design and planning processes and recognise also the value of the social practice placemaking practice in uncovering the lived expertism of residents (Till, 2014, pp.150–1), informed by a collaborative city habitus model of a holistic ecology of sense of place, locality, and social, environmental, cultural, economic and political factors. The outcome of arts in the public realm may be more politics with a small 'p', but no less significant for what can be seen in Dublin is the creation of Guattari's (2000) neighbourhood strategies where local groups at the grassroots are fundamental to transforming society. In this sense, the social practice placemaking of Dublin at this time, as exemplified by Art Tunnel Smithfield, was as much about a co-operational social structure as a critical spatial practice (Rendell, 2006) as material change in the urban form: the art installation of the garden presented itself as the object and site of experiencing as well as the means to start a process of reflection and tactical response. This process ran counter to Dublin's enclosure (Bresnihan and Byrne, 2014; Tonkiss, 2013, p.176) towards a commoning of vacant land (Tonkiss, ibid.) that worked to place the development agenda in the hands of the public (Roy, 2005, p.150). But this is a susceptible process and here lies the challenge inherent in the notion of Dublin's new urbanism. Smithfield may still be emerging as a liveable area of the city, a 'proto-urban space' (ibid., p.127) with increasing cultural programming, consumer activity and footfall, but it is still vulnerable, as the closure of Art Tunnel Smithfield and its return to dereliction attests.

7.1.3.4 The neoliberal dilemma

There is a dilemma when it comes to participation in social practice arts and social practice placemaking and the political. The expediency and relative low-budget cost of a social practice placemaking active citizenship sits in a conflicted position by its co-opting by city administrations, as exemplified by *Beta Projects* [n.19] in Dublin and the co-commissioning of arts activity by the Nine Elms developer, and in the wider placemaking sector by city administrations engaged in their own tactical urbanism projects and attracted by 'off the shelf' participatory placemaking and creative placemaking schema such as PPS's *Lighter, Quicker, Cheaper* (PPS [b]). Arguably, such schemes are counter to the social justice concern of social practice placemaking – where social practice placemaking works with a 'politics of difference' (Young, 1990, in Gieseking and Mangold, 2014, p.251); such schemes can, if deployed wholesale, result in homogenised placemaking strategies and outputs. The Drawing Shed was working ideologically to counter neoliberal planning and regeneration policy and schemes; Art Tunnel Smithfield and Big

Car were working to various degrees of ownership, with real estate, and in the latter's residential regard particularly, with wealth transfer (Roy, 2005, p.153). All though are examples of active citizenship indicating an increased confidence in the citizenry of itself to have a dynamic and efficacious role in 'microtopian politics' (Bishop, 2012; Rancière, 2004).

In the built environment context of urban design, planning and regeneration, participative arts practice is a symbolic violence, a neoliberal attempt to redress issues of exclusion; this cannot be successful in relation to goals of social change as it is a mechanism created by the dominant political structures to meet its own ends (Hoskyns, 2005, p.119). This participation is conceptual, scientific and managerial (Richardson and Connelly, 2005, p.78). As in the arts sector, participation in this context is not intended to fully engage the community and has become standardised and bureaucratised (Hou and Rios, 2003, p.20) – this is the participation of consultation, not of co-production. For both Bourriaud (1998/2006, p.165) and Bishop (2006 [a]), as soon as art comes close to activist, it sacrifices its integrity and legitimacy (Kester, 2011, p.33). This is where social practice placemaking can be accused of being deployed as a strategy of cultural politics where artists play a tactical role in gentrifying regeneration to reinforce an image of an area to create a new one (Lossau, 2006, p.47) supporting urban boosterism and city marketing (Hall and Smith, 2005, pp.175–6; Zebracki et al., 2010, pp.787–8), part of a 'new instrumentalism which seeks to use arts and culture as a tool for furthering a range of broader economic and social goals' (Gilmore, 2014, p.13) and the homogenising 'single logic' (Best, 2014, p.296) strategies of liveability aligned to Public Realm placemaking and creative placemaking. In this setting social practice placemaking becomes accountable against its function and but all too often ends up serving the existing forms of space occupation i.e. of received notions of work, rest and play and a civic beautification (Deutsche, 1991, in Miles, 1997, p.90; Kwon, 2004, p.65), at best 'diverting decoration' and at worst an 'empty trophy commemorating the powers and riches of the dominant class' (Kwon, 2004, p.65). Kester (2011, p.30) though views the social practice arts artist's role as to 'create alternative models of sociality to challenge the instrumentalizing of human social interaction characteristic of a post-industrial economic system'. Social practice arts has a 'key role to play in placing the arts at the centre of civil society', achieving this by an engagement grounded in a social and/or civic purpose and relational engagement practices (Froggett et al., 2011, p.102) by the inclusion of the community voice (Madyaningrum and Sonn, 2011, pp.359–60). This is a practice of participation established on Rancière's 'rupture' where social and aesthetic hierarchies are made horizontal (Jackson, 2011, p.52) and formal political participation is replaced by a more localised, directly relevant participation with more immediate and tangible results (Bishop and Williams, 2012, p.138).

The case studies illuminate the artists' conflicted position with regards to their relation to neoliberalism. The artists were aware they were operational in culturised (Zukin, 2010, p.xi), contested, increasingly privatised, exclusivised spaces of an invited pseudo-participation (Petrescu, 2006, p.83) that has interests in using the spectacle of art to distract and sanction a proscribed dissent and foster a passive citizenship (Bishop, 2004, in Whybrow, 2011, p.29; Fernando, 2007; Gilmore,

2004, in Gilmore, 2014, p.16; Ostwald, 2009, p.95; Tonkiss, 2013, p.104). Public and emplaced art, with social practice placemaking included in this, is co-opted in this scenario into neoliberalist urban policy to fulfil the remit of a de-centralised and disinvested state (Lydon and Garcia, 2015, pp.13–4) and to allocate resources through funders that are deemed deserving (Bresnihan and Byrne, 2014, p.3) – as seen explicitly in the case in Dublin and Dublin City Council/*Comhairle Cathrach Bhaile Átha Cliath* (DCC) [n.20] support of certain projects enacted by its public realm protocols. This co-opting by policymakers is a conflicted position for artists and the case studies themselves exemplify ways in which social practice placemaking can variously avoid, agitate or negotiate with policy: in Dublin, this was by artists setting the agenda for change; for Big Car, this was by taking the means of cultural production into their own ownership; for The Drawing Shed this was by subverting the message of policy into reflexive questioning. The research findings though accord with the discourse that art in the city is a site of exemplified, rhetorical critique and resistance to such neoliberal culturised processes (Debord, in Bishop, 2006 [a], p.12; Kester, 2011, p.226; McCormack, 2013, p.12; Zukin, 1995, p.264), social practice placemaking and its active citizenship being of Rancière's (in Jackson, 2011, p.52) 'rupture'. Where the data also points though is in the social practice placemaking artists' embracing of the limitations of the political agency of their practice, owning their positon as exemplifier and rejecting the notion that they have a responsibility to work on behalf of meso or macro bodies to resolve social, structural issues. The 'neoliberal dilemma' is complex, contextual, subjective and ambivalent – one person's arts-led regeneration is another's culturised creative placemaking (Markusen and Gadwa, 2010 [a, b]), based on that person's belief in the arts as intrinsic or instrumental (Froggett at el., 2011, p.9).

Learning from ambiguous urban experience thinking and ambivalent pragmatism, one can see how this is too complex a debate for an either/or response to the neoliberal dilemma: while social practice placemaking phenomenologically is situated in a political space, the degree to which this is explicit/implicit varies; projects can be willingly complicit, unknowingly co-opted and deliberately critical of and subverting of the neoliberal discourse for example. These positions may fluctuate in the course of a project or locale, or in their interpretation, regardless of intent. The research saw the actions of the social practice placemaking artist as consciously parasitic to the hegemonic function of space and cognizant of their subversive agency. None of the case studies, no matter how oppositional the project may have been to the macro or the meso, could avoid some level of interaction with the macro – the projects could not, and nor did they necessarily choose to, act in a political or funded isolation. It was also known that councils did not have the resources to act alone in the public realm and that, in their call to use the arts, the arts could also use this platform in an agitating way. In this, it was recognised that there was a need for inter-micro, meso, macro dialogue; social practice placemaking was seen as a means for doing this but only successful if its models of practice and process were initiated from the bottom up; if these interventions come from the top down, discursive agency is rendered redundant as part of a pseudo-participative (Petrescu, 2006, p.83) model.

7.2 Limitations and expansions

The following section details some of the limitations of the research and its text and questions that remain unanswered in this project. It goes on to offer thinking on the new knowledge produced from the research project and signpost implications of the findings for the placemaking and arts sectors. It closes with sections on directions and recommendations for future research emerging from this project.

7.2.1 Limitations of the text

This text holds a city-centric aspect to study, falling perhaps into the trap of 'urban triumphalism' (Schmid, 2014, p.4) that privileges a positive view of cities over suburban and rural, and it rests on the placemaking and art discourses of the same. The research project took place in the western, northern hemisphere, in developed nations, and whilst a degree of cross-cultural study was possible, it is noted that this is limited in comparison to the potential that a wider global siting could offer. It also uses the UK/European *social practice arts* term, not the US *socially engaged arts* term, and it references artists from both fields – the reasons for this have been articulated but this is still a conceptual limitation of the study, although it is anticipated that the thinking will still be able to travel across and through practice. As part of the nascent focus on arts knowledge in the placemaking sector, it is just as important to look at what concords as well as discords in thinking, and to look at a breadth of practices, which this research project has through including examples of social practice placemaking, creative placemaking and community art practices. Whilst this is the narrow focus of this research, attention in the wider placemaking content also needs to be given to other forms of placemaking, as identified in this research, such as Public Realm placemaking and participatory placemaking, and addressed via a similar lens for comparative analysis and learning.

The methodological depth of the project was limited by the restriction to duration adequacy, place embeddedness and circumstances in the moment that limited the social opportunities with subjects to enhance researcher–subject relatedness and social relationship understanding, overcome vernacular language barriers and verify inter-relational social consensus. Data collected was qualitative, not quantitative, partly due to best-fit meeting of the research aims but also the capacity of the researcher and the timeframe of the research project, and paucity of available quantitative data held within the projects themselves (though this is not to criticise the case studies for this, as only surface-level quantitative information will be asked for by funders and neither do they have the capacity to collect data outside of these requests). The research did not focus on data relating to the relative homogenous or heterogeneous making of the project demographic, for the same reasons. The research project was bounded durationally by the practical issues of levels of funding, and the duration of the PhD period itself. The closing of Art Tunnel Smithfield, whilst opening up a new angle of research and a valuable one for the generation of new knowledge vis-à-vis project failure and closure, limited more than with the other case studies the time that the researcher could spend on-site, as this occurred over the winter months with limited activity in the garden.

Gaps in the text are threefold. Firstly, it is known that the role and place of the architectural profession has been largely grouped under an urban professional class heading in this text [3.4], but that this is not representative of the special role, as creative urban professionals, that architects do and can play in placemaking as a whole, and social practice placemaking specifically. Architects played key roles in Art Tunnel Smithfield in particular and the researcher is aware of numerous examples of architects operating as social practice placemakers, from the research process of the contextual review, the field data collection, and their own professional practice and academic activity. Thus, whilst the distinct role of the architectural professional and the relation of architecture of placemaking was nascent in this text, this is a field deserving of its own study. Secondly, admitting that social practice arts and social practice placemaking are, on the whole, longer duration projects, creative practices most suited to longitudinal evaluation, the research suffers for the limit imposed by the academic and financial timeframe of the PhD endeavour and the comparative short amount of time for data collection in particular. Thirdly, the data collected was qualitative: there is a need for quantitative data pertaining to social practice arts and social practice placemaking to create a metrics-based evidence base for the practices, of use to the sector primarily for its funding and policy implications. Specific demographics of the participants were outside the scope of this study, but, as a practice concerned with marginalised voices and urban realm power dynamics, this is an omission from a study of social practice placemaking. Lastly, whilst other forms of placemaking have been presented in descriptive and critical form here, their in-depth study was outside the remit of this research – for a placemaking typology to be holistic and fully functional, similar research as undertaken here needs to take place around Public Realm placemaking, participatory placemaking and creative placemaking and their strategic, tactical and opportunistic modalities.

7.2.2 Unanswered and alternative research questions uncovered in the research project

As this is a text of western, northern hemisphere, and of developed world cities, and as differences in placemaking practices are known to occur outside of this global region (Friedmann, 2010, p.150), it was questioned during the research what these differences are, as specific creative practices and process. These differences have a value as studies in their own right and also as a cross-cultural study to inform the sector knowledge exchange practice and its professional development globally. This would also begin to answer questions around the demographics and power relations of types of placemakings. This scale of the city locales was chosen to meet the aims of the project, but there remains a question as to the scale of city-size and its density too, and how this affects social practice placemaking; a wider global context could also include suburban and rural areas. The significance of territories, especially in the US case study, was not anticipated, and thus, whilst these findings have been included and critiqued from a perspective of arts participation, a reading of this data through urban geography theory was omitted. It was keenly observed that the US case study was in an urban setting of large areas of unused to vacant land, at a scale that was absent totally in the UK, and whilst seen in Dublin on a city-wide scale, it was only as small pockets of land. It was mooted during the research if the scale of the arts-led regeneration in the US would be possible in the setting of the other case studies – and notionally considered not, for geographic, socio-economic and bureaucratic reasons. Less bureaucracy was observed in the US in terms of opening up space for use or purchase by the arts and its funding and philanthropic system appeared more positive to capital purchase than in the UK and Ireland case studies. The same question could of course be asked of non-Western placemaking scenarios.

With regards to an informal (visual) aesthetic that is seen across social practice placemaking projects globally, the relation to the normative culture and the homogenisation effect of prescriptive user-group needs and actions (Rappaport, in Fernando, 2007) is not yet addressed in placemaking literature; is the fact that so many urban interventions 'look the same' an indication of a common global movement or is it an indication of a neoliberal homogenising urbanisation? Looking from one placemaking project to another, one will see common tropes of tactical urban design. This though is also indicative of a global field of practice that can be made bespoke. The relative differences between social practice arts and socially engaged arts, their academy and practice differences, also raised speculative questions on their differentials to the social practice placemaking practice, though the researcher took their lead from the artists themselves in seeing a broad similarity of intent that cross-cut difference. The legacy of social practice placemaking projects was questioned by the interviewees and has been in this text; the text lacks the longitudinal duration to report on this, so this remains an unanswered question. Key themes relating to the case studies here, for a longitudinal study, pertain to individual and community transformation and enacted citizenship and its long-term effect on planning and policy.

The major question that has gone unanswered however, is that of placemaking as a social movement, as it is often and increasingly referred to in its popular media (PPS [a]). From one perspective, the concerns of social practice placemaking are too microtopian (Bishop, 2006 [b]) to effect wider social change; any 'trickle-up' (Burnham, 2010, p.139; Silberberg, 2013, p.10) to policy is limited; there is no established causal relation between aesthetics and political movements (Amin, 2008; Kaji-O'Grady, 2009; Kester, 2011, p.224; Ostwald, 2009, p.94); and the long-term legacy of the aesthetic dislocation is limited and any change affected can only be at the localised level and not of a wider citizenship (Gosling, 1996).

To take a positive view of citizen agency in the neoliberal city, through social movement theory, 'groups find ways of creating their own spaces of identity and solidarity through acts that both comply with and contradict certain neoliberal forces' (Long, 2013, p.55). Furthermore, and as shown above, emplaced arts are implicated in this: social practice arts is a means of mobilising social movements through a raised consciousness, to 'decode' the art process within its cultural loci (Larsen, 1999/2006, p.173), challenging the macro, dominant structures and negative social representations of the producing community (Murray, 2012, p.257).

The public also has an interest in arts in the public realm and an 'enthusiasm for community participation in city enhancement' triggered both by a feeling of disillusionment in the administration and/or by 'the growing realisation that government can't and perhaps shouldn't have to do it alone' (Legge, 2012, p.33) – a reiteration of the neoliberal dilemma itself of course. Here, Bishop's (2012, p.258) microtopian is the agency of social practice arts in the performative and the increase in expressive conscientisation and efficacy and in the social networks of the functional and elective communities (Gilmore, 2014, p.17; Nicholson, 1996, p.116) as prefigurative models of mobilisation (Tabb, 2012, p.202). Furthermore, placed in the right to the city (Lefebvre, 1984) discourse and the manifestation of authentic places, social practice placemaking could be said to connect the individual and the community to wider cultural forces. As a social practice arts informed practice, could it be said that social practice placemaking has the same potential?

To oppose this positivist stance though, Kester (2011, p.224) questions if such politics is that of 'pocket revolutions', 'isolated moments of transgression or resistance that will never coalesce into a coherent whole capable of toppling the vast apparatus of neoliberal capitalism' (ibid., p.224) based on a 'dialogical determinism', 'the naïve belief that all social conflicts can be resolved through the utopian power of free and open exchange' (ibid., p.182). This perspective leads one to question the potential of political agency as an outcome of emplaced arts as rather than side-stepping formal politics via flexible citizenship, it conversely perpetuates the urban discourse it set out to agitate state (Bishop and Williams, 2012, p.138). Kester (2011, p.207) again: 'The only way to avoid complicity with the existing (and implicitly monolithic) capitalist system is to abjure merely "cultural" interventions and instead pursue direct, revolutionary action. Anything less will only serve to legitimate the system'. Furthermore, and to undermine flexible citizenship further, 'The move toward localism is driven by expediency more than ideology' with business and city administrations instituting this as it's found to be cheaper and works better (Brown, 2014, p.176) and being proved correct in thinking it can appeal for urban citizen equitability without changing the structural status quo.

The findings of the research begin to signpost questions around social practice placemaking and social movement thinking. For example, whilst it was seen that the re-appropriation of space does not equal its control, as seen with Art Tunnel Smithfield in particular, it was also seen that for social practice placemaking to control land use is to take control of the means of production of regeneration and become the developer, as seen with Big Car. Any hyperlocalised microtopian concerns were also dissolved in the crossing of territories across a city. It was questioned in the field if the momentum and ethos of Big Car was a social movement of sorts for the city of Indianapolis; Big Car had a manifesto, modus operandi, incorporated multiple viewpoints and worked across multiple locations, engaged in collective (social) action, and was seen to be transforming the sociopolitical characteristics of Indianapolis. This is presented here as an unanswered question and the study of placemaking through the prism of social movement theory is required to answer it.

7.2.3 New knowledge contribution

The text has contributed new knowledge to the arts and placemaking sectors, principally through the investigation of social practice arts in place, and come to understand and term this social practice placemaking. Uniquely, this text conjoins in a transdisciplinary study arts theory and urban and placemaking theory. Thus, the text has addressed the identified gap in current research pertaining to the arts in placemaking and ways into placemaking participation and out through this to active citizenship via place attachment.

Through qualitative case study research the research project interrogated social practice placemaking projects as a placemaking practice informed by social practice arts and signals to the placemaking sector how arts practice can be understood in place as having a rich heritage of practice and a vital contemporary practice. The text gives a new lens into social practice arts, of the informal aesthetic, and uncovers the detailed arts practice and process of the identified social practice placemaking practice, in this process creating a placemaking typology [3.7.1] that delineated between more diverse forms of placemaking than the umbrella terms used to date of placemaking and creative placemaking. The political, economic, social and health impacts being asked of and accredited to placemaking (Silberberg, 2013, p.2) is resulting in a cumulative confusion augmented by the competing demands made and expectations of placemaking (Fleming, 2007; Markusen and Gadwa, 2012 [a], [b]); an understanding of the scope of each is essential to manage expectation and expedite clearer and more effective outcomes and outputs measuring.

As such, this text firstly aims to address the gap in knowledge in the placemaking sector pertaining to an understanding of the practice and process of a social practice arts–informed placemaking – *social practice placemaking* – which would notionally be otherwise termed creative placemaking, and in doing so, the text delineates between different types of placemakings, along art practice and participation lines. It also aims to fill the research gap identified by Chonody (2014, pp.30–1) of how mural art can be used in social practice, over that of community art practice. It secondly aims to bring a placemaking knowledge into the social practice arts and wider arts sector, to galvanise the social practice arts field of knowledge of working in place, with material processes and outcomes, to inform the wider arts sector of this ever-growing practice and to extend its knowledge of arts in place and in the public realm. Thirdly, the text aims to communicate to decision makers in city administrations, planners and funders the detail of the practice of social practice placemaking to which they may not be aware, to aid dialogue between parties and to help inform policy making and grant giving. In this latter regard, where the micro of the social practice placemaking project meets the macro of policy and the meso of city administrations and funders, the text also aims to contribute to debates on the role of art in place relating to an instrumentalised use of the arts in the public realm or as social work (Bishop, 2006 [a]; Froggett et al., 2011; Hamblen, 2014; Jackson, 2011; Miles, 1997) and its position vis-à-vis a neoliberal administration (Brown, 2014; McAllister, 2014; Sennett, 2012; Zukin, 1995). What this text terms the neoliberal dilemma of emplaced arts pertains to the neoliberal rhetoric of social inclusion and the co-opting of arts-in-place and emplaced arts to the ends of market forces and administrative institutions (Kwon, 2004). It also aims to address the gaps in knowledge pertaining to 'a host of overlapping and poorly defined terms' (Carmona et al., 2008, p.4) used in the arts and placemaking sectors by both interrogating those terms and presenting a placemaking typology.

7.2.4 Implications of the research

The implications of the research sit broadly as arts sector implications and urban and placemaking sector implications. Depending on the lens this research is read through, implications could be multiple; this section focuses on four core implications of the research, as grounded in the main themes of this text.

With regards to participation, the faceted model of participation and one scaled on degree of participation and extending out of it to co-production, is one that has learning potential, to both understand and develop creative and public engagement practices and processes. Understanding how people participate and barriers to participation then leads on to the composition of the participating group, their recruitment, and motivations to join, and the acknowledgement that this is an exclusory process often based on demographic assumption (Gosling, 1996, p.148). Thus, not only can this study benefit an understanding of the nature of participation, but also – and herein the second implication of the research – the understanding of the community as of varied form, as functional and neighbourhood-based (Nicholson, 1996, p.116), and further nuances in this as communities as of *mythic unity* (read, assumed intersectional accord), *sited* and both *temporary* and *ongoing invented communities* (Kastner, 1996, p.42). When professionals enter the 'community', it can be left to them to realise the types of communities found there – as seen in the case of Big Car's (2014) *Galería Magnifica* and The Drawing Shed on the Nine Elms estate, both funders working from a desire to 'work with the community' but not necessarily knowing at first who that community is or how it is formed and where. This places the artist or urban professional in a patronage role, ascribing categories from a pragmatic demand of the work in hand. An understanding of the complexity of communities would help before and during such situations, and also undermine the implicit advantage-position of the artist or urban professional knowing what is best for the community. Rather, the community would be viewed as expert in their own lives (Chonody, 2014, p.2; van Heeswijk, 2012) and capable of having a critical distance to their lived experience (Kastner, 1996, p.42). Furthermore, if social practice arts and social practice placemaking are to substantiate any claims to be a paradigm shift in arts practice, and take a place in the urban realm and planning (Mozes, 2011, p.19) and also avoid being subsumed into claims made of and from public art discourse (Kelly, 1984), a discourse needs to form that takes its cues from art, architecture, urban, sociological and psychological thinking. Another implication is that of the transformative potential of social practice arts or social practice placemaking. To validate and qualify the assertion that there is a transformative potential in these art forms, empirical, theoretical and cross-disciplinary study is necessary which positions both practices in relation to individuals, community and institutional protagonists, as well as wider social structures (Froggett et al., 2011, p.105).

A further implication of this research is that of evaluation of placemaking and the potential of this research to add to the creation of a 'public artscape' critical discourse (Mozes, 2011; Ostwald, 2009; Zebracki et al., 2010). Moss (in Mozes, 2011, p.19) states that artists are co-opted into projects as the shortcut to placemaking success, but that what this sector needs is an evidence-based evaluation of practice. If placemaking is to substantiate its claim to be the paradigm shift in the urban and planning realms, it has to avoid being subsumed into claims made of and from Public Art discourse. Thus, it requires its own substantive critical discourse, of which the placemaking typology here aims to be part of. There are implications of the placemaking typology for the urban sector, including for artists, architects, planners, policymakers and communities around how it can deepen an understanding of the relative agencies of the various modes of arts in place; how it may advance placemaking practice as a whole by its use to better understand differences and similarities between placemakings within the placemaking sector; and how it can be used by the placemaking sector to communicate its practices to constituent stakeholders in the community, arts, policy, urban design and masterplanning for example.

Furthermore, when it comes to evaluation of an exemplified practice, the measurement of metrics such as attendee numbers 'is to miss the point' (Froggett et al., 2011, p.103) as impact and legacy will be more profound and affective in individuals, groups and the space/place. Both research and evaluation need to focus on the effect of emplaced arts and inform planners and policy makers who are concerned to reposition the arts in relation to other cultural fields such as sports, education and health (ibid., p.105). The funded sector processes however do not lend themselves to such longitudinal, qualitative evidence collection. Measurements of success are more likely to be qualitative than quantitative which can pose a problem for the short-term funded, limited-capacity arts or community organisation or lone artist (Sutherland, 2010, p.179). Thus emplaced arts practice should not be measured by 'reductionist conceptions of "impact"' but by long-term and complex effects, such as the motivations of those in the project under question (Stern, in Lowe and Stern, in Finkelpearl, 2013, p.151) and if participants have changed, how so and how people have responded to the project (Chonody, 2014, p.202) as well as the process of personal and community transformation and influence on professional practice for example (Froggett et al., 2011, p.9). The art process and outcome can act as a method of evaluation too as inquiry research and interpretation (Chonody, 2014, pp.210–1). For Lowe, 'daily life is the art. … It is a project that actively tests hypotheses on the world' (Finkelpearl, 2013, p.353).

7.2.5 Directions for future research

Principally, as stated in 3.7.1, as the placemaking typology is presented as developmental, it is hoped that it will be reviewed in the field through its application in both arts and placemaking practices. It aims to be used by both academic and professional practitioners, and the non-artist/non-professional stakeholders from the micropublic that comprise project cohorts. It is suggested that case study and participant observation methodologies are developed further as a means to research social practice arts and social practice placemaking projects, but that added to this is the 'thinking-in-doing' (The Drawing Shed artist) aspect of this for the practitioner–researcher [1.6] in the field, as practice-based research, as nascent and informally explored in the researcher's *Thinker in Residence* role with The Drawing Shed and Big Car. The role encapsulated a co-production of knowledge in the moment, necessarily undermining and repositioning the externalist and monological or binary principle of the researcher in the field, conjoining theory and practice in the research endeavour. It is also suggested that a method such as Q methodology (Watts and Stenner, 2014) is also brought into the research-in-practice field as a way of introducing aspects of quantitative data collection and analysis, to the benefit of evaluative means and evidence development in the sector.

Emerging in particular from the research as a site worthy of special attention – but outside of the scope of this project per se – was that of social housing and the housing estate. It was recognised that the terroir of this setting was differentiated as a political and social locale, and the arts responses to it, equally specialised. Housing estate arts interventions have a history, such as with the *Gorbals Art Project*, Glasgow [n.21] and Mckenzie's *LUPA* [n.22], and both their threat, as with Zuloark [n.23] in Madrid, and their destruction, as with *Slipstream* (2010) by David Cottrell [n.24], are the concern of artists in global urban locations. As such sites become the frontline of gentrification – as seen here with the activity of The Drawing Shed, but also in London with the politically contentious redevelopment of Ernö Goldfinger's grade-II listed Balfron Tower in the London Borough of Tower Hamlets, where artists have been accused of political complicity by the social pressure group Balfron Social Club [n.25] – this spatial, artistic and political setting warrants deeper investigation.

It has been stated that placemaking is not a universally inclusive process and the demographic membership of participants in projects is a noted issue (Arts Council England, 2011; Hermansen, 2011; Mici, 2013; Nicholson, 1996; Puype, 2004, p.300; Roberts, 2009; Schneekloth and Shibley, 2000) – the 'low income placemaking' of the Placemaking Leadership Council [n.26] grouping accords with this view [3.3], also viewing large-scale public realm projects as excluding minority cultures and exacerbating cultural and social difference. Within the professional placemaking sector too, some may be excluded from types of practice. The experience of placemakings will not necessarily be homogenous – different actors will have different experiences (Zebracki et al., 2010, pp.790–1) and this is an urgent matter for research.

Furthermore with regards to the site of study, this text was of a western, northern hemisphere, and developed world cities setting; it is known by the researcher that practice differs in South America, Africa and Asia for example – in terms of sites of intervention, political discourse and creative responses – and these locations have informed the global placemaking sector, as in the case of Wikicuidad [n.27] for example, whose urban guerrilla interventions, such as painted zebra crossings, have had a global influence on tactical urbanist interventions. Standalone as well as cross-cultural research is a required future direction for the placemaking sector to expand sector knowledge and understanding as well as inform popular media on the same. For placemaking too, an investigation of its social movement agency is required. Social movement praxis was not the central concern of the text but as it emerged as a site of interest and proved a useful angle into the considerations of the text, it is thought to have further value for research. Furthermore, with the generally positive stance and reportage of placemaking in the popular and sector media, and with the 'growing sense of political renewal around the world', which 'animates the remarkable profusion of contemporary art practices concerned with collective action and civic engagement' (Kester, 2011, pp.6–7), the social movement claims of emplaced art and placemaking need to be interrogated.

Lastly, it is recommended that focus is given to the evaluative criteria and frameworks of social practice arts and social practice placemaking projects. With regards to evaluation, generic evaluation methods and questions are not able to capture and communicate the innovative nature of social practice placemaking. Such evaluation focuses on the features of a project, not its characteristics; does not take into account other factors attributable to social practice placemaking's success outside of itself, or subsumes them in the data; and does not generate enough data to create indicators that are place-specific. When any community-located arts-based placemaking is termed creative placemaking or community-driven placemaking, it is in danger of becoming the 'community arts' of the arts sector, looked down upon as a poor relation as it is lacking rigour (Schupback, 2012; Thompson, 2012, p.86) and quality. The essential next step for the placemaking sector is to have more peer-to-peer knowledge exchange; more dissemination of in-depth case studies; and longitudinal research (Hall and Smith, 2005, p.175; Schupback, 2012), challenges for many in an un/under-funded sector. What Markusen (2012) calls for is a longitudinal multivariate model of evaluation. A benchmark of best practice is needed (Ball and Essex, 2013), especially with regard to the social objectives of community-facing creative work and the common assertion of causality between urban public space, civic culture and political transformation (Amin, 2008, p.5). Increased knowledge exchange and evaluation is required in this endeavour.

Social practice placemaking practitioners such as Rick Lowe of Project Row Houses asks that their work be evaluated as something that has 'poetics of relationships' and as something that got a community talking, 'touching on people's curiosities and imaginations about what community building could be' but questions too that 'on the other side there's an issue of on what level did that impact really happen?' (Open Engagement, 2015, p.24). As Markusen (2012) states, the indicators of success for creative placemaking are 'fuzzy'; evaluation is not done in a place-specific way but by imposed indicators from funders from external data sources, referencing the NEA Arts Vibrancy Indicators [n.28] as a one-size-fits-all mechanism for evaluation and dissemination of findings, working in the absence of bespoke evaluation tools and measurements. The means and mode of measurement are both determined by their context and they also go on to shape the thinking underpinning city decision-making and policies (Gilmore, 2014, p.13). When qualitative data can be seen as 'fuzzy evidence' (Silberberg, 2013, p.53) there is a two-fold task ahead, to firstly prove the value of aesthetic and cultural experience, and secondly, to effectively measure the 'extrinsic, material impacts these experiences might bring, understood as cost saving or income compensation' (Gilmore, 2014, p.19). With regards the arts in particular, whilst some research has shown a link between the arts and a positive increase in economic and cultural assets, it has not answered the 'how' question of this. Research that specifically locates varying degrees of arts and cultural engagement may help locate to what degree the 'how' question can be answered at this time, and locate where this answer may come from.

A number of potential research themes arose during the research: the rural; intersectionality; and social movement. Firstly, a concern of rural placemaking and a challenge to the common impression of the rural as of a fixed, 'made' place identity, in comparison to the dynamic systems of the city (Sen and Silverman, 2014, p.5). Rather, such research would address the rural place as a dynamic social, politic and place identity sphere. Similar questions and aims from this research project could be asked of the rural place. Projects and practice-based research addressing this already include Deveron Arts [n.29], Northumbrian Exchanges [n.30], Rural Studio [n.31], M12 Studio [n.32] and The Wassaic Project [n.33]. In both the rural and the urban setting too, there is research potential as to the arts-based use of previous community service buildings – such as former pubs, post offices, churches and community halls for example, and how these are being used by artists as a base from which to question the neighbourhood community of its needs. Three examples of this are Deidre O'Mahony's *X-PO* (2008) [n.34], an artistic reimagining of a former post office in Kilnaboy in County Clare, Ireland; The Bevy [n.35], a former pub in Brighton, UK, now purposed as a community centre, and Homebaked [n.36] in Liverpool, UK, a bakery brought back into use by the local community, artists and Liverpool Biennial [n.37].

Secondly, the 'spatial knowing' (Thrift, 2005, p.43) of diverse elective communities in the urban realm, experienced variously as liberating, anonymising, inclusive or exclusive – informed by positive and adverse urban experience thinking demand an intersectional theory of space, drawing on space and urban theories, political theory, and queer and feminist theories. The role of arts in this as a means of expression is a further site of research. For example, as observed in this research, the co-produced social practice placemaking experience was an intersectional one. In the social aesthetic experience, there was a sharing of experience where difference was celebrated and explored. This created a shared space of a shared experience, with similarities between people recognised as common ground, the art experience recognised as not being the same or equal for all; this linked back to the degree of transformational potential: 'it varies as a shared experience…there isn't a shared experience as such' (The Drawing Shed artist).

Lastly, placemaking as a whole demands a social movement concept of framing. This text has seen social practice placemaking and art in place as a means of changing the nature of citizenship, and this has been seen in a political lens – the next step of this is to look at this as a form of activism to force overt political change through cultural activism and to investigate an analogous or literal understanding of placemaking as a social movement.

7.2.6 Recommendations of future actions

Firstly, the text asks that the placemaking typology [3.7] is interrogated through application in the sector. This demands it travel globally, and in this, begin to uncover cross-cultural and cross-sectoral differences and similarities of practice. With regards data collection, it is anticipated that Q Methodology could prove beneficial to a placemaking research project, as a tool to work systematically with both qualitative and quantitative data, as well as picture coding, as this subject matter lends itself to an experiential aesthetic interpretation. Leading on from this, the metrics for evaluating social practice need to be attended to: the network Americans for the Arts [n.38] has begun to ask what such objective metrics could be, who would create them, how they could be used, by whom, and to what ends (Open Engagement, 2015, p.23). Arguably, the arts and the urban need to join together in this as urban metrics can inform arts metrics development, extending from theories of dimensions of performance of urban form for example (Lynch, 1981, p.118).

A longitudinal study of the case studies was out of the scope of this project. However, all would warrant further research over time, as projects such as Big Car's Garfield Park arts-led regeneration development and Dublin's vacant land policy and activism around gentrification as in London evolve, and would answer some of the questions left unanswered here and in the academic and popular media pertaining to the social, political and economic legacy of such activity and the 'trickle-up' (Burnham, 2010, p.139; Silberberg, 2013, p.10) to planning and policy changes. Funding amounts have a major influence on project duration, levels effectively setting a limit on project duration and capacity; a longitudinal timeframe is out of the scope of most funders and this enforces on the projects four pragmatic and ethical concerns. Firstly, to manage the expectations of the participants that the project is of a short-term and/or may be stopped at short notice. Secondly, to manage their goal of a sustainable legacy whilst pitching from short-term project funding strand to short-term project funding strand. Thirdly, to programme in time for the devising and writing of long-term funding applications, whilst simultaneously working to capacity in the field. Fourthly, to maintain and keep extending long-term relations with funders whilst funded in the short-term. These four issues were all evidenced across the case studies and are known to be encountered across the arts and placemaking sectors, and as such are a site of potential further research as a bespoke project.

7.3 Closing remarks

Each of the case studies, in place, responded to the issues of that place. In Dublin, this was overwhelmingly 'the vacant land issue'; in Indianapolis, it was revitalising disinvested neighbourhoods through social practice arts as regeneration; the concern in London was with social housing and gentrification. Social practice arts or social practice placemaking activity in place was seen to have an iterative and subsequently generative place agency activation process – which could go on to activate place attachment. All case studies saw viewers through to protagonists as co-creators of a place. In Dublin this was seen through the journey of the Art Tunnel Smithfield volunteers becoming active in other area projects and politics. In the case of The Drawing Shed participation at all levels (Arnstein, 1969; Finkelpearl, 2013, p.1) created the art experience as one that was a 'visceral experience about what the place was' (The Drawing Shed participant). Through its 'dedicated presence and the encouragement of creativity' (Big Car artist) and its creative programming, Big Car worked through community arts to social practice placemaking to realise the cultural and economic assets of a city, unlocking and working with a latent agency of place to generate further place-based agency.

Social practice placemaking operates as a critique of the placemaking sector, and of itself, as a generative critical spatial practice (Rendell, 2006), breaking down linear and positivist correlations pertaining to people and place, preferring a faceted and pluralist approach that includes space for dissent and failure as essential to experimentation. As an artform influenced by social practice arts, the same critiques of practice can be levelled at it in relation to its calibre of creativity and political efficacy (Fletcher, in Bryan-Wilson, in Fletcher and July, 2014, p.145). However, social practice placemaking practitioners are aware that their role is not as social workers but as exemplifiers (McCormack, 2013, p.12) and do not claim wider or more significant creative or political objectives (Bryan-Wilson, in Fletcher and July, 2014, p.145). Social practice placemaking has aspects of shared practice with social practice arts, participatory arts and community arts, and of creative placemaking. But how it approaches participation, the role of the arts and of artists, and fiscal outcomes, differ and whilst it also works towards liveability metrics as other placemakings do (Gilmore, 2014; Knight Soul of the Community, 2010), it will approach these with a different intentionality, and many exclude some altogether – such as the fiscal imperative, which would be seen in forms of creative placemaking as central to the concern of practice but is weighted differently in the arts-led regeneration of Big Car and absent from the outcome priorities of Art Tunnel Smithfield and The Drawing Shed for example.

Place attachment facilitated in the social practice placemaking process is based on a performed identity, formed from gestural, ritual and navigated processes and mirrored in the performativity of elective belonging where living somewhere is an everyday practice (Lefebvre, 1984). Motivations to join social practice placemaking projects were articulated through Lippard's (1997) 'lure of the local', the desire for a re-connection to place and its people, a gravitational pull to local social networks from an elective belonging. Whilst causal links from social practice placemaking to place-attached behaviours were evidenced, they were seen more as faceted (Mihaylov and Perkins, 2014, p.69) and interactionist (Cozolino, 2006, p.127–8; Lewicka, 2014, p.51; Merleau-Ponty, 1962; Milligan, 1998, pp.1–2) than linear (Seamon, 2014, p.12); this is by virtue of the relational and dialogical aesthetic that is concordant with Kester's (2011, p.29) seven forms of intersubjective exchange, which includes a feedback loop that is of a faceted, not linear, course.

As a place-attached practice, social practice placemaking has an agency of active citizenship and civic participation; from this, at the level of the case studies as substantive projects, it was seen that their place-attached behaviour had an affect on the cities or boroughs they were working in, firstly, by changing the debate of arts in the placemaking context, and secondly, by affecting placemaking policy. By operating in the urban realm, the case studies, in various ways, had an interrelation with the macro structural forces of the city, that of planning and policy, and the administrations' funding streams, and also were interrelated with a meso strand of developers, philanthropists and other public or private funders. This was a complex relationship, played out through both a presence of the macro and meso in the structure in the project and also through their absence and opposition (Lilliendahl Larsen, 2014, p.330; Zukin, 1995, p.32). It did though form a 'public urbanity' (Lilliendahl Larsen, 2014, p.329) of an 'alternative grassroots political mobilisation' (Gosling, 1996, p.147–9) of urban co-creators, affecting a new arts-led political and planning paradigm based on a model of 'developmental relationships' (Benington, 1996, p.162). In the context of social practice placemaking, its everyday urbanism and re-appropriation of vacant or liminal space, in right to the city framing, represents the new production of space (Lilliendahl Larsen, 2014, p.321) that operates outside of and against what Bresnihan and Byrne (2014, p.13) term, of Dublin specifically, but arguably a term that can be applied elsewhere, 'enclosure'. In Dublin specifically, its re-appropriation of the urban commons is a 'refusal to embody the pathetic subjectivity of contemporary neoliberalism, a desire to be and do something else' (ibid., p.4) through a mobilisation structure of networks and collectives and social movement organisations (Hou and Rios, 2003, p.20).

Again, this places social practice placemaking in the situational and participatory arts field, located in social practice arts practice and process specifically. Performative (Froggett et al., 2011, p.97; Jackson, 2011, pp.33–4; Kester, 2004, p.90; McCormack, 2013, p.189; Rounthwaite, 2011, p.92) methods are employed by the artist to both mirror (Froggett et al., 2011, p.96) the issues of the locale back to it and to create the relational urban event space (Hannah, 2009, p.117) to enact possibilities of how spaces may be (Buser et al., 2013, p.624; Finkelpearl, 2013, p.49; Jackson, 2011, p.14), a 'reflection into action' (Kester, 2004, pp.90–101) of social practice placemaking materiality. As relational, social practice placemaking explores the uses and meaning of space through their multi-levelled drivers; as place-based, it examines the spatial and temporal relations inherent, as a spacetime (Munn, 1996, in Low, 2014, p.20), which conceives of placemaking as a material, territorial and networked practice (Foo et al., 2014, p.177). This conjoins urban design and liveability (Miles, 1997, p.2) and challenges the Public Art and new genre public art models of what art is in the public realm, acting as a 'critical instrument' towards social change (Moyersoen and Swyngedouw, 2013, p.149) – Rendell's (2006) critical spatial practice.

This text has placed social practice placemaking in a progressive politics of a collectivised, transducive, active citizenship emanating from its user-centred parasitic agency in the urban form (Klanten and Hübner, 2010, p.103). As a critical spatial practice and an 'embodied placemaking' (Sen and Silverman, 2014, p.4), social practice placemaking offers the potential of active citizenship as a radical discourse through its opening up of emancipatory space politics, as suggested by Lefebvre (1984). This is a human–material–place adaptive process of emergent experimentalism and assemblages of knowledge production from diverse constituents, that can be mobilised as such in the splash intervention to maintain the potential of aesthetic dislocation, or, over time, can become the modus operandi to identifying and working with local civic capacity. The alternative city form exemplified by social practice placemaking, and made material, is of an alternative political form too (Bresnihan and Byrne, 2014), with tools of critique and the offering of alternative situation creation and the situated, everyday practice of social and spatial reproduction with a critical awareness of its political agency vis-à-vis the formal structures of politic.

7.4 Bibliography

Ainsworth, M. D. (1991) 'Attachments and other affectional bonds across the lifecycle' in Parkes, C. M., Stevenson-Hinde, J. and Marris, P. (eds.) *Attachment across the life cycle*. London: Routledge.

Altman, I. and Low, S. M. (1992) *Place attachment*. New York: Plenum Press.

Amin, A. (2008) 'Collective culture and urban public life' in *City* 12, 1 (April).

Anderson, S. (1986) 'Studies towards an ecological model of urban environment' in Anderson, S. (ed.) *On streets*. Cambridge, MA: MIT Press.

Anderson, S. E. and Nielson, A. E. (2009) 'The city at stake: "stakeholder mapping" the city' in *Culture Unbound*, 1: 305–29.

Arnstein, S. R. (1969) 'A ladder of citizen participation' in *AIP Journal*, July, 216–23 [Online]. Available at: http://geography.sdsu.edu/People/Pages/jankowski/public_html/web780/Arnstein_ladder_1969.pdf (Accessed: 25th August 2015).

Arts Council England (2011) *Arts audiences: insight* [Online]. Available at: www.artscouncil.org.uk/media/uploads/pdf/arts_audience_insight_2011.pdf (Accessed: 20th December 2013).

Ball, S. and Essex, R. (2013) *A hidden economy: a critical view of Meanwhile Use in ixia*. Available at: www.publicartonline.org.uk/downloads/news/FINAL%20VERSION%20A%20hidden%20economy;%20a%20critical%20review%20of%20Meanwhile%20Use.pdf (Accessed: 2nd July 2013).

Beech, D. (2010) 'Don't look now! Art after the viewer and beyond participation' in Walwin, J. (ed.) *Searching for art's new publics*. Bristol: Intellect.

Benington, J. (1996) 'New paradigms and practices for local government capacity building within civil society' in Kraener, S. and Roberts, J. (eds.) *The politics of attachment: towards a secure society*. London: Free Association Books.

Best, U. (2014) 'The debate about Berlin Tempelhof Airport, or: a Lefebvrean critique of recent debates about affect in geography' in Stanek, Ł., Schmid, C. and Moravánszky, Á. (eds.) *Urban revolution now: Henri Lefebvre in social research and architecture*. Farnham: Ashgate Publishing Ltd.

Beyes, T. (2010) 'Uncontained: the art and politics of reconfiguring urban space' in *Culture and Organisation* 16, 3: 229–46.

Bishop, C. (2006) [a] 'Introduction – viewers as participants' in Bishop, C. (ed.) *Participation*. London: Whitechapel Gallery and The MIT Press.

Bishop, C. (2006) [b] 'The social turn: collaboration and its discontents' in *Artforum*, 178–83 [Online]. Available at: www.gc.cuny.edu/CUNY_GC/media/CUNY-

Graduate-Center/PDF/Art%20History/Claire%20Bishop/Social-Turn.pdf (Accessed: 28th November 2015).
Bishop, C. (2012) *Artificial hells: participatory art and the politics of spectatorship*. London: Verso.
Bishop, P. and Williams, L. (2012) *The temporary city*. Abingdon: Routledge.
Bonjour, L. (2011) 'A priori knowledge' in Bernecker, S. and Pritchard, D. (eds.) *The routledge companion to epistemology*. New York: Routledge.
Bosch, S. and Theis, A. (eds.) (2012) Connection: artists in communication. *Interface: centre for research in art, technologies, and design*. Belfast: Dorman and Sons Ltd.
Bourriaud, N. (1998/2006) 'Relational aesthetics' (1998) in Bishop, C. (ed.) *Participation*. London: Whitechapel Gallery and The MIT Press.
Bowlby, J. (1979/2010) *The making and breaking of affectional bonds*. Abingdon: Routledge.
Boyer, M. C. (2014) 'Reconstructing New Orleans and the right to the city' in Stanek, Ł., Schmid, C. and Moravánszky, Á. (eds.) *Urban revolution now: Henri Lefebvre in social research and architecture*. Farnham: Ashgate Publishing Ltd.
Bresnihan, P. and Byrne, M. (2014) 'Escape into the city: everyday practices of commoning and the production of urban space in Dublin' in *Antipode* 47, 1: 36–54.
Brown, A. (2012) 'All the world's a stage: venues, settings and the role they play in shaping patterns of arts participation' in *Perspectives on non-profit strategies* [n.p.]: Wolfbrown.
Brown, B. (2014) 'The rise of localist politics' in McClay, W. M. and McAllister, T. V. (eds.) *Why place matters: geography, identity, and civic life in modern America*. New York: New Atlantis Books.
Burnham, S. (2010) 'Scenes and sounds: the call and response of street art and the city' in *City* 14, 1–2 (February–April).
Buser, M., Bonura, C., Fannin, M. and Boyer, K. (2013) 'Cultural activism and the politics of place-making' in *City* 17, 5: 606–27.
Carmona, M., de Magalhães, C., and Hammond, L. (2008) *Public space: the management dimension*. London: Routledge.
Carrus, G., Scopelliti, M., Fornara, F., Bonnes, M. and Bonaiuto, M. (2014) 'Place attachment, community identification, and pro-environmental engagement' in Manzo, L. C. and Devine-Wright, P. (eds.) *Place attachment: advances in theory, methods and applications*. Abingdon: Routledge.
Castells, M. (1983) *The city and the grassroots: a cross-cultural theory of urban movements*. London: Edward Arnold.
Chonody, J. M. (2014) 'Approaches to evaluation: how to measure change when utilizing creative approaches' in Chonody, J. M. (ed.) *Community art: creative approaches to practice*. Champaign, IL: Common Ground.
Cleveland, W. (2001) 'Trials and triumphs: arts-based community development' in *Public Art Review* Fall/Winter 2001, 17–23.
Colombo, M., Mosso, C. and de Piccoli, N. (2001) 'Sense of community and participation in urban contexts' in *Journal of Community and Applied Social Psychology* 11: 457–64.
Courage, C. (2015) 'Planning as a social art form' in *New Planner*, September 2015.
Cozolino, L. (2006) *The neuroscience of human relationships: attachment and the developing social brain*. New York: W. W. Norton and Company.
Crawford, M. (1999) 'Blurring the boundaries: public space and private life' in Chase, J., Crawford, M. and Kalishi, J. (eds.) *Everyday urbanism*. New York: The Monacelli Press.
Critical Art Ensemble (1998) 'Observations on collective cultural action' in *Art Journal* 57, 2 (Summer 1998).
Debord, G. (1958) 'Theory of the Dérive and definitions' in Gieseking, J. and Mangold, W. (eds.) *The people, place, and space reader*. New York: Routledge.
Debord, G. (2006) 'Towards a situationist international' in Bishop, C. (ed.) *Participation*. London: Whitechapel Gallery and The MIT Press.
Deutsche, R. (1992) 'Public art and its uses' in Seine, H F. and Webster, S. (eds.) *Public art: content, context and controversy*. New York: Iconeditions.

Dewey, J. (1958) *Art as experience*. New York: Putnam.
Dixon, J. and Durrheim, K. (2000) 'Displacing place-identity: a discursive approach to locating self and other' in *British Journal of Social Phycology* 39, 27–44.
Doherty, C., Eeg-Tverbakk, P. G., Fite-Wassilak, C., Lucchetti, M., Malm, M. and Zimberg, A. (2015) 'Foreword' in Doherty, C. (ed.) *Out of time, out of place: public art (now)*. London: ART/BOOKS with Situations and Public Art Agency Sweden.
Durkheim, E. (1893) *The division of labour in society reprint*. Basingstoke: Palgrave Macmillan, 2013.
Fernando, N. A. (2007) 'Open-ended space: urban streets in different cultural contexts' in Franck, K. A. and Stevens, Q. (eds.) *Loose space: possibility and diversity in urban life*. Abingdon: Routledge.
Finkelpearl, T. (2013) *What we made: conversations on art and social cooperation*. Durham: Duke University Press.
Fischer, C. S. (1975) 'Toward a subcultural theory of urbanism' in *American Journal of Sociology* 80, 6: 1319–41.
Fleming, R. L. (2007) *The art of placemaking: interpreting community through public art and urban design*. London: Merrell Publishers Ltd.
Fletcher, H. and July, M. (2014) 'Hello' in Fletcher, H. and July, M. (eds.) *Learning to love you more*. Munich: Prestel.
Foo, K., Martin, D., Wool, C. and Polsky, C. (2014) 'Reprint of "The production of urban vacant land: relational placemaking in Boston, MA neighborhoods"' in *Cities* 40: 175–82.
Franck, K. A. and Stevens, Q. (2007) 'Patterns of the unplanned: urban catalyst' in Franck, K. A. and Stevens, Q. (eds.) *Loose space: possibility and diversity in urban life*. Abingdon: Routledge.
Friedmann, J. (2010) 'Place and place-making in cities: a global perspective' in *Planning Theory and Practice* 11, 2: 149–65.
Froggett, L., Little, R., Roy, A. and Whitaker, L. (2011) New model of visual arts organisations and social engagement, *University of Central Lancashire Psychosocial Research Unit* [Online]. Available at: http://clok.uclan.ac.uk/3024/1/WzW-NMI_Report%5B1%5D.pdf (Accessed: 12th August 2015).
Gablik, S. (1992) 'Connective aesthetics' in *American Art* 6, 2: 2–7.
Giddens, A. (1992) *The transformation of intimacy*. Cambridge: Polity Press.
Gieseking, J. and Mangold, W. (eds.) (2014) *The people, place, and space reader*. New York: Routledge.
Gilmore, A. (2014) 'Raising our quality of life: the importance of investment in arts and culture' *Centre for Labour and Social Studies* [Online]. Available at: http://classonline.org.uk/docs/2014_Policy_Paper_-_investment_in_the_arts_-_Abi_Gilmore.pdf (Accessed: 12th August 2015).
Giuliani, M. (2003) 'Theory of attachment and place attachment' in Bonnes, M., Lee, T. and Bonainto, M. (eds.) *Psychological theories for environmental issues*. Aldershot: Ashgate.
Gosling, J. (1996) 'The business of "community"' in Kraener, S. and Roberts, J. (eds.) *The politics of attachment: towards a secure society*. London: Free Association Press.
Grodach, C. (2010) 'Art spaces, public space and the link to community "development"' in *Community Development Journal* 45, 4: 474–93.
Guattari, F. (2006) 'Chaosmosis: an ethico-aesthetic paradigm' in Bishop, C. (ed.) *Participation*. London: Whitechapel Gallery and The MIT Press.
Hall, T. and Smith, C. (2005) 'Public art in the city: meanings, values, attitudes and roles' in Miles, M. and Hall, T. (eds.) *Interventions: advances in art and urban futures (Vol 4)*. Bristol: Intellect Books.
Hamblen, M. (2014) 'The city and the changing economy' in Quick, C., Speight, E. and van Noord, G. (eds.) *Subplots to a city: ten years of In Certain Places*. Preston: In Certain Places.

Hannah, D. (2009) 'Cities event space: defying all calculation' in Lehmann, S. (ed.) *Back to the city: strategies for informal intervention*. Ostfildern: Hatje Cantz.
Haynes, J. (1984) *Thanks for coming!* London: Faber and Faber.
Hermansen, C. (2011) 'Social creativity', January 2009. *SCRIBE: Scarcity and Creativity in the Built Environment working paper no 3* [Online]. Available at: www.scibe.eu/wp-content/uploads/2010/11/03-CH.pdf (Accessed: 2nd July 2013).
Holmes, J. (1993) *John Bowlby and attachment theory*. London: Routledge.
Hope, S. (2010) 'Who speaks? Who listens? Het Reservaat and critical friends' in Walwin, J. (ed.) *Searching for art's new publics*. Bristol: Intellect.
Hoskins, M. W. (1999) 'Opening the door for people's participation' in White, S. (ed.) *The art of facilitating participation: releasing the power of grassroots communication*. New Delhi: Sage Publications.
Hoskyns, T. (2005) 'City/democracy: retrieving citizenship' in Till, J., Blundell-Jones, P. and Petrescu, D. (eds.) *Architecture and participation*. London: Spon.
Hou, J. (2010) *Insurgent public space: guerrilla urbanism and the remaking of contemporary cities*. London: Routledge.
Hou, J. and Rios, M. (2003) 'Community-driven place making: the social practice of participatory design in the making of Union Point Park' in *Journal of Architectural Education* 57, 1: 19–27.
Jackson, S. (2011) *Social works: performing art, supporting publics*. Abingdon: Routledge.
Kaji-O'Grady, S. (2009) 'Public art and audience reception: theatricality and fiction' in Lehmann, S. (ed.) *Back to the city: strategies for informal urban interventions*. Ostfildern: Hatje Cantz.
Kaprow, A. (1996/2006) 'Notes on the elimination of the audience' (1966) in Bishop, C. (ed.) *Participation*. London: Whitechapel Gallery and The MIT Press.
Kastner, J. (1996) 'Mary Jane Jacobs: an interview with Jeffrey Kasther' in Crabtree, A. (guest ed.) *Public art, art and design profile no 46*. London: Academy Group Ltd.
Kearns, P. and Ruimy, M. (2014) *Beyond Pebbledash and the puzzle of Dublin*. Kinsale: Gandon Editions Kinsale.
Kelly, O. (1984) *Community, art and the state* [n.p.]: Comedia.
Kester, G. H. (2004) *Conversation pieces: community and communication in modern art*. Berkeley: University of California Press.
Kester, G. H. (2011) *The one and the many: contemporary collaborative art in a global context*. Durham: Duke University Press.
Klanten, R. and Hübner, M. (2010) *Urban interventions: personal projects in public spaces*. Berlin: Gestalten.
Klanten, R., Ehmann, S., Borges, S, Hübner, M and Feireiss L. (2012) *Going public: public architecture, urbanism and interventions*. Berlin: Gestalten.
Knight Communities Overall Knight Soul of the Community (2010) *Where people love where they live and why it matters: a national perspective* [Online]. Available at: www.neindiana.com/docs/national-research/soul-of-the-community---overall.pdf?sfvrsn=4 (Accessed: 13th August 2015).
Kravagna, C. (2012) 'Working on the community: models of participatory practice' in Deleuze, A. (ed.) *The 'do-it-yourself' artwork: participation from Fluxus to new media*. Manchester: Manchester University Press.
Krupat, E. (1985) *People in cities: the urban environment and its effects*. Cambridge: Cambridge University Press.
Kuskins, J. (2013) 'Love or hate it, user-generated urbanism may be the future of cities', 23rd September 2013, *Urbanism*. Available at: http://gizmodo.com/love-it-or-hate-it-user-generated-urbanism-may-be-the-1344794381 (Accessed: 4th December 2013).
Kwon, M. (2004) *One place after another: site-specific art and located identity*. Cambridge, MA: The MIT Press.
Landesman, C. (1997) *An introduction to epistemology*. Cambridge, MA: Blackwell Publishing.

Larsen, L. B. (1999/2006) 'Social aesthetics' in Bishop, C. (ed.) *Participation*. London: Whitechapel Gallery and The MIT Press.
Lefebvre, H. (1984) *The production of space*. Translated, Nicholson-Smith, D. Malden, MA: Blackwell Publishing.
Legge, K. (2012) *Doing it differently*. Sydney: Place Partners.
Legge, K. (2013) *Future city solutions*. Sydney: Place Partners.
Lehmann, S. (2009) *Back to the city: strategies for informal intervention*. Ostfildern: Hatje Cantz.
Lepofsky, J. and Fraser, J. C. (2003) 'Building community citizens: claiming the right to place-making in the city' in *Urban Studies* 40, 1: 127–42.
Lewicka, M. (2014) 'In search of roots: memory as enabler of place attachment' in Manzo, L. C. and Devine-Wright, P. (eds.) *Place attachment: advances in theory, methods and applications*. Abingdon: Routledge.
Lilliendahl Larsen, J. (2014) 'Lefebvrean vagueness: going beyond diversion in the production of new spaces' in Stanek, Ł., Schmid, C. and Moravánszky, Á. (eds.) *Urban revolution now: Henri Lefebvre in social research and architecture*. Farnham: Ashgate Publishing Ltd.
Lippard, L. (1997) *The lure of the local: senses of place in a multi-centered society*. New York: New Press.
Livingston, M., Bailey, N. and Kearns, A. (2008) *People's attachment to place: the influence of neighbourhood deprivation*. Glasgow: Glasgow University.
Long, P. (2013) 'Sense of place and place-based activism in the neoliberal city: the case of "weird" resistance' in *City* 17, 1: 52–67.
Lossau, J. (2006) 'The gatekeeper – urban landscape and public art' in Warwick, R. (ed.) *Arcade: artists and placemaking*. London: Black Dog Publishing.
Low, S. (2014) 'Spatializing culture: an engaged anthropological approach to space and place (2014)' in Gieseking, J. and Mangold, W. (eds.) *The people, place, and space reader*. New York: Routledge.
Lydon, M. and Garcia, A. (2015) *Tactical urbanism: short-term action for long-term change*. Washington: Island Press.
Lynch, K. (1981) *A theory of good city form*. Cambridge, MA: MIT Press.
Madyaningrum, M. E. and Sonn, C. (2011) 'Exploring the meaning of participation in a community art project: a case study on the Seeming Project' in *Journal of Community and Applied Social Psychology* 21: 358–70.
Mancilles, A. (1998) 'The citizen artist' in Frye Burnham, L. and Durland, S. (eds.) *The artist as citizen: 20 years of art in the public arena*. New York: Critical Press.
Markusen, A. (2012). OTIS Report on the Creative Economy [Online]. Available at: www.otis.edu/otis-report-creative-economy (Accessed: 13th July 2016).
Markusen, A. and Gadwa, A. (2010) [a] *Creative placemaking white paper* [Online]. Available at: www.nea.gov/pub/CreativePlacemaking-Paper.pdf (Accessed: 5th October 2013).
Markusen, A. and Gadwa, A. (2010) [b] *Creative placemaking white paper executive summary* [Online]. Available at: www.nea.gov/pub/CreativePlacemaking-Paper.pdf (Accessed: 5th October 2013).
Marris, P. (1991) 'The social construction of uncertainty' in Parkes, C. M., Stevenson-Hinde, J. and Marris, P. (eds.) *Attachment across the life cycle*. London: Routledge.
Martin, D. G. (2013) 'Place frames: analysing practice and production of place in contentious politics' in Nicholls, W., Miller, B. and Beaumont, J. (eds.) *Spaces of contention: spatialities and social movements*. Farnham: Ashgate.
McAllister, T. V. (2014) 'Making American places: civic engagement rightly understood' in McClay, W. M. and McAllister, T. V. (eds.) *Why place matters: geography, identity, and civic life in modern America*. New York: New Atlantis Books.
McCormack, D. P. (2013) *Refrains for moving bodies: experience and experiment in affective spaces*. Durham: Duke University Press.
McGonagle, D. (2007) 'Foreword' in Butler, D. and Reiss, V. (eds.) *Art of negotiation*. Manchester: Cornerhouse Publications.

Merleau-Ponty, M. (1962) *The phenomenology of perception*. New York: Humanities Press.
Mici, B. (2013) 'Participation makes for successful placemaking', 8th March 2013. *Planetizen* [Online]. Available at: www.planetizen.com/node/61068 (Accessed: 4th December 2013).
Mihaylov, N. and Perkins, D. D. (2014) 'Community place attachment and its role in social capital development' in Manzo, L. C. and Devine-Wright, P. (eds.) *Place attachment: advances in theory, methods and applications*. Abingdon: Routledge.
Miles, M. (1997) *Art, space and the city: public art and urban futures*. London: Routledge.
Miles, M. (2008) 'Critical spaces: monuments and changes' in Cartiere, C. and Willis, S. (eds.) *The practice of public art*. New York: Routledge.
Milligan, M. J. (1998) 'Interactional past and potential: the social construction of place attachment' in *Symbolic Interaction* 21, 1: 1–33.
Minton, A. (2009) *Ground control: fear and happiness in the twenty-first century city*. London: Penguin Books.
Mitchell, D. (2003) *The right to the city*. New York: The Guildford Press.
Moyersoen, M. and Swyngedouw, E. (2013) 'LimiteLimite: cracks in the city, brokering scales, and pioneering a new urbanity' in Nicholls, W., Miller, B. and Beaumont, J. (eds.) *Spaces of contention: spatialities and social movements*. Farnham: Ashgate.
Mozes, J. (2011) 'Public space as battlefield' in Fernández Per, A. and Mozes, J. (eds.) *A+T Strategy and Tactics in Public Space, Independent Magazine of Architecture and Technology*, Autumn 2011, Issue 38.
Murray, M. (2012) 'Art, social action and social change' in Walker, C., Johnson, K. and Cunningham, L. (eds.) *Community psychology and the socio-economics of mental distress*. Basingstoke: Palgrave Macmillan.
Nancy, J. L. (1986/2006) 'The inoperative community' in Bishop, C. (ed.) *Participation*. London: Whitechapel Gallery and The MIT Press.
Nicholson, G. (1996) 'Place and local identity' in Kraener, S. and Roberts, J. (eds.) *The politics of place attachment: towards a secure society*. London: Free Association Press.
Nowak, J. (2007) *A Summary of Creativity and Neighbourhood Development: Strategies for Community Investment TRFund* [Online]. Available at: www.sp2.upenn.edu/siap/docs/cultural_and_community_revitalization/creativity_and_neighborhood_development.pdf (Accessed: 4th December 2013).
Obrist, H. U. (2015) *Ways of curating*. London: Penguin Books.
O'Neill, P. (2014) 'The curatorial constellation – durational public art, cohabitattion time and attentiveness' in Quick, C., Speight, E. and van Noord, G. (eds.) *Subplots to a city: ten years of In Certain Places*. Preston: In Certain Places.
Open Engagement (2015) 'In conversation: Rick Lowe and Lisa Lee' in *Place and Revolution, Open Engagement in Print* 003 [n.p.].
Ostwald, M. J. (2009) 'Public space and public art: the metapolitics of aesthetics' in Lehmann, S. (ed.) *Back to the city: strategies for informal intervention*. Ostfildern: Hatje Cantz.
Pahl, A. (1996) 'Friendly society' in Kraener, S. and Roberts, J. (eds.) *The politics of attachment: towards a secure society*. London: Free Association Books.
Petrescu, D. (2006) 'Working with uncertainty towards a real public space' in *If you can't find it, give us a ring: public works*. [n.p.]: Article Press/ixia.
Project for Public Spaces [a] (2015) *The year in placemaking*. Available at: www.pps.org/blog/2015-the-year-in-placemaking/ (Accessed: 13th February 2016)
Project for Public Spaces [b] *The lighter, quicker, cheaper transformation of public spaces*. Available at: www.pps.org/reference/lighter-quicker-cheaper/ (Accessed: 13th February 2016).
Puype, D. (2004) 'Arts and culture as experimental spaces in the city' in *City* 8, 2: 295–301.
Rancière, J. (2004/2006) 'Problems and transformations in critical art' in Bishop, C. (ed.) *Participation*. London: Whitechapel Gallery and The MIT Press.
Rendell, J. (2006) *Art and architecture: a place between*. London: I. B. Tauris.

Richardson, T. and Connelly, S. (2005) 'Reinventing public participation in the age of consensus' in Till, J., Blundell-Jones, P. and Petrescu, D. (eds.) *Architecture and participation*. London: Spon.

Roberts, P. (2009) 'Shaping, making and managing places: creating and maintaining sustainable communities through the delivery of enhanced skills and knowledge' in *Town Planning Review* 80, 4–5.

Rounthwaite, A. (2011) 'Cultural participation by group material: between the ontology and the history of the participatory art event' in *Performance Research: A Journal of the Performing Arts* 16, 4: 92–6.

Rowles, G. D. (1983) 'Place and personal identity in old age: observations from Appalachia' in *Journal of Environmental Psychology* 3, 4: 299–313.

Roy, A. (2005) 'Urban informality: toward an epistemology of planning' in *Journal of the American Planning Association* 71, 2 (Spring 2005).

Rubinstein, R. and Parmelee, O. (1992) 'Attachment to place and the representation of the life course by elderly' in Altman, I. and Low, S. (eds.) *Place attachment*. London: Plenum Press.

Saegert, S. (2014) 'Restoring meaningful subjects and "democratic hope" to psychology (2014)' in Gieseking, J. and Mangold, W. (eds.) *The people, place, and space reader*. New York: Routledge.

Scannell, L. and Gifford, R. (2010) 'Comparing the theories of interpersonal and place attachment' in Manzo, L. C. and Devine-Wright, P. (eds.) *Place attachment: advances in theory, methods and applications*. Abingdon, Oxon: Routledge.

Schmid, C. (2014) 'The trouble with Henri: urban research and the theory of the production of space' in Stanek, Ł., Schmid, C. and Moravánszky, Á. (eds.) *Urban revolution now: Henri Lefebvre in social research and architecture*. Farnham: Ashgate.

Schneekloth, L. H. and Shibley, R. G. (2000) 'Implacing architecture into the practice of placemaking' in *Journal of Architectural Education* 53, 3: 130–40.

Schupback, J. (2012) 'Defining creative placemaking: a talk with Ann Markusen and Anne Gadwa Nicodemus' in *NEA Arts Magazine* [Online] 2012, no 3. Available at: www.arts.gov/about/NEARTS/storyNew.php?id=01_defining&issue=2012_v3 (Accessed: 3rd July 2013).

Seamon, D. (2014) 'Place attachment and phenomenology: the synergistic dynamism of place' in Manzo, L. C. and Devine-Wright, P. (eds.) *Place attachment: advances in theory, methods and applications*. Abingdon: Routledge.

Sen, A. and Silverman, L. (2014) 'Introduction – embodied placemaking: an important category of critical analysis' in Sen, A. and Silverman, L. (eds.) *Making place: space and embodiment in the city*. Bloomington: Indiana University Press.

Sennett, R. (1970) *The uses of disorder: personal identity and city life*. New Haven: Yale University Press.

Sennett, R. (2012) *Together: the rituals, pleasures and politics of cooperation*. London: Allen Lane.

Sepe, M. (2013) *Planning and place in the city: mapping place identity*. Abingdon: Routledge.

Serres, M. (1980/2007) *The parasite*. Minneapolis: University of Minnesota Press.

Sherlock, M. (1998) 'Postscript – no loitering: art as social practice' in Harper, G. (ed.) *Interventions and provocations: conversations on art, culture and resistance*. Albany: State University of New York Press.

Shirky, C. (2008) *Here comes everybody: how change happens when people come together*. London: Penguin Books.

Shumaker, S. and Taylor, R. (1983) 'Toward a clarification of people-place relationships: a model of attachment to place' in Feimer, N. and Geller, E. (eds.) *Environmental psychology*. New York: Praeger.

Silberberg, S. (2013) Places in the making: how placemaking builds places and communities. *MIT Department of Urban Studies and Planning* [Online]. Available at: http://dusp.mit.edu/cdd/project/placemaking (Accessed: 13th August 2015).

Sime, J. D. (1986) 'Creating places or designing spaces?' in *Journal of Environmental Psychology* 6: 49–63.
Sixsmith, J. (1986) 'The meaning of home: an exploratory study of environmental experience' in *Journal of Psychology* 6, 4: 281–98
Sorenson, A. (2009) 'Neighbourhood streets as meaningful spaces: claiming rights to shared spaces in Tokyo' in *City and Society* 21, 2: 207–29.
Sosa, D. (2011) 'Perceptual knowledge' in Bernecker, S. and Pritchard, D. (eds.) *The Routledge companion to epistemology*. New York: Routledge.
Suderberg, E. (2000) 'Introduction: on installation and site specificity' in Suderberg, E. (ed.) *Space, site and intervention: situating installation art*. Minneapolis: University of Minnesota Press.
Sutherland, A. (2010) 'An audience with…' in Walwin, J. (ed.) *Searching for art's new publics*. Bristol: Intellect.
Tabb, W. (2012) 'Beyond "cities for people, not for profit": cities for people and people for systemic change' in *City* 16, 1–2 (February–April 2012).
The Drawing Shed (2015) [a] *Manual for possible projects on the horizon: that things fall apart…and…thinking through making*. The Drawing Shed: Some[w]Here. [n.p.]
Thompson, N. (2012) 'Socially engaged art is a mess worth making' in *Architect*, August 2012, 86–7.
Thrift, N. (2005) 'Driving in the city' in Featherstone, M., Thrift, N. and Urry, J. (eds.) *Automobilities*. London: Sage.
Till, K. E. (2014) '"Art, memory, and the city" in Bogotá: Mapa Teatro's artistic encounters with inhabited places' in Sen, A. and Silverman, L. (eds.) *Making place: space and embodiment in the city*. Bloomington: Indiana University Press.
Tonkiss, F. (2013) *Cities by design: the social life of urban form*. Cambridge: Polity.
Tuan, Y. (1974) *Topophilia: a study of environmental perception, attitudes, and values*. New York: Columbia University Press.
Tuan, Y. (2014) 'Place/space, ethnicity/cosmos: how to be more fully human' in McClay, W. M. and McAllister, T. V. (eds.) *Why place matters: geography, identity, and civic life in modern America*. New York: New Atlantis Books.
Uitermark, J., Nichols, W. and Loopmans, M. (2012) 'Cities and social movements: theorizing beyond the right to the city' in *Environment and Planning A* 44: 2546–54.
van Heeswijk, J. (2012) *Public art and self-organisation, London* (conference), August 2012. Available at: http://ixia-info.com/events/next-events/public-art-and-self-organisation-london/ (Accessed: 15th January 2016).
van Hoven, B. and Douma, L. (2012) 'We make ourselves at home wherever we are' – older people's placemaking in Newton Hall' in *European Spatial Research and Policy* 19, 1: 65–79.
Watson, S. (2006) *City publics: the (dis)enchantments of urban encounters*. Abingdon, Oxon: Routledge.
Watts, S. and Stenner, P. (2014) *Doing Q methodological research: theory, method and interpretation*. London: Sage Publications Ltd.
White, S. (1999) 'Participation: walk the talk' in White, S. (ed.) *The art of facilitating participation: releasing the power of grassroots communication*. New Delhi: Sage Publications.
Whitehead, F. (2006) *What do artists know* [Online]. Available at: http://embeddedartistproject.com/whatdoartistsknow.html (Accessed: 16th June 2015).
Whybrow, N. (2011) *Art and the city*. London: I. B. Tauris.
Wilmott, P. (1987) *Friendship networks and social support*. London: Policy Studies Institute.
Zebracki, M. Van Der Vaart, R. and Van Aalst, I. (2010) 'Deconstructing public artopia: situating public-art claims within practice' in *Geoforum* 41: 786–95.
Zukin, S. (1995). *The cultures of cities*. Malden, MA: Blackwell Publishers.
Zukin, S. (2010) *Naked city: The death and life of authentic urban place*. Oxford: Oxford University Press.

7.5 List of URLs

1. The Drawing Shed: www.thedrawingshed.org/
2. Allan Kaprow: www.allankaprow.com/
3. Big Car: www.bigcar.org/
4. *IdeasFromElse[W]here*: www.thedrawingshed.org/ideasfromelsewhere
5. Harrell Fletcher: www.harrellfletcher.com/
6. Project Row Houses: http://projectrowhouses.org/
7. Rebuild Foundation: https://rebuild-foundation.org
8. Art Tunnel Smithfield: http://arttunnelsmithfield.com/
9. *Some[w]HereResearch*: www.thedrawingshed.org/projects-2014/somewhere-research
10. Art in Odd Places: http://recall.artinoddplaces.org/
11. Art in Odd Places Indianapolis: www.artinoddplaces.org/indianapolis/
12. Thinker in Residence – The Drawing Shed: https://ideasfromelsewhere.wordpress.com/2014/06/11/ideasfromelsewhere-thinker-in-residence
13. Thinker in Residence – Big Car: www.bigcar.org/space-and-community/
14. Thinker in Residence – Art in Odd Places: www.artinoddplaces.org/aiop-2015-recall-thinker-in-residence-quinn-dukes/
15. *LiveElse[W]here*: www.thedrawingshed.org/live-elsewhere
16. Big Car, Service Center: www.bigcar.org/project/service-center-for-culture-and-community/
17. Garfield Park Creative Community: www.bigcar.org/project/garfield/
18. *Live Lunch*: www.thedrawingshed.org/liveelsewhere-live-lunch-may-31-2014
19. *Beta Project*: https://dubcitybeta.wordpress.com
20. Dublin City Council/*Comhairle Cathrach Bhaile Átha Cliath*: www.dublincity.ie
21. Gorbals Art Project: www.gorbalsartsproject.co.uk/
22. LUPA (2011–13): www.facebook.com/LUPA.E2/info
23. Zuloark: www.zuloark.com/
24. *Slipstream* (2010), David Cottrell: www.cotterrell.com/projects/4492/slipstream-v/
25. Balfron Social Club: http://50percentbalfron.tumblr.com/
26. Placemaking Leadership Council: www.pps.org/about/leadership-council/
27. Wikicuidad: http://thisbigcity.net/wikicity-citizens-improve-cities/
28. NEA Arts & Livability Indicators: Assessing Outcomes of Interest to Creative Placemaking Projects: www.arts.gov/artistic-fields/research-analysis/arts-data-profiles/arts-data-profile-8/arts-data-profile-8
29. Deveron Arts: http://deveron-arts.com/home/
30. Northumbrian Exchanges: www.n-ex.org.uk/
31. Rural Studio: www.ruralstudio.org/
32. M12 Studio: http://m12studio.org/
33. The Wassaic Project: http://wassaicproject.org/
34. *X-PO* (2008), Deidre O'Mahony: www.deirdre-omahony.ie/public-art-projects/x-po.html
35. The Bevy: www.thebevy.co.uk/
36. Homebaked: http://homebaked.org.uk/
37. Liverpool Biennial: www.biennial.com/
38. Americans for the Arts: www.americansforthearts.org/

Appendix
Demographic and statistical information

A.1 Global demographic information

In 2011 it was estimated that 3.5bn people lived in urban areas (Legge, 2012, p.5). In 2013, the world's population stood at 7,172,872,424 and had a current yearly growth of 1.10 per cent (Geohive [c]), with a total of 633 countries in 2010 with populations of over 750,000 (Geohive [a]). The top 25 cities had populations ranging from c8M to c36M, this group predicted to have populations ranging from 12M to 38M in 2015 (Geohive [b]). In these countries urban population percentages in 2010 were 61.8 for Ireland; 82.3 for the USA; and 79.6 for the UK (Geohive [d]).

A.2 Dublin, Republic of Ireland

Historically, Ireland's population entered decline in 1841 to the 1960s, from 6.5M to 2.8M at its lowest in 1961. The years 2002–2006 however saw an 'unparalleled' annual population increase of 80K or close to 2 per cent and at the highest variant predictions this rate is set to continue to 2021 (Central Statistics Office [b], pp.6–8) and 2026 (Central Statistics Office [e], p.1). The city saw a population percentage change during 2002–2006 of 4.2 per cent; its regional area saw a change of 7.2 per cent, with a total population in 2011 of 1,273,069 (Central Statistics Office [c], pp.9–10) and 97.8 per cent of the population in the aggregate town area (Central Statistics Office [c], p.14). To place this in context, the other four large city and suburban areas of Cork, Limerick, Galway and Waterford had a combined population of 403,083 in 2006, with a 3.8 per cent increase to 418,333 in 2011 (Central Statistics Office [c], p.117). The 2011 population figure is 527,612 in Dublin City and 1,273,069 in the wider area (Central Statistics Office [d]). The Dublin City area has 208,716 households, the wider Dublin region 468,122 (Central Statistics Office Ireland [a]). Thirty-two per cent of the inner city population is aged 20–34 years, compared to the national average of 24 per cent (Cudden, 2013, p.51). Smithfield is located north of the River Liffey/*An Life* – 'the north side' as it is called in the vernacular – an area of the city where 70 per cent of the population of the electoral division are born outside Ireland (Duncan, 2012). Smithfield is located in and around the Arran

Quay administrative area of inner city Dublin; this had a total population in 2011 of 15,841, the area experiencing an overall population increase (Cudden, 2013, p.75). The Smithfield median property value is €145,521, compared to Dublin's at €265,000 and the rest of Ireland at €175,000; median rental prices are €1,037, €1,018 and €993 respectively (RateMyArea.com).

A.3 London, UK

London's population stood at 8.3M mid-2012, 13 per cent of the total UK population (Office for National Statistics). The Drawing Shed's first project was in Waltham Forest, London's second most densely populated borough, its population at 265,800 people living in approximately 96,900 households (Waltham Forest Borough Council). The ethnic demographic of the area has changed from a white population of 74.4 per cent in 1991 to 50 per cent, with the largest ethnic minority groups being Black Caribbeans, Pakistanis and Black Africans; it has the eighth largest Muslim population in England and the fourth largest in London. It is a 'deprived borough': whilst its unemployment rate is 9.1 per cent, below London's average, the proportion of its 'Lower Super Output Areas' in the 30 per cent most deprived in England is 81 per cent (Trust for London [a]). The area's average age of residents is 34.4 years, compared to the UK average of 40 years, and average earnings are £24,200, slightly lower than the London average; the borough ranked the fifteenth most deprived borough nationally, and London's sixth most deprived, according to the 2010 Index of Multiple Deprivation. The average house price in the borough at June 2014 was £333,300, up 28 per cent from 2013 (Waltham Forest Borough Council). Those without educational qualifications stood at 20.8 per cent; 49.9 per cent of its population own or have a mortgage on residential property and 10.7 per cent live in social housing (I Live Here). The Drawing Shed's second project took place in Wandsworth, in London's southwest. The 1960s saw slum clearance and industrial decline and consequent movement of population to other London boroughs and suburbs; the area is subject to regeneration at the time of writing however. The population was at 307,000 in 2011; it has a population density of 90 persons per hectare, higher than the Greater London average though comparatively low for inner London. Seventy-one per cent of residents are white, compared to 60 per cent in London; it has a high proportion of Black or Ethnic Minority (BME) residents compared to the national average although it has the second lowest proportion of BME residents in Inner London. Recent population growth has been attributed to young professional workers, who, whilst bringing wealth into the area, are transient with career progression (Trust for London [b]). The median household income in Wandsworth is above the London median but there are wards in the borough where 1 in 4 households earn under £15,000 and approximately 50 per cent of all households with children in Wandsworth are in receipt of Child Tax Credits. Over half the borough population is single or cohabiting, amongst the highest rates in the country (Wandsworth Borough Council).

A.4 Indianapolis, Indiana, USA

In the USA, 80 per cent of Americans live in urbanised areas (Lydon and Garcia, 2015, p.67). Indianapolis has a population estimate of 843,393 for 2014, an estimated increase of 1.7 per cent since the previous census. Its population is 86.1 per cent white, compared to 77.4 per cent for the US nationally, with 4.7 per cent 'foreign born', compared to 12.9 per cent nationally. Its median household income is $48,248, with 70 per cent home ownership rate of a median $122,800 property value, and 15.3 per cent living below the poverty line (United States Census Bureau State and County QuickFacts). The city is ranked the US's ninth poorest city, with 29.1 per cent of households with income below $25,000, and one of the largest national increases in child poverty and income inequality rates (Kennedy, 2015 – who also states the city's population to be 828,841). Indianapolis is 78th on the Creative Class list in the US, the creative class having a 33 per cent share of the working population (compared to 48.8 per cent of Durham, ranked first), with 43.6 per cent in the service industries (Florida, 2011, pp.404–8) and is in the bottom ten of the 'Openness-to-Experience' ranking, a personality inclination that favours innovation and creativity (ibid., p.250).

A.5 Bibliography

Central Statistics Office (Ireland) [a] *Number of private households and persons in private households in each province, county and city*. Available at: www.cso.ie/quicktables/GetQuickTables.aspx?FileName=CNA33.asp&TableName=Number+of+private+households+and+persons+in+private+households+in+each+Province+,+County+and+City&StatisticalProduct=DB_CN (Accessed: 20th August 2013).

Central Statistics Office (Ireland) [b] *Population and labour force predictions 2011–2041*. Available at: www.cso.ie/en/media/csoie/releasespublications/documents/population/2008/poplabfor_2011-2041.pdf (Accessed: 20th August 2013).

Central Statistics Office (Ireland) [c] *Population classified by area*. Available at: www.cso.ie/en/media/csoie/census/documents/census2011vol1andprofile1/Census,2011,-,Population,Classified,by,Area.pdf (Accessed: 20th August 2013).

Central Statistics Office (Ireland) [d] *Population of each province, county and city, 2011*. Available at: www.cso.ie/en/statistics/population/populationofeachprovincecountyandcity2011/ (Accessed: 20th August 2013).

Central Statistics Office (Ireland) [e] *Regional population projections*. Available at: www.cso.ie/en/media/csoie/releasespublications/documents/population/current/poppro.pdf (Accessed: 20th August 2013).

Cudden, J. (2013) *Dublin demographics 2013*. Available at: www.slideshare.net/jcudden/dublin-demographics-2013 (Accessed: 23rd September 2015).

Duncan, P. (2012) 'Dublin holds most "people born abroad"' 14th May 2012, *The Irish Times* [Online]. Available at: www.irishtimes.com/news/dublin-holds-most-people-born-abroad-1.714646 (Accessed: 23rd September 2015).

Florida, R. (2011) *The rise of the creative class revisited*. New York: Basic Books.

Geohive [a] *Agglomerations over 750,000 (with 2010 ranking)*. Available at: www.geohive.com/earth/cy_aggmillion2.aspx (Accessed: 20th August 2013).

Geohive [b] *Agglomerations: top 25 for 2015*. Available at: www.geohive.com/earth/cy_agg2025.aspx (Accessed: 20th August 2013).

Geohive [c] *Current world population (ranked)*. Available at: www.geohive.com/earth/population_now.aspx (Accessed: 20th August 2013).

Geohive [d] *Urban/rural division of countries for the year 2010*. Available at: www.geohive.com/earth/pop_urban1.aspx (Accessed: 20th August 2013).

I Live Here: Britain's Worst Places To Live (2015) *Walthamstow, Waltham Forest*. Available at: www.ilivehere.co.uk/statistics-walthamstow-waltham-forest-40928.html (Accessed: 23rd September 2015).

Kennedy, B. (2015) 'Indianapolis' in 'America's 11 poorest cities', 8th February 2015, *CBS Moneywatch* [Online]. Available at: www.cbsnews.com/media/americas-11-poorest-cities/4/ (Accessed: 23rd September 2015).

Legge, K. (2012) *Doing it differently*. Sydney: Place Partners.

Lydon, M. and Garcia, A. (2015) *Tactical urbanism: short-term action for long-term change*. Washington: Island Press.

Office for National Statistics. *London's population was increasing the fastest among the regions in 2012*. Available at: www.ons.gov.uk/ons/rel/regional-trends/region-and-country-profiles/region-and-country-profiles---key-statistics-and-profiles--october-2013/key-statistics-and-profiles---london--october-2013.html (Accessed: 23rd September 2015).

RateMyArea.com. Smithfield, Dublin, Ireland. Available at: http://dublin.ratemyarea.com/areas/smithfield-162 (Accessed: 23rd September 2015).

Trust for London [a] *Waltham Forest*. Available at: www.londonspovertyprofile.org.uk/indicators/boroughs/waltham-forest/ (Accessed: 23rd September 2015).

Trust for London [b] *Wandsworth*. Available at: www.londonspovertyprofile.org.uk/indicators/boroughs/wandsworth/ (Accessed: 23rd September 2015).

United States Census Bureau State and County QuickFacts. *Indiana*. Available at: http://quickfacts.census.gov/qfd/states/18000.html (Accessed: 23rd September 2015).

Waltham Forest Data and Statistics. Available at: www.walthamforest.gov.uk/Pages/Services/statistics-economic-information-and-analysis.aspx (Accessed: 23rd September 2015).

Wandsworth Borough Council. *Statistics and census information*. Available at: www.wandsworth.gov.uk/downloads/200088/statistics_and_census_information (Accessed: 23rd September 2015).

Index

abstracted participation 113, 130
active citizenship 138–40, 143, 164, 191–7, 211
aesthetic dislocation 38–9, 41, 65, 66, 83, 95, 96, 119, 158, 176, 181–2, 185, 186, 200, 211
aesthetics: dialogical 35–7, 40, 65, 95–6, 189; informal 94, 127–9, 155, 157, 177, 183–4, 200, 202; political movements and 78; relational 35–7, 40; social 40, 41–2, 66, 68, 76, 81; temporary 155–6
aesthetic third 39, 43, 68, 81, 83, 95, 182, 183
affective spaces 7, 46, 93, 147
agonism 79–80
agonistic agency 185
ambivalent pragmatism 81–3
anomie 16, 17, 159, 166
antagonism 79
architects 199
architecture 58–60, 67
art, symbolic economy of 3
art experience, participatory 29
Art in Odd Places Indianapolis (AiOPIndy) 156, 169–70, 179
artist(s) 29, 30, 31, 33; as catalysts 41–2; disposition of 156–7; as facilitators 33, 34; as leaders 153–4, 177–8; as placemakers 169; position of 100–5; role of 38, 40–3, 69–70, 103–4, 129–30, 145–6, 156–7, 168–70; special dispensation of 100–1; staff artist 153–4
artists/non-artists binary 34–5
art objects 30, 104, 128–9; re-materialised 45–6
ArtPlace 55
art(s): access to 156; agency of 78; at Art Tunnel Smithfield 126–38; community 17, 33, 34, 129, 174–5;
emplaced 17, 37–9, 47, 77–8, 173–81; instrumentalisation of 2, 47, 55–6, 82, 149, 196, 202; meaning of 32; Modernist 32; mural 38, 128, 158, 174–5; new genre public art 31–2; participatory xvii, 29–30, 32–5, 37–8, 67, 83; place attachment and 17, 45, 77, 148–9, 160, 166, 192–4, 209–10; place of 105–12; as political 31; process 45; Public Art 31, 36, 37, 58, 204; in public realm 47–8; relational 35–7; site-specific 30, 32, 44; social practice xvii, 1, 40–7, 78–9, 198, 202–4; in urban realm 29–32
arts-based placemaking 2
arts in place 6, 29–52, 77, 203
arts-led regeneration 164–70
arts organisations, role of 168–70
Art Tunnel Smithfield 4, 125–52, 195–6, 198–9, 209; active citizenship and 138–40; art practice and process of 126–38; introduction to 125–6; new urbanism and 140–9; participation issues 129–31, 179; place attachment and loss 131–5, 148–9; social horticulture 127–9; space/place thresholds and boundary crossing 135–8; temporary land use 140–9
'attached person' 14
attachment theory 13
audience 29, 30, 32, 42, 70, 130–1
authenticity 17, 167

beautification 126–9
belongingness 148, 159, 164, 182, 183, 209, *see also* place attachment
Beta Projects 131, 143, 144, 145, 195
Big Car 4, 153–72, 174–5, 195–6, 209; artists at 153–4, 156–7, 168–70; arts-led regeneration and 164–70; arts practice

and process of 154–7; collaborative participation 155–6; introduction to 153; place identity and 160–4; social cohesion and 157–60; social justice and 156; social practice placemaking 153–72; temporary aesthetic 155–6
boundary crossing 135–8, 160–4
bricolage 39
built environment 4, 44, 53, 54, 67, 149, 196

capitalist spectacle 78
case studies 4–5; Art Tunnel Smithfield 125–52, 179, 195–6, 198–9, 209; Big Car 4, 153–72, 174–5, 195–6, 209; demographic context of 17–20; The Drawing Shed 93–124, 175, 177, 179, 195, 209
cities 1; culturalisation of 3, 5–6; experience of 3–4, 9–11; interactions in 11; livability of 5; ludic city 44; right to the city discourse 6, 79–80; spaces 6–9
citizens 68, 78
citizenship 2, 3, 5, 8, 16, 46, 77, 80–1, 138–40, 143, 164, 191–7, 207–8, 211
city-making 1, 8
civic culture 21
civic participation 3, 77, 78, 138–40
collaboration 34–5
collaborative participation 155–6
communal conservatism 77
community art 17, 33, 34, 129, 174–5
community capital 71
community conscientisation 70–1
community driven placemaking 56–7, 61
community(ies) 3–4, 67, 175; experimental 46; functional 175; micro 181; ongoing invented 32; place-attached 16; sense of 70–1; sited 32; temporary invented 32
co-produced participation 114
co-production 8–9, 43–46, 67–70, 114, 178, 187–8
countersites 6, 9, 157, 184, 185
creative class 19, 55
creative placemaking 2, 55–8, 65, 143, 153, 167–8, 178, 190, 195–9, 202, 206, 209
creativity 13, 31, 68
critical barometer 47
critical materialism 35–6, 68, 128
critical spatial practices 8, 9
cultural gatekeepers 40

culturalisation 3, 5–6, 13
cultural policy 146
cultural production 67

Dadaism 30
democracy 80–1
demographic context 17–20
density-intensity hypothesis 11
depopulation 17
dialogical aesthetics 35–7, 40, 65, 95–6, 189
dialogic discussion 45–6
dissensus 185
diverted space 43
The Drawing Shed 4, 19–20, 93–124, 174, 175, 195, 209; degrees of participation 112–18; dialogical aesthetic 95–6; duration 99–100; gallery 105–8; housing estate 108–11; informal aesthetic 94, 177; introduction to 93–4; location 105–12; participation 179; performativity 96–8; position of the artist 100–5; public space 111–12; reflexive and transformative outcomes 118–20; relational objects 94–5, 97, 104, 116–17; social practice placemaking and 94–100
driving 160
Dublin 18–19, 82, 184; Art Tunnel Smithfield 125–52; land access and private property in 136–7; new urbanism 19, 140–9; praxis of the commons 126; public space in 18–19, 136–8; social practice placemaking 143–9; urban regeneration 144; vacant land issue 138–49, 208; vacant space in 126
Dublin City Council (DCC) 18, 125, 126, 128, 131, 133, 134, 137–49, 197
duration 44, 99–100, 176–7, 181, 185–6

education theory 72
embedded artist position 101–3, 110
embodied placemaking 67, 211
empathetic identification 36
emplaced arts 2–3, 6, 17, 37–9, 47, 77–8; aesthetic dislocation of 38–9; murals as 38; social practice placemaking and 173–81
engaged participation 113–14
exclusion 16–17
exemplification 46–7, 182–5
experimental communities 46
expert/amateur dichotomy 8–9, 35, 58–9, 67–8

Fletcher, Harrell 156, 174
flexible citizenship 81
framing 207–8
functional community 175
functional site 30

Garfield Park Creative Community 165–8
gated communities 17
generative planning 194–5
gentrification 6, 17, 62–3, 76, 78, 103–4, 109–10, 165, 180, 182, 187, 188, 205, 208
global context 1–28
grassroots activism 138–40, 144, 146–8
group membership 11

'habit-memory' 12
habitus 3, 72
Haynes, Jim 174
Hope, Sophie 34

IdeasFromElse[W]here 94–7, 100–1, 103, 105–6, 108, 111, 113–17, 120, 174, 190
identity 5, 12; group 10; individual 10; neighbourhood 5–6, 160–1, 163–4, 169, 175; place 12, 14–17, 159–64
Indianapolis 19, 143; Big Car 153–72; territoriality 160–4; travel and the car 160; urban regeneration 144, 164–70, 208
informal aesthetic 94, 127–9, 155, 157, 177, 183–4, 200, 202
informal spaces 8
instrumentalisation 2, 47, 55–6, 82, 149, 196, 202
interconnectedness 40
intersectionality 45, 102, 207
intersubjectivity 29, 40
Ireland 18–19, 136–7, 138–9, *see also* Dublin

Kaprow, Allan 29, 174

Ladder of Participation 37
land access 136
Lighter, Quicker, Cheaper 63–5, 195
liveability 5, 19
LiveElse[W]here 97–8, 99
Live Lunch 97–8, 99
located co-production 187–8
London 19–20; The Drawing Shed 93–124, 120; urban regeneration 144, 208
loneliness 16

'loose space' 8
Lowe, Rick 41, 45, 174, 206
ludic city 44

market forces 3
Mary's Abbey 144–5
material change 188–91
meaning-making 6, 32, 45, 54, 68
mediated participation 113
micro-community 181
micropublics 4, 8, 15, 16, 36–7, 45, 81, 132, 159, 162, 181, 192
mixophilia 10
mixophobia 10
Modernist art 32
murals 38, 128, 158, 174–5

National Endowment for the Arts (NEA) 55
neighbourhoods 3–6, 160–1, 163–4, 169, 175
neoliberal dilemma 2–3, 77–9, 195–7
neoliberal urban economics 5–6, 80
new genre placemaking 61
new genre public art 31–2
new situationism 68, 100, 157
New Situationists 29
new urbanism 10, 16, 19, 57, 140–9
non-artist participation 113–14, 179

objects: art 104, 128–9; relational 94–5, 97, 104, 116–17, 188–9
Obrist, Hans Ulrich 174
Office for Metropolitan Architecture 58
ongoing invented communities 32
On The Rack 97

Park(ing) Day 64
participant-as-sole-artist 114
participation 178–80, 203; abstracted 113, 130; active citizenship 138–40, 143; Art Tunnel Smithfield 129–31; barriers to 112, 115, 116, 125, 130–1, 179, 184, 189, 198, 203; collaborative 155–6; co-produced 114; degress of, at Drawing Shed 112–18; engaged 113–14; issues of and around 129–31; mediated 113; non-artist 113–14, 179; prolematising 115–18; transformation and 119
participatory art xvii, 29–30, 32–5, 37–8, 43, 67, 83
participatory placemaking 62–5
Percent for Art 31

performative event space 67
performative principle 41
performativity 96–8, 175–6
place attachment 2, 4, 13–17, 45, 77, 148–9, 160, 166, 192–4, 209–10; arts and 17; Art Tunnel Smithfield 131–5; components of 14–15; defined 13–14; performed 182–5
place change 186–7
place efficacy 186–7
'place-faking' 56
place identity 14–17, 159, 160–4
place intensification 15
place loss 131–5, 187
placemaking: aims and benefits of 56–8; architecture and 58–60; community driven 56–7, 61; concept of 53–4, 60–1; creative 2, 55–8, 65, 143, 153, 167–8, 178, 190, 195–9, 202, 206, 209; embodied 67, 211; expanding understanding of 60–5; grassroots-led 138–40, 144, 146–8; new genre 61; participatory 62–5; vs. placeshaping 104; Public Realm 61–2, 65, 73, 198; rural 207; as social movement 200–1, *see also* social practice placemaking
Placemaking Leadership Council (PLC) 57
placemaking typology 2, 3, 37, 47–8, 53–92, 204–5, 208; illustration of 74–5; need for 72–3; purpose of 73
place politics 191–2
place realisation 15
place release 15
place(s): of art 105–12; authentic 17, 167; dimensions of 13; identity and 12; sense of 12–13; shaping 11–13, *see also* space(s)
placeshaping 104
pluralist place politics 191–2
pocket revolutions 201
political participation 31
politics: neoliberal dilemma 2–3, 77–9, 195–7; pluralist place 191–2; progressive 79–80; of social practice placemaking 76–83
positionality, of the artist 100–5
pragmatism 81–3
private property 136
problem-posing pedagogy 30
progressive politics 79–80
project closure 133–4
Project for Public Spaces (PPS) 54
Project Row Houses 206

pseudo-participation 69
pseudopublic spaces 62–3
psychosocial distancing 10
Public Art 31, 36, 37, 58, 204
public-authority interaction 16
Public Realm placemaking 61–2, 65, 73, 198
public space 8, 10, 18–19, 54, 77, 78; The Drawing Shed 111–12; Dublin 18–19, 136–8; re-imagining 53
public urbanity 8, 9

reflexivity 46, 68, 76, 120, 175, 183
relational aesthetics 35–7, 40
relational objects 94–5, 97, 104, 116–17, 188–9
relative expertism 9, 42–3, 154, 168–9, 177, 180, 184, 186, 195
relocalism 21
re-materialisation 45–6
resingularisation 41
right to the city discourse 6, 79–80
role dissonance 42
rural placemaking 207

self-criticality 44–5
Sennett, Richard 42, 43, 45, 174
shock 38
sited communities 32
site-specific art 30, 32, 44
situated context 42–3
social activism 47, 71
social aesthetics 37, 40, 41–2, 66, 68, 76, 81
social capital 15–16, 17, 70, 158, 159–60
social cohesion 3, 15–16, 77, 149, 157–60, 191–7
social horticulture 127–9
Social Impact of the Arts Project (SIAP) 54
social inclusion 203
sociality 10
social justice 156, 195
socially engaged art xvii, 198
social movements 42, 200–1, 207
social participation 78
social practice art(s) xvii, 1, 40–7, 78–9, 198, 202–4; duration 44; as gestural 45; heritage 174–5; outcomes 46–7; practice 41–3; process 43–5; as social work 47
social practice placemaking xiii–xiv, 6, 65–72, 128, 173–211; aims and outcomes of 70–2; ambivalent

pragmatism and 81–3; as arts-led regeneration 164–70; Art Tunnel Smithfield 132, 138–40, 148–9; Big Car 153–72; case studies 4–5; citizenship and 80–1; concept of 1; The Drawing Shed and 94–100; Dublin 143–9; emplaced arts and 173–81; evaluation of 206; in global context 1–28; neoliberal dilemma and 77–9; place identity and 160–4; political implications of 76–83; practice 66–8, 173–81; process 68–70, 173–81; right to the city discourse and 79–80; social cohesion and 157–60, 191–7; theory of 20–1; transformative outcomes 180–1, 203–4; urban public realm and 181–91
social relations 12
social work 47
socio-spatial positionality 68
socio-spatial production 7–8
Some[w]Here Research 95–7, 101–3, 120
space/place thresholds 135–8
space(s): affective 7, 46, 93, 147; diverted 43; informal 8; 'loose space' 8; performative event 67; production of 11–12; pseudopublic 62–3; public 8, 10, 18–19, 53, 54, 77, 78, 111–12, 136–8; re-appropriation of 7–8, 9, 76, 80, 125, 126, 138, 147, 148–9, 191, 201, 210; shaping 11–13; vacant 7–9, 18, 126, 138–49, 200
spacetimes 7, 9, 43, 147
SPARK: Monument Circle 153, 169
spatial boundaries 182
splash interventions 157
staff artist 153–4
storytelling 97–8, 114, 118
symbolic interactionism 189

tactical urbanism 63–5
temporary aesthetic 155–6
temporary invented communities 32
temporary land use 140–9
territoriality 112, 160–4
territory crossing 160–4
theoretical framework 3–4
threshold crossing 135–8
topophillic connections 14
tradition 39
transformation 180–1, 203–4
travel 160
The Tube 165–6, 167, 169

unitary urbanism 35, 66
United Kingdom 19–20
United States 19
urban commoning praxis 19
urban creatives 69
urban design 6, 8–9, 57, 64, 146, 196
urban development 3, 21
urban environment 4, 5–6
urbanisation 6, 77–8
urbanism: new 10, 16, 19, 57, 140–9; tactical 63–5; unitary 66
urban lived experience 1, 3–4, 9–11, 20–1, 69, 166, 190–1
urban planning 9, 13, 80, 142–3, 146, 194–5
urban population 1–2
urban public realm 181–91
urban regeneration 54, 69, 103–4, 144, 164–70, 196, 208
urban spaces 6–9; art in 29–32; shaping 11–13
urban studies 5
urban thinking/theory 2, 9–10
urban triumphalism 198

vacant space 7–9, 18, 138–49, 200
Vitality Indices 13

Walker, Jim 154, 156–7
Waltham Forest 19–20
Wandsworth 20